Springer Series on Environmental Management

Robert S. DeSanto, Series Editor

William J. Petak
Arthur A. Atkisson

Natural Hazard Risk Assessment and Public Policy

Anticipating the Unexpected

With 89 Figures

Springer-Verlag
New York Heidelberg Berlin

William J. Petak
University of Southern California
School of Public Administration
University Park
Los Angeles, CA 90007 U.S.A.

Arthur A. Atkisson
University of Wisconsin—Green Bay
Green Bay, WI 54302 U.S.A.

On the front cover: A visible spectrum image, taken at 2200 Greenwich Mean Time (4 PM Eastern Daylight Standard Time) on August 7, 1976 from NOAA's Eastern Geostationary Satellite (GOES-EAST) located at 75° W longitude, 36,000 km above the Equator. Hurricane Belle is located at approximately 27° N and 74° W, northeast of the Bahama Islands. This low sun angle image clearly shows the internal structure of the clouds arcing the hurricane. Photo courtesy of the United States Department of Commerce, National Oceanographic and Atmospheric Administration.

Library of Congress Cataloging in Publication Data
Petak, William J.
Natural hazard risk assessment and public policy.
(Springer series on environmental management)
Bibliography: p.
Includes index.
1. Natural disasters. I. Atkisson, Arthur A. II. Title. III. Series.
GB5014.P47 1982 363.3′49 82-10468

Softcover reprint of the hardcover 1st edition 1982

 This book was prepared with support of NSF Grant ERP-75-09998. However, any opinions, findings, conclusions, and/or recommendations herein are those of the authors and do not necessarily reflect the views of NSF.

Typeset by Ms Associates, Champaign, Illinois

9 8 7 6 5 4 3 2 1

ISBN-13: 978-1-4612-5688-5 e-ISBN-13: 978-1-4612-5686-1
DOI: 10.1007/978-1-4612-5686-1

To Ramona and Jean

Series Preface

This series is dedicated to serving the growing community of scholars and practitioners concerned with the principles and applications of environmental management. Each volume is a thorough treatment of a specific topic of importance for proper management practices. A fundamental objective of these books is to help the reader discern and implement man's stewardship of our environment and the world's renewable resources. For we must strive to understand the relationship between man and nature, act to bring harmony to it, and nurture an environment that is both stable and productive.

These objectives have often eluded us because the pursuit of other individual and societal goals has diverted us from a course of living in balance with the environment. At times, therefore, the environmental manager may have to exert restrictive control, which is usually best applied to man, not nature. Attempts to alter or harness nature have often failed or backfired, as exemplified by the results of imprudent use of herbicides, fertilizers, water, and other agents.

Each book in this series will shed light on the fundamental and applied aspects of environmental management. It is hoped that each will help solve a practical and serious environmental problem.

Robert S. DeSanto
East Lyme, Connecticut

Preface

Throughout history, individuals and governments have sought means for limiting the adverse impacts associated with the exposure of people and property to the hazards of the natural environment. In the United States, the hazards of major concern have included riverine flood, coastal storm surge, tsunami, earthquake, expansive soil, landslide, hurricane wind, and severe wind.

In attempts to mitigate the effects of exposures to these hazards, rivers have been dammed, deepened, and diked; coastlines have been equipped with seawalls; storm cellars have been dug in backyards; buildings have been elevated above the level of expected flood heights; and structures have been strengthened to reduce their vulnerability to these forces. Numerous types of building strengthening, area protection, site development, and other technologies are available to reduce the risks associated with exposure to natural hazards, and their mandatory application can be forced through adoption and enforcement of a wide variety of federal, state, and local public policies. Also, the risk of loss may be spread through use of insurance schemes, and the impact of catastrophic hazardous occurrences on exposed populations may be reduced through disaster relief and recovery measures financed by nonimpacted parties.

However, every public and private response that can be made to the risks presented by natural hazard exposures imposes costs on someone, somewhere, at some time. In some cases, the costs of such ventures may exceed the value of the risk reduction produced by the purchased mitigation. In still other cases, the use of the mitigation may engender a false sense of public security and lure additional numbers of people into contact with hazardous areas, thereby increasing the total losses associated with such exposures. What to do about the continuing exposure of people and property to natural hazards is, therefore, both a question of considerable complexity as well as one of increasing importance to public policy makers at federal, state, and local levels. Should building code requirements be strengthened? Should governmentally enforced restrictions be imposed on the use of such hazardous areas as floodplains, earthquake-prone sites, and steep hillsides subject to land slippage? What public problems are posed by the voluntary and involuntary exposure of people and property to natural hazards, and what problems might be produced by public effort to control such exposures?

It is to these and other important questions that this book is addressed. Based primarily on a project funded by the National Science Foundation under Grant ERP-75-09998, the book also contains material derived from a project conducted for the Federal Emergency Management Agency and another funded by the Federal Insurance Administration. In none of these studies, nor in this book, has it been our intention to tell policy makers what solutions should be applied to hazard induced public problems. Rather, we have sought to present data and conclusions that may aid policy makers in performing this task. Specifically, we identify: the various stakeholder groups whose interests are bound up in hazard exposure and hazard mitigation situations; candidate lists of possible public problems; the range of technologic and policy options that may be appropriate to solving each listed candidate problem. Further, we identify and assess the more important costs and benefits associated with each policy option, and we demonstrate how risk analysis can be applied in defining public problems and assisting in the policy-making process.

Our research used risk analysis techniques that resulted in: (1) the generation of annual expected natural hazard loss estimates for 1970 and 2000; (2) the identification of specific strategies and technologies theoretically capable of reducing such losses; (3) the identification of the amortized annual costs associated with selected mitigation strategies; (4) the identification of the candidate public problems and stakeholder groups associated with natural hazard exposures and alternative technology-forcing policy options; (5) the identification and critical evaluation of past and current public policies, institutional arrangements, and administrative practices aimed at mitigation of natural hazards losses; (6) the identification and assessment of the contemporary social, technical, administrative, political, legal, and economic constraints on natural hazards policy making operations; and (7) the development and assessment of policy options appropriate for coping with hazard-related public problems between 1970 and 2000.

Although we accept full responsibility for the conclusions and approaches that appear in this book, we acknowledge the formidable debt we owe the individual members of the several teams of investigators participating in the total effort. Their patient understanding and their help in educating us about previously unknown subjects and techniques made our work possible. Specifically, we wish to acknowledge Jon Chrostowski, Ronald Eguchi, Dr. Peter Gordon, Dr. Gary Hart, Joseph Hirschberg, James Krohn, Larry Lee, Dr. James Slosson, and Dr. John Wiggins. Their contributions are all well documented in the reports listed in the bibliographies.

Dr. Daniel Alesch and Dr. Michael Kraft of the University of Wisconsin-Green Bay deserve special acknowledgment for their contributions to the policy analysis recommendations.

Special thanks are due Shirley Kay, Chris Burch, Marge Judge, Nancy Mulford, Jean Atkisson, and Beth Petak, who were largely responsible for manuscript preparation. Also, special thanks are due Ramona Petak for her able assistance in preparation of the book index. Their tasks proved to be the most laborious. We wish also to acknowledge the very large extent to which the professionally challenging environment of the J. H. Wiggins Company contributed to the completion of the several federally funded studies that are the real parents of this work. Through the support of the company, engineers, economists, management scientists, geoscientists, policy scientists, and computer system specialists were assembled and fused into effective working teams.

Finally, special acknowledgment is due Dr. Charles Thiel, formerly a member of the research staff of the National Science Foundation and the Federal Emergency Management Agency, and Dr. G. Patrick Johnson, National Science Foundation, for their consistent support and patience while we were completing this very difficult project.

William J. Petak
Arthur A. Atkisson
Los Angeles, California
Green Bay, Wisconsin
August 16, 1982

Contents

PART I

ANTICIPATING THE UNEXPECTED AS A FOCUS OF PUBLIC POLICY

Public policies are the authoritative *megadecisions* or choices made by, or on behalf of, legally recognized groups of people that guide future subsidiary decisions. As authoritative decision-constraining guides, public policies may deal with the end states to be achieved by future decisions, with the means to be used or avoided in reaching those ends, with other special conditions to be observed in making and implementing such decisions, and with various combinations of the above.

Public policies are embedded in such documents as statutes, ordinances, constitutions, executive orders, regulatory rules and regulations, and opinions of judges.

Typically, public policies are directed at the elimination or reduction of the *public problems* associated with some set of *social impacts*, at the distribution of the benefits of problem-solving activity to one or more stakeholder groups, and at the imposition of problem-solving costs on some such groups. Social impacts consist of the observable primary, secondary, and higher-order consequences associated with some set of causal phenomena, and are raised to the status of public problems when they produce socially unwanted gaps between the states of affairs that are socially desired, or expected, and the states of affairs that are socially perceived to be extant.

Throughout history, such socially unwanted gaps have been created by a rich variety of causes, including the exposure of people and property to hazardous natural events. Volcanic eruptions, riverine and coastal flooding, forest fires, and many other natural events have produced consequences of such magnitude in some communities and societies that they have become public problems.

In terms of those potential problem-causing situations in which the probabilities of event occurrences and their consequences may vary widely from year to year, considerable attention is now being directed to anticipating the unexpected, that is, to the application of risk analysis methods to policymaking. As a formal system for aiding policy makers in reaching rational decisions, the methods of risk analysis may be considered as consisting of those aimed at (1) the identification of risk, (2) the estimation of risk, and (3) the evaluation of risk. Risk estimation is concerned with determining the probabilities that some series of events of various magnitudes will occur in the future, specifying the associated consequences, and identifying the populations exposed to such risks. On the other hand, risk evaluation is concerned with assessing the importance of the identified risk factors to the various stakeholders whose interests are embedded in the candidate problem situations.

In the three chapters included in this section, we examine the attributes of nine natural hazards, the range of primary, secondary, and high-order effects produced, and the role of risk analysis in defining and assessing the associated candidate problems. The characteristics of the American public policy system are examined as are the roles of risk analysis, problem analysis, policy analysis, and technology assessment in the making of major policy decisions. In this section we examine the past and current public policies which have been employed in the United States for the management of natural hazards, and present the concepts which are necessary to understanding the use of the findings presented in later sections of the book.

1
Natural Hazards as Public Policy Problems

Introduction

Throughout history public policy makers have sought to anticipate the unexpected in order to reduce the risk to human life and safety posed by intermittently occuring hazardous natural and man-generated events. The Code of Hammurabi (circa 1950 B.C.) decreed that should a house collapse and kill the occupant, the builder of the house must forfeit his own life. Etched on the cliffs of ancient Babylon, the code was designed to serve the same people who were troubled not only by hazardous structures created by other people, but also by the intermittent flooding of their fertile plains, which lay between the Tigris and Euphrates Rivers. Prompted by the need to record and predict such flooding, the Babylonians developed crude systems of mathematics, laid the intellectual groundwork for the modern field of hydrology, and developed government financed systems for flood control and irrigation.

However much the pre-Hammurabi Babylonians may have viewed building collapse and community flooding as the acts of some malevolent God, as unexpected events to be borne as the natural concomitant of human existence, it is clear that succeeding generations comprehended that the periodicity of floods can be predicted, their causes understood, and that corrective loss bearing measures can be applied by government. They believed buildings collapsed not because of some natural law, but due to carelessness and oversight by their builders. At least these threats to human life and safety could be controlled through the design and enforcement of appropriate public policies.

The lesson is being relearned by contemporary decision makers worldwide. Government sponsored risk management programs now attract increased attention by policy makers at the highest levels of government and the electorate. No matter what the source of risk to human life, safety, and well-being, these electorates now demand that government reduce the uncertainties associated with hazardous events and design and enforce policies to mitigate their effects.

Natural Hazards as Candidate Problems

Natural hazards consist of such phenomena as avalanches; earthquakes; expansive soils; forest, field, marsh, and prairie fires of natural origin; hailstorms; hurricane winds; landslide; lightning; riverine flooding; coastal storm flooding; urban headwater flooding; severe rain, snow, and ice storms; land subsidence; tornadoes; tsunamis; and volcanic eruptions. Each has caused much human suffering and economic loss in earlier periods of history and each continues to do so in the modern world.

There are 516 active volcanoes from which eruptions occur approximately once every 15 days. Monitoring instruments currently record approximately 2000 tremors beneath the crest of the earth's surface each day, and almost twice each day earthquakes of sufficient magnitude to damage buildings and other structures occur. Quakes of sufficient strength to produce widespread damage and death occur 15 to 20 times each year. Above the surface of the earth 1800 orbiting thunderstorms can be observed at any given time, and lightning strikes the planet's outer skin at the rate of 100 times per second.

In late summer, 50 or more hurricanes can be observed forming somewhere in the world, and during approximately the same season of the year from 600 to 1000 tornadoes strike somewhere within the United States at a rate of 4 or more strikes per day. Nearly one half billion members of the planet's total population now reside in riverine and coastal floodplains where they produce one-third of the world's food production; and, on any given day, some fraction of these plains are covered by floodwaters.

Almost no portion of the earth's surface is free from the risks produced by hazardous natural events. Some of these occur infrequently, but with catastrophic results. For example, over the past 10,000 years, Mt. Ranier in the northwestern corner of the United States has produced 12 volcanic ash eruptions, one or more lava flows, several hot avalanches of rock debris, and at least 55 large mudflows. As noted by Foster (1980) one of the largest of these deposited 2.5 billion cubic yards of material over approximately 125 square miles in the now populated lowlands of Puget Sound.

Most of the coastal Minoan cities on the island of Crete were destroyed by an enormous tsunami (giant seismic sea wave) that struck the island between 1450 and 1480 B.C. The tsunami apparently was generated by numerous violent earthquakes accompanying a major volcanic eruption on the island of Santorin.

Twenty-one years before the adoption of the Declaration of Independence, earthquakes shattered Massachusetts, and during the War of 1812, the highest magnitude earthquake in the history of the United States left parts of Missouri and Arkansas permanently sunken. In the immediate post-Civil War years another devastating earthquake struck South Carolina, and in 1871 a forest fire consumed Peshtigo, Wisconsin, and caused 1182 deaths. In 1889, flood waters claimed 2209 lives in Johnstown, Pennsylvania, on a single day, and 11 years later (September 8, 1900) the largest civil disaster in U.S. history occurred when a "great" hurricane pushed the waters of a storm surge over Galveston, Texas, and caused 6000 deaths. Only two years later, but at another point on the globe, a volcanic eruption from Mt. Pelee produced a flowing avalanche that struck the city of St. Pierre and killed 30,000. On April 18, 1906 a "great" earthquake rocked San Francisco and together with the resulting fires caused 500–700 deaths and more than $374 million in property damage. In September 1928, a Florida hurricane caused 1833 deaths over a two-day period, while the previous March a California dam collapse sent a wall of water over an unwary populace and swept 450 to their deaths. Many of these natural disasters prior to 1930 rank among the most severe civil calamities in U.S. history, having claimed more lives per event than the sinking of the Titanic in 1912 (1517 dead); the Texas City ship explosion of 1947 (561 dead); the Coconut Grove Night Club fire of 1942 (492 dead); the Monongah, West Virginia, coal mine explosion of 1970 (361 dead), and the worst air traffic accident in history in March 1977 (581 dead).

In more recent years, the Palm Sunday tornadoes of 1965 claimed 271 lives in five states; hurricane Camille (1969) destroyed over $1.4 billion in property and caused 256 deaths; the South Dakota flash flood of 1972 killed 236; the Alaska earthquake (1964) claimed 131; and Agnes, the hurricane and tropical storm of 1972, caused 118 deaths and the loss of more than $3.1 billion in property. Also, on a single day in 1974, tornadoes caused 318 deaths in several southern and midwestern states. Two years later, a major earthquake occurred in Tangshan, China, and resulted in an estimated 750,000 deaths (Frazier, 1979).

Although less dramatic in their impact, a variety of other natural hazards during these same time periods also produced considerable property damage and substantial annual economic losses. These hazards include expansive soils, land subsidence, landslides, the erosion of river and shore banks, droughts, and storms of hail, ice, snow, and rain.

In our efforts to mitigate the effects of exposures to natural hazards, population warning systems have been placed in operation, and rivers have been dammed, deepened, and diked. Coastlines have been equipped with seawalls; storm cellars have been dug in backyards; buildings have been elevated above the level of expected flood heights; and a variety of means have been employed to strengthen structures and thereby reduce their vulnerability to the forces exerted by winds, land movement, and other natural hazards. Sadly, however, these efforts have produced less than satisfactory results.

Construction of flood control facilities has seemed to prompt heavy migration into flood-prone areas and has thereby increased the real costs of flood exposures. Governmental provision of disaster relief, low cost loans, and subsidized insurance has seemed to encourage, rather than inhibit, private risk-taking activity. A public unwillingness to acknowledge the threat of future loss-producing occurrences in high-hazard areas, and an accompanying faith that government will somehow protect them, have contributed to a continuing population movement into such high-hazard areas as the hurricane and flood-prone belts of water-adjacent areas along the Gulf Coast and South Atlantic. Similar population movements have occurred in seismically active areas and along the shores of rivers and lakes subject to periodic flooding. As a result, the United States now faces the probability that one or more major community catastrophes, each far greater in loss of life and property than any previous one, may occur during the next several decades. But similarly, we also face the risk of overreacting to the threats posed by natural hazards, through implementation of public policies that may produce costs far in excess of their potential benefits.

Numerous types of building strengthening, area protection, site development, and other technologies are available to reduce the risks associated with exposure to natural hazards, and the mandatory application of these technologies can be forced through federal, state, and local public policies.

Hazard-mitigating amendments to building codes, subdivision standards, and land use regulations can be enacted. Hazard zone identification measures can be adopted, and legally sanctioned systems for avoidance of development in such areas can be inaugurated. The risk of loss may be spread through use of insurance schemes, and the impact of catastrophic hazardous occurrences on exposed populations may be reduced through community safety plans, disaster relief, and recovery measures financed by nonimpacted parties.

What mix of these measures to employ, when, where, at what cost, and to whom, has therefore become a major public policy question, influenced greatly by the public attitudes and emotions of the moment. Public and governmental responses to hazardous natural occurrences vary. Studies performed by other researchers have shown that the immediate aftermath of a severe, disaster-scale event is characterized by high levels of public support for governmental action to avert future disasters. Such support may be exhibited both by hazard area residents and people far removed from the disaster site. Indeed, media exaggerations of disaster impacts increase with distance from the disaster site, and "outsiders," in general, tend both to overestimate the recovery needs and exaggerate the real impacts on the affected communities. Those who did not experience the disaster tend to deny that they too may be threatened by a similar future occurrence on their home turfs, but for a while tend to support governmental action to aid those less fortunately located.

Within the hazard area, the immediate time period following the occurrence of a severe natural event may be characterized by a high level of public interest in ways to avert future disasters and a flurry of public policy proposals may be generated. However, between natural disasters, there is generally a lack of public interest in hazard-related legislation and a corresponding lack of legislative activity. In contrast, during the emotional period immediately following a disaster, many hazard-targeted legislative proposals may be advanced and enacted into law, many of which may be imprudently drafted. Moreover, the magnitude and social impacts of the event may be grossly exaggerated.

Depending on the nature of the hazard, the immediate postevent period may see a variety of public policy actions. Hillside lots may be zoned to limit the use of slide-prone areas by all but the most carefully engineered structures; building code standards may be changed to require new buildings to be resistant to the damage-producing force of winds and earthquakes; hazard warning systems may be inaugurated along with the development of community safety or population evacuation plans; and public investments may be poured into such area protection facilities as dams, seawalls, levees, channel improvements, and revetments. Public funds may be appropriated to aid communities and property owners in restoring their properties after damage-producing natural events have occurred; low cost publicly subsidized loans and insurance schemes may be provided; and attention may even focus on the strengthening or removal of damage-prone structures in high-hazard areas.

Interestingly, however, residents in areas that have experienced hazardous natural events tend to deny that such events will recur and believe they will not personally, in the future, experience the adverse impacts of such occurrences. Even new migrants into a high-hazard area exhibit this propensity. Although somewhat more fearful than long-time residents, they nonetheless deny that future hazardous events will affect them. Even in areas that have experienced particularly severe disasters, and where population evacuation systems are the primary means for averting future loss of life, voluntary compliance with evacuation orders tends to drop off as a function of the length of time since the last severe disaster. Individuals seem to assume either that disastrous events will not recur, or that the impacts of such events will somehow be ameliorated or eliminated by governmental action, that the individual will somehow escape the impacts of such events, and that any personal losses will be reimbursed by some governmental or insuring institution.

However, every public and private response that can be made to the risks presented by natural hazard exposures imposes costs on someone, somewhere, at some time. In some cases, the costs of such ventures may exceed the value of the risk reduction produced by the purchased mitigation; in still other cases, the use of the mitigation may engender a false sense of security and lure additional people into hazardous areas and thereby increase the total losses associated with such exposures. What to do about the continuing exposure of people and property to natural hazards is, therefore, both a question of considerable complexity as well as one of increasing importance to public policy makers at federal, state, and local levels. Should building code requirements be strengthened? Should governmentally enforced restrictions be imposed on the use of such hazardous areas as floodplains, earthquake-prone sites, and steep hillsides subject to land slippage? Should communities be required to adopt community safety plans? What public problems are posed by the voluntary and involuntary exposure of people and property to natural hazards, and what problems might be produced by public efforts to control such exposures? To what extent should those who benefit from governmental investments in hazard-mitigating facilities be expected to bear the costs of those facilities?

Questions such as these now confront policy makers within the United States and throughout the world. They are not easy to answer, since few of the situations viewed as constituting major public problems can be said to stand alone. Instead, each is inextricably related to other "good" and "bad" features of contemporary life, and action taken to address any one perceived problem may produce waves of desirable and undesirable effects throughout the networks to which the focal problem is linked. The boundaries between "problems" and "nonproblems" are frequently obscured to even the most well informed and well intentioned; what is "right to do" may be enormously difficult to determine; and even the most expert technicians may experience difficulty in determining the efficacy and cost of alternative problem solutions. If, as Sir Edmund Burke once noted, "The public interest requires that we do today what men of intelligence and good will would wish—five or ten years hence—had been done" to cope with public problems, it is difficult for those who are both well informed and well meaning to determine where the public interest lies. As John Gardner (1970) observed, today's public policy battles typically are fought in milieus in which "issues are confused, where you are never sure you're right, where good and bad are inextricably fused with the partly good and the partly bad, where often you can't do one worthy thing without endangering some other worthy thing." It is to be the dilemmas, uncertainties, and complexities of modern public problem solving that the policy-making system is directed. Through this system, the ends and framework for natural hazards management are fashioned.

Characteristics of the Public Policy System

In order to understand how natural hazards are or might be examined by the actors and entities within the public policy system it is first necessary to understand the characteristics of the system and how it functions in respect to the formulation, evaluation, and revision of public policy decisions. From the perspective of Anderson (1975) and Jones (1977) the process through which public policies are framed, adopted, implemented, and amended is characterized by a critical sequence of events, all depicted in Figure 1-1.

Within the limits of the Anderson-Jones perspective, the public policy process begins when some state of affairs is perceived as being intrinsically or instrumentally unsatisfactory by an element of society and is perceived by it to qualify as a "public problem." The perceived existence of this state of affairs may subsequently generate a public state of dissatisfaction and lead to demands on the public policy system. Under appropriate circumstances these demands may result in policy maker dissatisfaction and the subsequent inclusion of policy-making demands on the public policy agenda. At this stage in the process, a "public policy problem" may be said to have been identified. Thereafter, the inclusion of an item on the public policy agenda leads to the making of one or more public policy decisions, to the output of governmentally generated services or products, and to the generation of "policy output effects."

If decisions have been appropriately framed and decision-implementing activities appropriately conducted, the supposition is that these policy output effects will include a resolution or amelioration of the "public problem" that activated the entire sequence of events; and that no new "public problems" will result

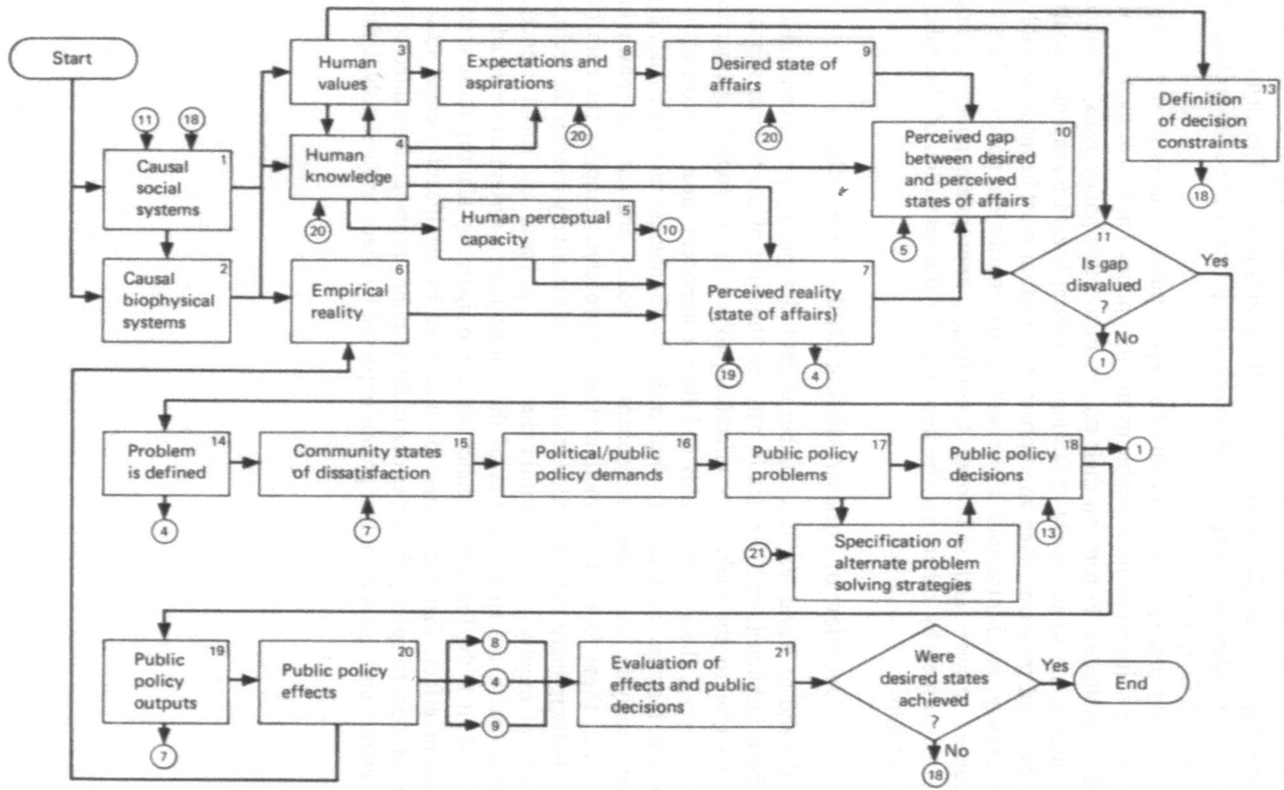

Figure 1-1. The public problem-defining and problem-solving process.

from the decision-making process, the decisions themselves, or the decision-implementing activities that were a part of the policy process. The plight of the public policy maker in dealing with seemingly straightfoward "public problems" is discussed in Chapter 8.

Few would disagree that rationally conceived public policies should be targeted on the solution of these problems and that policy-implementing programs should be evaluated, in part, on their success in meeting policy goals. If public policy makers are to act wisely in fashioning natural hazard management goals, objectives, and programs, then appropriate information must be made available. Unfortunately, good intentions, alone, will not necessarily lead policy makers to the correct answers and solutions. Accordingly, appropriate policy-assisting studies performed by others may be necessary for the resolution of the policy maker's plight.

Overview of Policy Assisting Study Approaches

Since the end of World War II a number of multidisciplinary professional fields have been fashioned, and justified in part, on the basis of the rational support they can give to the making of effective public policies. Joined together as *policy sciences*, these several fields and their component areas of professional operations may be grouped into four major methodologic sets: risk analysis, problem analysis, technology assessment, and policy analysis. As shown in Figure 1-2, these four sets of methodologic approaches comprise the core of the multidisciplinary field of policy science and substantially overlap the technical fields of inquiry included within the natural, physical, and social sciences. Consistent with the characteristics of the public policy system as depicted in Figure 1-2, it is the function of policy science to rigorously identify those social impacts or effects that are or may be unwanted by significant societal stakeholder groups, to specify the types of candidate public problems associated with those effects and the stakeholder groups impacted by them, to rigorously define

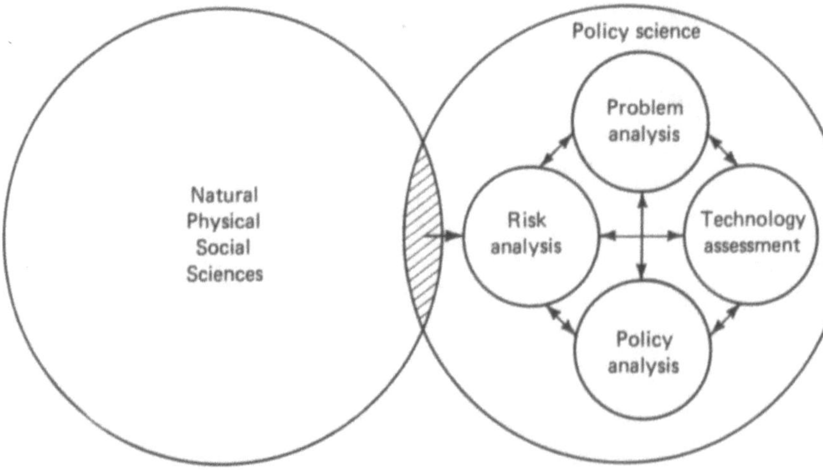

Figure 1-2. Interrelationship between risk analysis and all other fields.

the networks of causes preceding the occurrence of these problems, to identify and test alternative strategies for problem solving, and to measure the payoffs associated with these candidate problem solving solutions. In our view, the four major components of policy science are as follows.

Risk Analysis

Risk analysis, or risk assessment, typically is defined as including three related operational elements: risk identification, risk estimation, and risk evaluation. The identification of a societal risk may target on the description of a "candidate problem," system side effect, or "impact," while risk estimation targets on the quantitative description of the population at risk, the identification of the impact-causing events, the specification of the probability that such events will occur, the elucidation of the consequences associated with the various magnitudes of expected events, and the integration of these magnitude/event probability calculations into a quantitative measure of risk. As a concept, risk typically is viewed as being a function of two major factors: (a) the probability that an event, or series of events of various magnitudes, will occur, and (b) the consequences of the event(s). Hence, risk estimation is aimed at identifying (a) the probability that particular events of specified intensities will occur over some indicated time frame; (b) the area and/or population that will be exposed to the events (e.g., the population at risk); (c) the vulnerability of the area or exposed population to effects associated with the events; and (d) the consequences to the populations at risk of exposure to the expected series and intensities of the events over some specified time frame. The outputs of a risk estimation are in the form of probabilistically derived estimates of the consequences to the population at risk resulting from exposure to the hazard, typically in the form of annual expected, most likely, and maximum probable quantitative estimates. The third operational element in risk analysis is termed "risk evaluation." Activities performed in the risk evaluation phase of a study are intended to help answer the question, "How safe is safe enough?" Risk evaluation relies heavily on providing the policy maker with referents against which the possible significance of the focal risk may be weighed. In concept, at least, the analytic activities associated with risk evaluation are intended to result in conceptually sound bases for answering the central question and to provide considerable openness to this aspect of policy making. As observed by Fischoff et al. (1980), "by detailing every step of their work, from problem definition through value and fact assessment to bottom-line calculations, good analysts maximize the possibility for peer review and political review."

Technology Assessment

Joseph Coates (1976) has suggested that:

> technology assessment is the name for a class of policy studies which attempt to look at the widest possible scope of impacts in the society of the introduction of a new technology or the extension of the established technology in new and different ways. Its goal is to inform the

> policy process by putting before the decision maker an analyzed set of options, alternatives and consequences . . . [it] is extremely wide-sweeping in scope; it is not the decision process itself but only one input into that process.

As observed by the Office of Technology Assessment of the United States Congress (1977), "The objective of each assessment is to provide an early appraisal of the probable impacts and uncertainties of technology programs, so that both beneficial and adverse factors can be identified and considered in the legislative planning process. Both near-term and longer-term effects, whether intended or unintended, are examined, as are the diverse interests and viewpoints of the many different stakeholders who will foreseeably be affected by the technology."

D. E. Kash (1977) has observed that "technology assessments are a class of policy studies which assume that technologies cause or have significant influence on social change." Kash suggests that four important questions are asked by assessors of technology: (1) What are the consequences resulting from the use of the technology, as distinguished from the impacts that are used to justify its development? (2) Do the stakeholders see themselves as affected beneficially or adversely, and how intensely? (3) What can be done to enhance beneficial impacts? (4) What are the sociotechnological options that can modify impacts in the direction desired by policy makers and affected parties?

Armstrong and Harmon (1977) reviewed numerous definitions of technology assessment and concluded that all the examined definitions arise from two basic assumptions: (1) that the implementation of a new technology or the expansion of an old one is (or should be) a conscious societal choice, and (2) that it is not technology itself that is inherently harmful to society, but rather the consequences of the decisions concerning the ways in which the technology will be applied and managed. They classified these assessments into two major types: problem oriented and technology oriented. These authors further stated that a problem-oriented technology assessment focuses on one or more central problems and then considers alternative candidate solutions to each problem. In contrast, a technology-oriented assessment focuses on the future impacts of a single technology along alternative paths and seeks to determine associated side effects. According to these authors an assessment of this type tends to focus on one of three classes of technology: a *physical technology*, a *management technology*, or a *combination of the two*. The technologic focus may be further characterized by examination of an existing or emerging technology.

Whatever the focus of a technology assessment, it seems clear that its primary purpose is to provide informed, comprehensive inputs to the public policy system while making explicit the value orientations within which decisions are to be made.

Problem Analysis

This special type of analysis may be viewed as the interface between technology assessments, public policy analyses, and the acts of public decision making that are the raison d'être for the entire set of activities. From our perspective, problem analyses identify (a) problem states of affairs; (b) networks of causes and effects to which these states are linked; (c) underlying values associated with the states; (d) value holders and problem-experiencing populations and areas, and (e) candidate targets and methods of intervention.

As a formal set of methods, problem analysis is based on the view that the states of affairs conventionally referred to as "problems" are as much a product of human values and perceptions as they are empirical reality.

Because these states of affairs require human valuing, they must be viewed as being potentially different from the states of affairs that may be identified objectively as "impacts," "consequences," or "effects." As the term is used here, an *impact* is an empirically verifiable consequence of a defined cause. An impact is inherently neither good nor bad. In contrast, a *problem* is a situation in which one or more attributes of an observer's perceived reality are seen as being incongruent with certain aspirations and/or values.

To the extent that *problem analysis* includes the rigorous specification of stakeholder groups and the group-related values that might lead them to define some aspects of a situation as constituting "problems," we have little doubt that problem analysis is not yet a well recognized or practiced area of professional specialization. Instead, what passes for problem analysis is the cluster of professional activities targeted on identification of health, safety, and other *impacts* or *effects* that occur in specific situations and the explication of the networks of causes to which these states are linked. The identification and description of the populations at risk of experiencing such effects have also been a part of such activities. Epidemiologists, human ecologists, and others have been foremost in this area.

Of course, the preparation of problem statements is the function of many specialists. Most analysts, no matter what their disciplinary orientation, have found the act of problem analysis to be a necessary tool. As Martino (1972) has noted:

> The almost universal experience of operations researchers, management analysts, and forecasters of all varieties, when called in to assist in solving some problem, is that the initial statement of the problem is at best incomplete and often completely erroneous. Thus, it becomes their first task to take the initial statements of the problem and find out what the real problem is. Upon reflection, it will be seen that the situation is almost inevitable. If the nature of a problem is obvious, the services of an analyst are not needed. Hence, the practitioner of any variety of rational analysis must expect that the problems he is presented with will be poorly structured, incompletely defined, and difficult to interpret. Thus, one of the functions he performs for the decision-maker is to clarify the true nature of the problem.

Policy Analysis

In its contemporary form, policy analysis is considered to be a component of the broader field of policy sciences. It is concerned both with the performance of *descriptive* and *substantive* studies. The former are aimed at identifying the policies and policy structures or systems that are already in place and may influence the success or failure of some focal problem-solving effort, and further are targeted on interpreting legislative policies as a necessary first step toward the design of their implementing systems.

Heavily based in law, political science, and administrative systems analysis, descriptive policy studies are targeted both on process and "what is." *Substantive policy analysis*, on the other hand, is oriented to the content of proposed or existing public policies, their efficacy in dealing with specified problems or in achieving specified goals, and with the efficiency they exhibit in terms of such goal-accomplishing activities. Dror (1971) has suggested that policy analysis involves "an integration between . . . political science and public administration, on the one hand, and systems analysis, decision theory, and economic theory, on the other hand, within the broad framework of policy sciences." He suggests that a main emphasis of policy analysis is its attention to the political aspects of policy making and to the creative search for new policy alternatives utilizing both quantitative data as well as: "Gestalt images, qualitative models, and qualitative methods. . . . This involves imaginative thinking, systematic integration of trained intuition into policy analysis (e.g., through the Delphi method), development of qualitative tools (such as metaphor construction, scenarios, counterfactual thinking) and construction of broad qualitative models of complex issues . . . futures thinking with long-range predictions, [specification of] alternative states of the future, and speculative thinking on the future as essential background for current policy-making."

Policy analysis *begins* once problems have been specified. Focusing on the possible targets of intervention in problem-causing systems, policy analyses identify (a) alternative problem-solving strategies and policy choices; (b) criteria for the selection of policy alternatives; (c) decision-influencing constraints; (d) estimates of the payoffs, outcomes, and side effects that may be expected from each policy alternative, and (e) issues of fact and value, the protagonists and antagonists associated with each, and the networks of arguments and assumptions advanced by each.

An Integrative Model

We make no claim, of course, that any one of these areas of professional application need "do it all," or that any one professional within any of these areas has expertise in handling all the operations associated with his own area of professional practice. What we do argue, however, is that the central actors in each of these areas must understand the ultimate, *total system* uses to which the products of their labors will be put, the potential linkages between their own activities and those of other specialists, and the modifications in output formats and language (whether symbolic, mathematical, or in conventional narrative form) compatible with the needs and understanding of other actors within the total system. Further, we suggest that the sum of the outputs from the four areas must ultimately be subjected to the test of whether they have met their essential justification for existence, that is, the removal of ambiguity, uncertainty, and irrationality from the policy-making and public problem-solving system.

To this end, it is useful to employ a *normative model* that identifies both the major chunks of activities and the major types of products that can usefully be built into policy-assisting activities. A simple version of the model is presented in Figure 1-3. Consisting of twenty-seven identified operations, all but two of the model elements (23 and 26) involve information gathering, synthesis, and analytic contributions from one or more types of professional supporters of the policy-making system.

As an intendedly value neutral policy-assisting system designed to provide all participants in a policy-making process with the underlying assumptions, conclusions, and value propositions for any proposed problem-solving solutions, the operations depicted in Figure 1-3 can be initiated at any one of a number of starting points. Although the design suggests that the full sweep of operations begins with the performance of Model Elements 1–3, virtually any point in the system can mark the beginning of such a study cycle. Some such cycles are inaugurated on the basis of tentative definitions of suspected problems, others on the basis of concern over some perceived change in the quality of the environment, and still others by concern for anticipated changes in technology. Also, some operations may be and usually are performed at different times in the sequence than others, and any given set of such operations may further be performed by individuals and research teams unknown to each other. The major features of the model follow.

Risk Analysis

The completion of the risk estimation sequence of operations begins with the identification of some factor or variable of interest to the analyst or client. Identified in the model as the risk identification operation, this step typically focuses on some hazard (the occurrence of an earthquake or flood), some potential risk-producing activity (cigarette smoking, emission of sulfur oxides from coal-fired steam power plants, etc.), or some pathophysiologic or psychosocial state of health. The latter usually are expressed in the form of a rate of occurrence greater than some specified norm or as a trend in such states (rates of injury and morbidity in specified occupations, escalations in the trend of lung cancer death rates, the occurrence of escalated leukemia death rates among workers exposed to benzene in the workplace, etc.). The suggested operations then proceed to analysis of the biophysical and socio/cultural causal system networks leading to the occurrence of the identified risk variable, to a comparable analysis of the network of effects on individuals, groups, and land areas suspected to be associated with the focal subject variable, and to an impact analysis that identifies the specific areas and populations at risk of experiencing the effects associated with the variable.

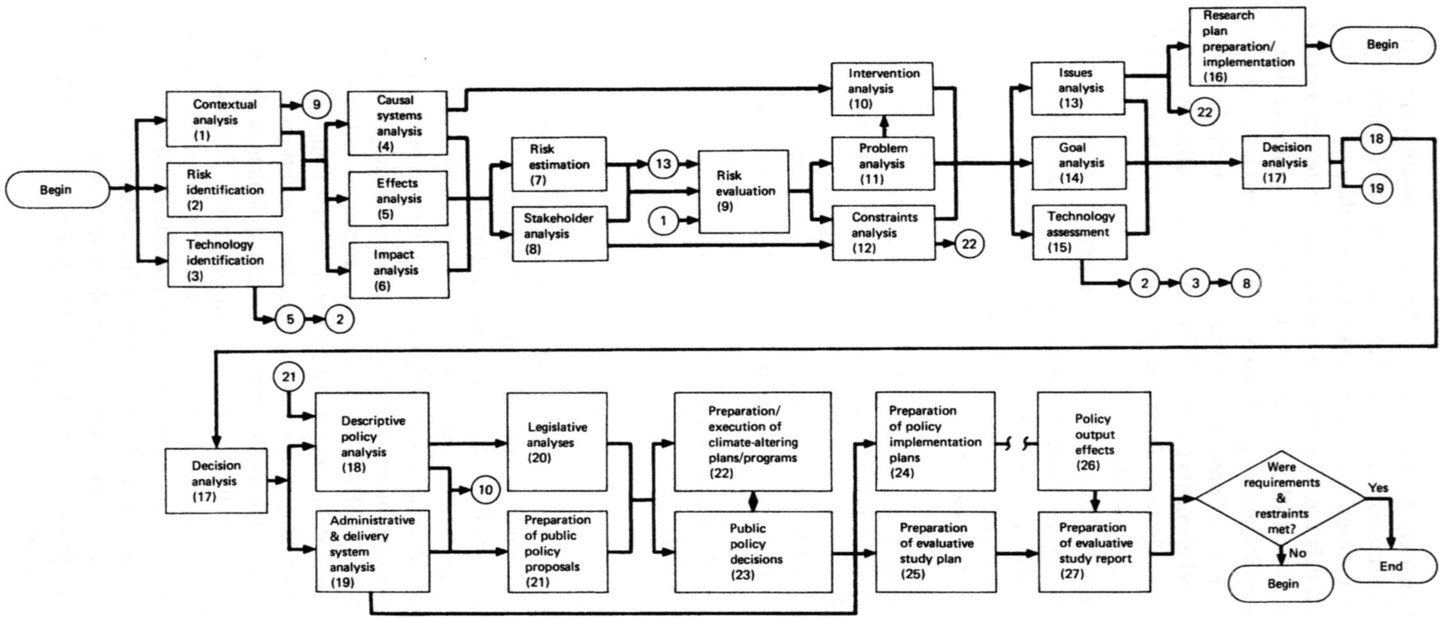

Figure 1-3. Model of value neutral policy-assisting system.

It is helpful to use diagrams to chart both *causal* and *effects* networks. By footnoting the connecting lines on the resulting diagrams, models of suspected causal and effects-producing systems can be prepared in a concise form, instantly exposing the assumptions of the analyst to review by other concerned experts. Logic errors, oversights, and feedback relationships become clarified by such techniques. An example of such a flow chart is presented in Figure 1-4. In addition, the resulting models are in convenient form for further, more quantitative, and computer-based analysis. The balance of the operations associated with the Risk Estimation cycle conform to standard contemporary professional practice, as described above.

Risk estimation can be, and frequently is, performed as a comparatively free-standing set of activities disconnected in time from other operations within the model. Thus, current federal air quality management legislation, which includes elements of statutes adopted in the 1960s, requires federal agencies to develop *air quality criteria* that, in practice, are defined as predictive or descriptive statements whose purpose is to describe the relationship between specified ambient burdens of air pollutants and the occurrence of specified effects on human beings, physical facilities, growing plants, and other elements of the human ecosystem (Faith and Atkisson, 1972). To our knowledge, the first air quality criteria were adopted by the California State Department of Public Health under a legislative mandate deriving from a policy request first made to the legislature by Governor Edmund G. Brown in the early 1960s. In respect to the criteria issued by the department to deal with the phenomenon of photochemical smog, the criteria were expressed in probabilistic terms and targeted on identifying the size and characteristics of the population groups within the state that would be affected by varying levels of photochemical smog products, and on identifying the frequency with which such exposures would occur.

This historic effort was not, however, mimicked by the air quality criteria later adopted by the U.S. Environmental Protection Agency and its predecessor entities with the U.S. Department of Health, Education and Welfare. Most air quality criteria developed by the federal government simply state the levels of specific pollutants at which certain population segments will begin to experience some specified effects. Little effort was directed at the time of their promulgation, or in subsequent periods, to a more solidly risk-oriented approach to specification of the impacts anticipated from the exposure of population sets of varying sizes to pollutant levels of varying magnitudes and frequencies. Similar disabilities are to be found in many of the criteria documents issued by the National Institute of Occupational Safety and Health. Thus, we define risk estimation as including both those operations targeted on the immediate resolution of some public problem or public policy question as well as those targeted on the development of environmental health, occupational health, and other criteria statements, which may or may not lead to direct problem-solving and/or regulatory activity within the immediate future.

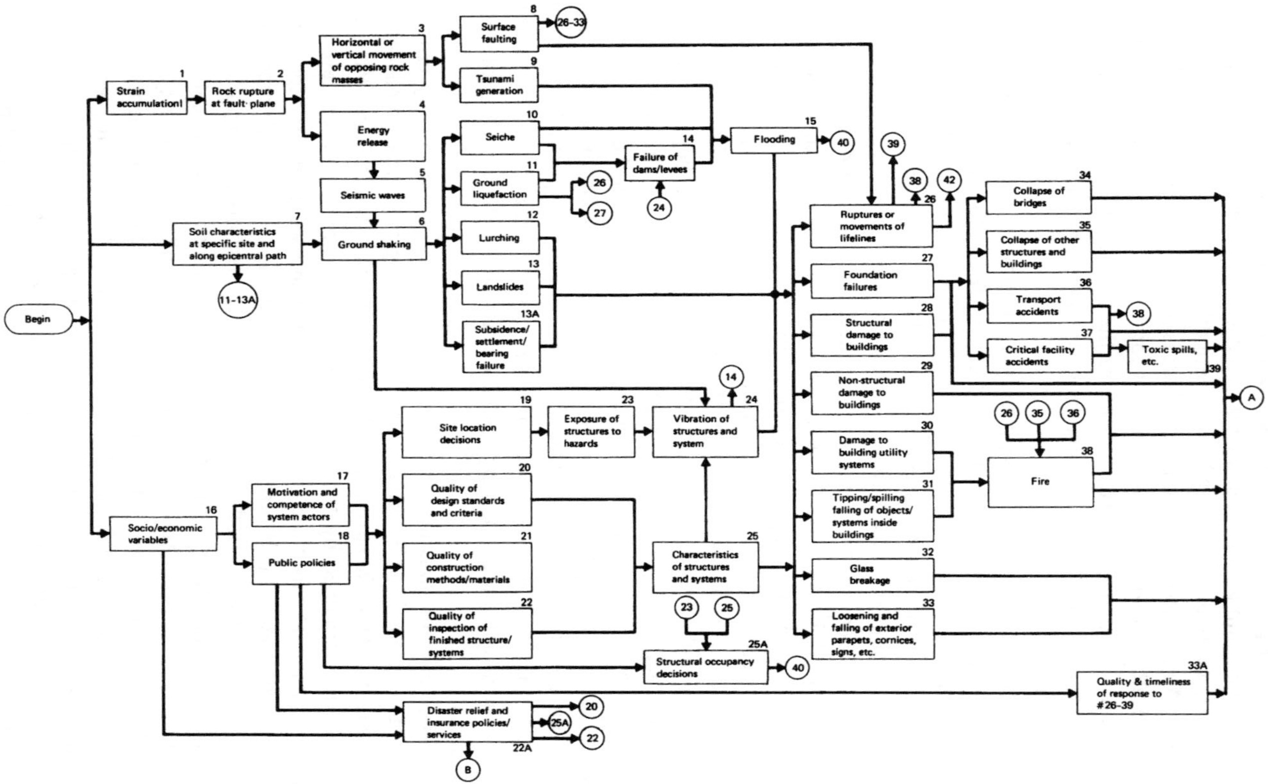

Figure 1-4. Illustrative network of earthquake-related events.

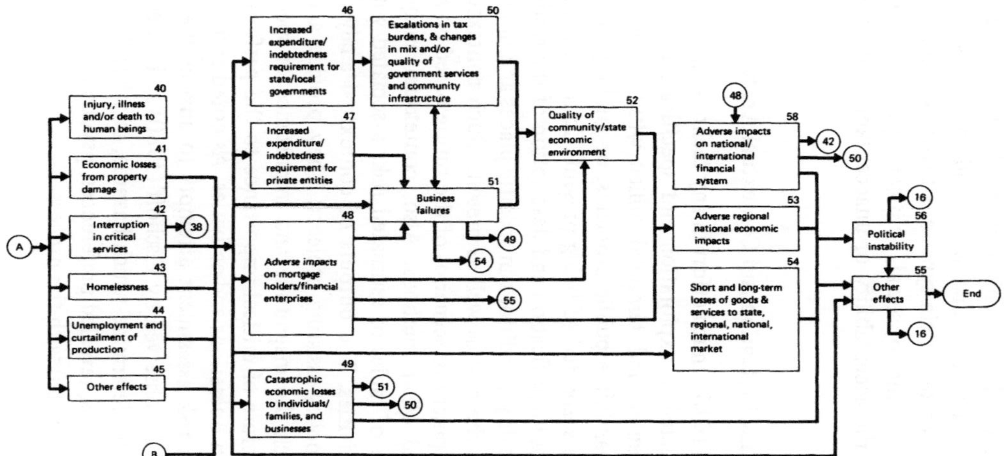

Figure 1-4. (continued)

Stakeholder and Constraints Analysis

This portion of the model is intended to represent those activities targeted on identification of the stakeholder groups whose interests are involved in situations in which some set of impacts or effects occur or in which the effects, costs, and benefits of efforts to control the occurrence and distribution of such impacts or effects will be felt. These operations elements are further intended to focus on identification of the value commitments of these several groups, the identification of group-related factors that would aid in determining the extent to which each would designate certain types of situations as constituting problems, and an assessment of the range of social, political, legal, and economic constraints that might influence the acceptability or unacceptability of proposed problem-solving strategies.

Risk Evaluation and Contextual Analysis

According to Coates (1976), much effort in any technology assessment or policy analysis must be directed at exploration and specification of the attributes of the social, environmental, and temporal milieu within which a particular set of impacts, effects, candidate public problems, or public policy questions have been identified or are suspected of being present. Contextual analyses therefore extend to the rigorous identification and description of the reality of morbidity, mortality, and life quality conditions in specific places and in respect to specific population subsets in the past as well as present. Such contextual analyses may also extend to development of scenarios of the future based on statistical examination of trends and other methods. Also, contextual analyses include the development and use of tables of estimated risks associated with various kinds of exposures and activities. The outputs of contextual analyses can serve the needs of numerous policy-assisting projects: They provide a framework, or a baseline, against which the magnitude and importance of specific risks can be assessed and are therefore of critical importance to the performance of project-related risk evaluation operations and activities. As noted earlier, the risk evaluation component of a risk assessment attempts to answer the question, "How safe is safe enough?" As observed by Fischoff (1980) and others, the evaluation component attempts to determine which hazards, in what form, society should tolerate. Referent analysis, revealed preferences, psychometric studies, and other techniques may be associated with the process.

Problem Analysis and Intervention Analysis

The several operations and activities represented by this section of the model are those that result in the preparation of formal problem statements that identify the several candidate intrinsic and instrumental problems associated with the focal set of risk factors, variables, or impacts, and that further relate each candidate problem to the interests of each of the identified groups of stakeholders. Such statements also are intended to contain a summary of both the *causal* and *effects* system networks revealed by earlier phases of the analysis, and, in reference to the output of these analyses, to identify the problem-causing factors that are the candidate targets for intervention in the problem-causing system. Typically, problem analyses should further extend to at least a preliminary identification of the candidate list of methods appropriate for dealing with each of the identified intervention targets. The importance of this set of operations cannot be overemphasized. It is intended to establish the framework within which all subsequent activities will be conducted and to guide the design and conduct of all prior information collection and analysis activities. This set of operations should lead to a companion set of documents describing the central issues of fact, value, and law surrounding the identified problems, and the candidate list of goals from which the subsequent goal plan is to be developed.

Issues Analyses and Research Plan Preparation/Implementation

Only rarely does the policy-making system generate formal documents that crisply identify the issues of fact, value, and law germane to resolution of the policy questions; and even more rarely are formal documents prepared that identify antagonists and protagonists associated with each issue and the arguments and assumptions on which their respective positions are based. The test to which any issue analysis should be put is whether it serves equally well the interests of both the protagonists and antagonists whose views are embedded in an identified issue. Moreover, such analyses should contain extensive and critical reviews of the literature on which the disputants rely. Recognizing that issues of fact can only be resolved through the use of the hypotheses being questioned, and through the generation of information necessary to test such hypotheses, issues analyses also should include at least summary descriptions of the research plans whose implementation is necessary to the resolution of the focal issues of fact. Also, since the exigencies of the public policy system may require policy makers to act before the scientific community has resolved the most pressing issues of fact, a formal issues analysis can sometimes profitably include decision trees whose design will permit policy makers to trace the factual outcomes associated with the acceptance or rejection of network assumptions. Similarly, although science cannot resolve issues of value, issues analyses may nonetheless identify value conflicts associated with the arguments of each party involved in an issue; and may further, through the use of acceptable facts and scientific procedures, state the probable consequences or effects that can result from either an affirmative or negative resolution of value issue.

Finally, the formats should require that all issues statements be composed as "yes/no" questions. Issues of fact are those that ask "is," "will be," or "does" questions. On the other hand, issues of value ask "ought" questions. Issues of law are a special subset of issues of fact since they focus on predictions on how the courts will decide.

Goal Analysis

In terms of the relationships between the policy-assisting professional and the policy maker or policy-influencing client, the goal analysis serves a critical function in stimulating the development of a goal plan that fixes the directions in which policy decisions are to move the system. The goal analysis, therefore, is envisioned as involving specification of alternative sets of goal statements, each set being distinguishable from others by its representational focus, viz., on the basis of which stakeholder interests and which ascendant values are to be served by the selected problem solution, and which are to be ignored or rejected. Using outputs of the constraints analysis and other earlier phases of the total inquiry, the goal analysis should specify the probable consequences associated with the selection of one goal set over another and of individual goals within subsets.

Both goal analyses and goal plans may be disconnected from the performance of a specific problem-targeted project. Like risk estimation, when cast in the form of criteria statements, goal analyses and plans may be comparatively free-standing elements intended to contribute to the shaping of a total system.

Decision Analyses

Decision analyses are studies that (1) begin with reference to some formal problem statement(s) and some agreed-on goal plan(s); (2) rigorously define (through development of a strategy plan) the alternative targets and methods for intervening in problem-causing systems; (3) state the criteria through which the acceptability of alternative strategies will be judged; (4) test the sensitivities of system outcomes to implementations of one or more of the selected decision alternatives, and (5) further weigh the benefits and costs of the alternative choices under examination. Model building, operations and systems analysis, benefit/cost analysis, sensitivity analysis, multiattribute analysis, and the other conventional decision-assisting methods are subsumed within this function.

Typical textbook discussions of decision-assisting methods begin with a statement that such methods are intended to assist individuals in making optimal choices from a set of competing alternatives. Rarely do such discussions specify where these alternatives come from. Thus, we have provided a partial answer to this question. That is, the problem analysis phase of the total operation will itself lead to at least gross specification of a range of candidate targets and methods for intervention in a problem-causing system. However, the decision analysis phase of the total cycle of operations should initially concentrate on development of a rather wide-ranging candidate strategy plan built around a normative model.

A normative or prescriptive model is a representation of what ought to be. Such a model is developed without attention to the constraining influences of politics, economics, technologic feasibility, and other factors. In effect, such a model sets up a "straw man" approach to the solution of a specified set of problems and serves the purpose of leading the analyst away from conventional wisdom to more constraint-free specification of alternatives. Such models should stand at the core of the strategy plans developed by analysts, and provide, in turn, the alternatives examined through the use of formal decision-assisting methods.

Technology Assessment

The broad definition of technology assessment implies that it includes all the elements in the model presented as Figure 1-2. Although we do not disagree with such a definition, as used in this model the term is more narrowly focused on a set of studies aimed either at identification and assessment of the technologies whose use may lead to the occurrence of candidate public problems, or to the identification of technologies (present and future) that may be incorporated into candidate problem-solving strategies. In the latter sense, such assessments obviously should focus on identification of the side effects that may be associated with use of the selected technology.

Administrative, Descriptive Policy, and Legislative Analysis

According to Coates (1976) the performance of technologically based policy analyses requires the identification of the decision systems whose outcomes either influence the occurrence of the focal public problem(s), or influence the design and implementation of optimal solutions. Descriptive policy analyses are aimed at meeting this need, and are concerned with the identification of those existing organizational structures and public policies which define the boundaries of problem relevant reality. Such analyses are aimed at identifying the current set of legally sanctified goals, policies, implementing programs, and related administrative or regulatory institutions germane to the focal inquiry. Such analyses identify the legally required decision routines and processes associated with such policies. On the other hand, administrative and delivery systems analyses are focused on identification of the networks of administrative institutions whose activities and legal responsibilities derive from the public policies identified in the descriptive policy analysis, and whose activities, orientations, prejudices, representational foci, and expertise may be germane either to the occurrence of the problem states or to the implementation of solutions.

Such analyses are pointed also at identifying the internal information collection, processing, and analysis routines of these institutions as well as the central characteristics of their internal decision processes. Typically, such analyses result in matrices linking specific types of activities, decision processes, and legal responsibilities germane to the inquiry to the appropriate administrative institution or regulatory agency. These may be referred to as functional-organizational or functional-authority matrices.

At the stage in the policy-making process in which legislative proposals actually are drafted, submitted to appropriate committees, and acted on, another special type of descriptive policy analysis is required also. This has been described as legislative analysis. Legislative analysis examines the content of *competing* legislative proposals. Through the use of cross-impact matrices and other methods, such analyses determine the commonalities and distinctions between competing policy proposals, and examine the consistency of each to some overarching set of decision-influencing criteria.

Policy Implementation Planning and Program Evaluation Planning

This stage in the policy-making process is devoted to objective assessment of the legislature's policy intentions, as revealed in statutes, court opinions, and their predecessor documents. Such efforts also make use of analyses of administrative and delivery systems. They result in the assignment of organizational/institutional responsibilities as necessary for the implementation of legislative intent, in the modification or invention of policy-implementing administrative and information-development routines, in the fashioning of interorganizational coordination plans, and in the development of resource allocation (budgetary, staffing, and logistics support) requests as necessary to secure the timely and effective accomplishment of legislative ends. It is at this stage in a rationally conceived policy-making and policy-implementing system that criteria also are developed to guide subsequent evaluation of the total sweep of policy-implementing activities. Where possible, formal evaluative research plans should be developed to guide such efforts.

Preparation of Climate-Altering Plans/Programs

In the earlier description of the attributes of those states referred to as "problems," it was noted that problems are as much a product of values and reality perceptions as they are of empirical reality. Public misperceptions of "what is" frequently leads policy makers into situations in which they are asked to deal with problems that may result more from myth than reality. When such situations occur, the educational mission of legislators becomes of ascendant importance. Further, the need to alter the "political climate" within which these institutions and individuals operate becomes particularly acute. Accordingly, the development and implementation of climate-altering plans/programs/strategies may occupy the center stage of a legislator's concern and require the types of information and risk assessment outputs generated earlier in the cycle.

Conclusion

The foregoing systems model was presented as an approach to identifying the problems encountered in arriving at a level of acceptable risk. Evaluation of risk is, of course, only part of the decision-making process when setting natural hazard policies. A rational decision maker must, as part of the process, evaluate the benefits in terms of changes in risk/or regulatory impacts and weigh the benefits against the costs, while fully recognizing the increasing degree of uncertainty associated with highly complex sociotechnical processes. Thus, the process is often more intuitive than explicit. The point to be emphasized here is that risk analysis is only a part and not the totality of the decision-making process. Considering all the problems of risk analysis, it should be understood that it is not necessarily the most difficult part of the policy-making process.

Risk analysis is that part of the process that many characterize as objective. It must be recognized that objective measurements of risk must, as with all intellectual endeavors, involve some element of subjectivity. The very choice of what questions to ask and issues to consider, as well as methods to be employed, involves judgment, especially in the face of large uncertainties. This choice is, therefore, subjective and in part based on the analyst's judgment of what may be the most significant factors. Problems encountered in the application of risk analysis have frequently been attributed to attempts to reduce risk analysis to a mechanistic application of methods or techniques. The purpose of the foregoing systems approach has been to clearly demonstrate that an application of purely mechanistic techniques is not satisfactory in the formation of public policy regarding the acceptability of risk.

There are a number of factors that must be considered in the establishment of acceptable risk criteria and standards. First, any approach should be evaluated in terms of its effectiveness and efficiency in achieving the desired reductions in mortality and morbidity and impacts on the general environment. Second, there must be a concern shown for certain kinds of deaths such as those viewed as extraordinary, frightening, or painful. Third, there must be concern for political and public perceptions of risk, as well as issues of equity. Specifically, the manager needs to address the questions of who pays, who benefits, who is risked, and who volunteers to take selected risks. The normative model presented here as an approach to achieving rational environmental risk management has been designed to consider these factors and bring them to focus within an overall decision-making system.

We have attempted to apply the basic elements of this systems model in performing this analysis. Thus, Chapters 2 and 3 discuss the characteristics of the natural hazard problem and an analysis of the policy approaches applied to the management of the hazards. Part II provides the results of a risk analysis, and a mitigation cost and impact analysis. Part III focuses on stakeholders, problems, and constraints on the decision-making system, and finally presents a set of recommended policy options.

References

Anderson, J. E. 1975. Public Policy Making. Praeger Publishers, New York.

Armstrong, J. E., and W. W. Harmon. 1977. Strategies for Conducting Technology Assessments. Department of Engineering–Economic Systems, Stanford University, Stanford, California.

Coates, J. F. 1976. The role of formal models in technology assessment. Technol. Forecast. Soc. Change. 9(1/2):139–190.

Dror, Y. 1971. Design for Policy Sciences. Elsevier, New York.

Faith, W. L., and A. A. Atkisson. 1972. Air Pollution. Wiley, New York.

Fischoff, B., S. Lichtenstein, P. Slovic, S. Derby, and R. Keeney. 1980. Approaches to Acceptable Risk (report to the Nuclear Regulatory Commission). Decision Research, A Branch of Perceptronics, Eugene, Oregon.

Foster, H. D. 1980. Disaster Mitigation for Planners; The Preservation of Life and Property. Springer-Verlag, New York.

Frazier, K. 1979. The Violent Face of Nature: Severe Phenomenon and Natural Disasters. Wm. Morrow & Co., Inc., New York.

Gardner, J. W. 1970. Recovery of Confidence. W. W. Norton, New York.

Jones, C. O. 1977. An Introduction to the Study of Public Policy. Duxbury Press, North Scituate, Massachusetts.

Kash, D. E. 1977. Testimony before the U.S. Department of Internal Hearings on Technology Assessment, June 8, 9, 10, 14, 1976. *Cited in* Armstrong and Harmon.

Martino, J. P. 1972. Technological Forecasting for Decision-Making. Elsevier, New York.

U.S. Congress. 1977. Technology Assessment in Business and Government. Joint Office of Technology Assessment. U.S. Govt. Printing Office, Washington D.C.

2

Natural Hazards Characteristics and Mitigations

Natural hazards are defined as those events that occur in nature and are capable of producing injury or death to people and/or damage to property. Human activities may alter the frequency with which such events occur, increase or reduce their severity or intensity, alter the size of the area(s) impacted, influence the rate of exposure of persons and/or property, and influence the vulnerability of hazard-exposed persons and property. Vulnerability-altering activities always are referred to as "mitigations," but this term also applies to all other activities intended to alter hazard exposures, frequencies, intensities, and impacts.

Earthquake

Earthquakes are natural events that involve the moving or shaking of the earth's crust, and are believed to be caused by the release of stresses accumulated as a result of rock rupture along opposing fault planes in the earth's outer crust. One source of such ruptures is the continuous collisions between the ten major and several minor plates comprising the earth's 10- to 50-mile thick outer and floating crust. Earthquakes are measured in terms of their magnitude or intensity, and by such specific physical parameters as ground acceleration, ground velocity, duration of shaking, and the interval between seismic waves.

The most common measure of earthquake magnitude, the Richter Scale, employs a nine-point logarithmic scale to present seismographic recorded vibrational amplitude and energy release. Each of the nine magnitude levels corresponds to a tenfold change in vibrational amplitude and a 31.5-fold change in energy release. In general, the greater the magnitude of an earthquake, the greater its damage potential. The quantities of TNT necessary to produce releases of energy comparable to that of various Richter magnitude earthquakes are depicted in Table 2-1.

The direct and indirect effects caused by an earthquake are referred to as "seismic effects" and reflect the quake's intensity. The most common system used to describe quake intensity is the Modified Mercalli Intensity (MMI) Scale. Structured around twelve levels of severity, the MMI Scale assigns a value of I to a barely perceptible quake, a value of VI to a quake in which some momentary measured damage may occur, and a value of XII to one in which virtually total destruction may occur. In its current form, the MMI Scale (Table 2-2) contains descriptions of effects associated with various intensity quakes.

Seismic effects from earthquakes include ground displacement resulting from surface faulting or other abrupt earthquake-related land level changes; damage from ground shaking; ground failure such as lurching, soil liquefaction, differen-

Table 2-1. Energies of Earthquakes (Richter Magnitude 1.0–9.0)

Richter earthquake magnitude	Approximate earthquake energy (T.N.T.)
1.0	6 ounces
1.5	2 pounds
2.0	13 pounds
2.5	63 pounds
3.0	397 pounds
3.5	1,990 pounds
4.0	6 tons
4.5	32 tons
5.0	199 tons
5.5	1,000 tons
6.0	6,270 tons
6.5	31,550 tons
7.0	100,000 tons
7.5	1,000,000 tons
8.0	6,270,000 tons
8.5	31,550,000 tons
9.0	199,000,000 tons

Energies of some major earthquakes

Location	Date	Energy (tons Richter Magnitude of T.N.T.)	
Anchorage, Alaska	1964	8.5	31,550,000
San Francisco, California	1906	8.2	12,550,000
Kern County, California	1952	7.7	1,990,000
El Centro, California	1940	7.1	250,500
Long Beach, California	1933	6.3	15,800
San Francisco, California	1957	5.3	500

Source: AIA Research Corporation (1975).

Table 2-2. MMI Scale of 1931, 1956 Version

I.	Not felt. Marginal and long-period effects of large earthquakes.
II.	Felt by persons at rest, on upper floors, or favorably placed.
III.	Felt indoors. Hanging objects swing. Vibration like passing of light trucks. May not be recognized as an earthquake.
IV.	Hanging objects swing. Vibration like passing of heavy trucks; or sensation of a jolt like a heavy ball striking the walls. Standing motor cars rock. Windows, dishes, doors rattle. Glasses clink. Crockery clashes. In the upper range of IV, wooden walls and frames creek.
V.	Felt outdoors; direction estimated. Sleepers awakened. Liquids disturbed, some spilled. Small unstable objects displaced or upset. Doors swing, close, open. Shutters, pictures move. Pendulum clocks stop, start, change rate.
VI.	Felt by all. Many frightened and run outdoors. Persons walk unsteadily. Windows, dishes, glassware broken. Knickknacks, books, etc. fall off shelves. Pictures fall off walls. Furniture moved or overturned. Weak plaster and masonry D (weak materials) cracked. Small bells ring (church, school). Trees, bushes shaken (visibly) or heard to rustle.
VII.	Difficult to stand. Noticed by drivers of motor cars. Hanging objects quiver. Funiture broken. Damage to masonry D, including cracks. Weak chimneys broken at roof line. Fall of plaster, loose bricks, stones, tiles, cornices. Some cracks in masonry C (ordinary workmanship and mortar, not reinforced). Waves on ponds; water turbid with mud. Small slides and caving along sand and gravel banks. Large bells ring. Concrete irrigation ditches damaged.
VIII.	Steering of motor cars affected. Damage to masonry C; partial collapse. Some damage to masonry B (good workmanship, mortar and design; reinforced, designed to resist lateral forces). Fall of stucco and some masonry walls. Twisting, fall of chimneys, factory stacks, monuments, towers, elevated tanks. Frame houses moved on foundations if not bolted down; loose panel walls thrown out. Decayed piling broken off. Branches broken from trees. Changes in flow or temperature of springs and wells. Cracks in wet ground and on steep slopes.
IX.	General panic. Masonry D destroyed; masonry C heavily damaged, sometimes with complete collapse; masonry B seriously damaged. Frame structures, if not bolted, shifted off foundations. Frames rocked. Serious damage to reservoirs. Underground pipes broken. Conspicuous cracks in ground. In alluviated areas, sand and mud ejected, earthquake fountains, sand craters.
X.	Most masonry and frame structures destroyed with their foundations. Some well-built wooden structures and bridges destroyed. Serious damage to dams, dikes, embankments. Large landslides. Water thrown on banks of canals, rivers, lakes, etc. Sand and mud shifted horizontally on beaches and flat land. Rails bent slightly.
XI.	Rails bent greatly. Underground pipelines completely out of service.
XII.	Damage nearly total. Large rock masses displaced. Lines of sight and level distorted. Objects thrown into the air.

Source: Richter (1958).

tial settlement, landsliding, and abnormal water-wave action (i.e., seiche in the case of inland lakes or tsunami in the case of ocean tidal action).

In general, the potential for damage from earthquakes in any given area is proportional to the actual energy released (magnitude of the quake), the distance of the area from the source of the energy release (epicenter), the local rock and soil characteristics of sites within the area, and the vulnerability of property to the resulting quake intensity. The earthquake "hazard," therefore, is based on complex interrelationships between earthquake faults, weak geologic materials, and human activity in the earthquake-prone region.

When an earthquake occurs, the initial source of damage to buildings and other structures is a product of ground movements that cause structures to vibrate. During moderate ground shaking, the vibrations may be in the building's elastic range and exhibit no damaging amplitudes. During strong ground shaking episodes, however, building structural members may undergo plastic strains and the movement may therefore induce some cracking of these members. If severe enough, such cracking can result in the collapse of whole buildings and/or their major components. Also, wood frame buildings that are not securely anchored to their foundations may be jolted off their mounts.

In almost all major U.S. earthquakes, structures most seriously damaged by ground movements have been unreinforced structures built from sand-lime-mortar brick load-bearing walls. Multistory steel frame buildings have fared comparatively well, although extensive nonstructural damage has sometimes occurred.

Mixes of avoidance and building strengthening strategies may be used to reduce the risk of building damage and death in seismically active areas. Except for necessary utility and transportation lines, avoidance strategies call for *no* construction across fault lines and immediately adjacent areas. In such areas, land use should be confined to agriculture or recreation. Outside such areas, building strengthening strategies may be employed.

In the design of important structures, such as 50-story buildings, nuclear power plants, large dams, long suspension bridges, and offshore drilling platforms, complex dynamic analyses are required, which involve the specifications of the design spectrum, the damping to be used, and the allowable design stresses. Standards specifying the required resistance of buildings to quake-induced lateral forces have been included in the Uniform Building Code and in state and local building regulations. Construction techniques appropriate for meeting such standards have been discussed by Whitman et al. (1973), Culver et al. (1975), and others.

Landslides

A landslide is an event in which surface masses of slope-forming earth move outward and downward from their underlying and stable floors in response to the force of gravity (Flawn, 1970). Such movements include *falls, creeps, flows,* and *slides* and may be triggered by various factors (Table 2-3).

In any of these events, the movement of the sliding, creeping, falling, or flowing mass of earth occurs whenever the "shear resistance" of the mass is exceeded by the "shear stress." The former term refers to the friction-induced binding of the mass to its floor, while the latter refers to the pull of gravity on the mass. Shear stress may be heightened on a mass by steepening its slope or by increasing its weight, while shear resistance may be lowered by lubricating the interface between the mass and its floor.

Transitional and *rotational* slides are mass movements of earth that involve a distinct surface of rupture or zone of weakness separating the earth slide material from a more stable underlying material. Generally, slides of this type occur under conditions where (1) slopes are sufficiently steep to allow for lateral down-slope movement of materials in response to gravity; (2) there is an underlying zone of weakness at a potential surface of rupture; and (3) there is an introduction of a natural or man-made disturbing factor sufficient to initiate instability and movement. Rapidly moving large slides of this type have the capacity to completely destroy buildings, roads, bridges, and other man-made structures, and cause deaths if they occur in highly urbanized or developed areas.

Unstable or potentially unstable slopes are those areas susceptible to slides, falls, creeps, or flows. Topographic, climatologic, geologic, and hydrologic factors contribute to slope instability. Of these, the geologic properties of slope-forming materials are a primary factor determining the susceptibility of a slope to earth movements. Although such movements can occur in any type of rock material, certain bedrock formations or rock types exhibit a high susceptibility to such movement:

1. Many of the younger (Mesozoic and Cenozoic) igneous (granitic) and metamorphic rocks found in the western United States have undergone intense fracturing and subsequent weathering. Therefore, these younger rocks generally have a greater propensity toward landsliding than the older, less fractured igneous rocks commonly found in the eastern portion of the country (i.e., New England).
2. Mesozoic and Cenozoic sedimentary rocks generally contain large amounts of clay, especially montmorillonite. The presence of the clay material has a definite deleterious influence on slope stability.
3. Many of the Cenozoic volcanic rocks in the western portion of the country appear landslide prone in that they contain zones of montmorillonite (altered volcanic ash), in addition to being highly fractured and, in some cases, weathered.
4. Serpentines consist essentially of secondary minerals normally derived by alteration of magnesium-rich silicate minerals. These materials are frequently susceptible to landsliding due to inherent weaknesses.

According to Terzaghi (1950), slides are caused both by internal and external factors. Internal causes lead to sliding without any change in surface condition, and result primarily from an increase in pore water pressure and a concurrent decrease in cohesion. External causes include oversteepening of the slopes, addition of weight from materials placed along the upper portions of slopes,

Table 2-3 Factors and Agents Influencing the Occurence of Landslides

Name of agent	Event or process that brings agent into action	Mode of action of agent
Transporting agent	Constriction operations or erosion	1. Increase of height or rise of slope
Tectonic stresses	Tectonic movements	2. Large-scale deformations of earth crust
Tectonic stresses or explosives	Earthquakes or blasting	3. High-frequency vibrations
Weight of slope-forming material	Process that created the the slope	4. Creep on slope
		5. Creep in weak stratum below foot of slope
Water	Rains or melting snow	6. Displacement of air in voids
		7. Displacement of air in open joints
		8. Reduction of capillary pressure associated with swelling
		9. Chemical weathering
	Frost	10. Expansion of water due to freezing
		11. Formation and subsequent melting of ice layers
	Dry spell	12. Shrinkage
	Rapid drawdown	13. Produces seepage toward foot of
	Rapid change of elevation of water table	14. Initiates rearrangement of grains
	Rise of water table in distant aquifer	15. Causes a rise of piezometric surface in slope forming

Slope materials most sensitive to action	Physical nature of significant actions of agent	Effects on equilibrium conditions of slope
Every material	Changes state of stress in slope-forming material	Increases shearing stresses
Stiff, fissured clay, shale	Changes state of stress and causes opening of joints	Increases shearing stresses and initiates process 8
Every material	Increases slope angle	Increase of shearing stresses
Every material	Produces transitory change of stress	
Loess, slightly cemented sand, and gravel	Damages intergranular bonds	Decrease of cohesion and increase of shearing stresses
Medium or fine loose sand in saturated state	Initiates rearrangement of grains	Spontaneous lique-faction
Stiff, fissured clay, shale remnants of old slides	Opens up closed joints, produces new ones	Reduces cohesion, ac-celerates process 8
Rigid materials resting on plastic ones		
Moist sand	Increases pore-water pressure	Decrease of frictional resistance
Jointed rock, shale		
Stiff, fissured clay and some shales	Causes swelling	Decrease of cohesion
Rock of any kind	Weakens intergranular bonds (chemical weathering)	
Jointed rock	Widens existing joints, produces new ones	
Silt and silty sand	Increases water content of soil in frozen toplayer	Decrease of frictional resistance
Clay	Produces shrinkage cracks	Decrease of cohesion
Fine sand, silt, previ-ously drained	Produces excess pore-water pressure	Decrease of frictional resistance
Medium or fine loose sand in saturated state	Spontaneous increase of pore-water pressure	Spontaneous lique-faction
Silt or sand layers be-tween or below clay layers	Increases pore-water pressure	Decrease of frictional resistance

Table 2-3 (continued)

Name of agent	Event or process that brings agent into action	Mode of action of agent
Water (continued)	Seepage from artificial source of water (resevoir or canal)	16. Seepage toward slope 17. Displaces air in the voids 18. Removes soluble binder 19. Subsurface erosion

Source: Broms (1975).

added weight from increased moisture content, and seismic or man-induced vibrations.

Legal designation of these landslide-prone areas as greenbelts, open space, or agricultural lands provides for non-conflicting uses. If land development, however, is necessary, a careful evaluation of the seriousness and causes of geologic problems and a strict adherence to engineered design and construction procedures must be employed to secure correction of adverse conditions.

Human activities, of course, can lead to the occurrence of these slide-inducing external and internal causal factors. Such activities can be controlled by subdivision and building codes containing stringent soil testing, site grading, and site avoidance requirements.

Falls are common where there are cliffs or massive broken, faulted, or jointed bedrock; or where steep ledges are undercut by natural processes or human activities. Falls, then, refer to the falling of soil or rock masses where a sliding surface does not occur.

The most effective and least expensive way to mitigate the fall hazard is to avoid such areas entirely, since there is no way to completely eliminate possible damage. Thus, human occupancy in an active fall area is incompatible with the idea of risk reduction. Certain techniques can be employed to decrease the hazard; however, complete removal of all potentially unstable rock or material usually is not possible.

Flows are characterized by a situation where the surface material breaks up and moves down a slope and flows as a viscous fluid. Such types of flows are earth-flows, mudflows, debris flows, flow-slides, and spontaneous liquefaction. The difference between the various types is the rate of movement.

Flows become serious threats only when people choose to live in active flow areas. Often there is little opportunity to provide warning, since flows can occur as fast as a rising storm or a rapid increase in springtime temperature. The consequences of improper utilization of flow areas range from occasional inconvenience to death and total destruction of buildings. Few flow-prone areas

Slope materials most sensitive to action	Physical nature of significant actions of agent	Effects on equilibrium conditions of slope
Saturated silt	Increases pore-water pressure	Decrease of frictional resistance
Moist, fine sand	Eliminates surface tension	Decrease of cohesion
Loess	Destroys intergranular bond	
Fine sand or silt	Undermines the slope	Increase of shearing stresses

are suitable sites for permanent structures. The unpredictable and often rapid movement of a flow makes even the location of a semipermanent structure, such as a mobile home, hazardous.

Flows can be channeled, diverted, or dammed. Generally, the cost will be high relative to the amount of protection afforded. The principal difficulties associated with engineered structures to control mudflows are related to the great mass of material contained in the flow, thereby requiring strong, and consequently, expensive structures.

Creeps are slow, down-slope movements of an earth mass. Although not dangerous to life, creeps may signal the onset of a potentially dangerous slope condition and may require engineering control to prevent the earth mass from displacing structures or moving over transport corridors.

Expansive Soil

Although all soils experience some shrinking and swelling as a function of variations in the moisture content of their environments, excessive shrink-swell behavior can produce the cracking and/or failure of building walls, foundations, slabs, and veneers.

Expansive soil generally refers to soil and rock containing clay materials with the capacity of undergoing significant volumetric changes when subjected to variances in water content.

The degree of shrink-swell capacity of any given soil is related to the proportion of clay contained in the soil and more specifically to the proportion of active clay materials such as montmorillonite (betonite).

One measure of the shrink/swell potential of soils, the coefficient of linear extensibility (COLE) has been developed by the U.S. Soil Conservation Service. The coefficient represents an estimate of the vertical component of swelling in a natural soil clod. Soils with coefficients lower than 3% are judged to rank low in soil expansion capability; those between 3.0 and 6.0% are judged to be moderate; and those with coefficients greater than 6.0% are judged to rank high in this capability. The higher the coefficient, the more severe the shrink/swell behavior of a soil is likely to be; the greater the shrink/swell behavior, the greater the damage potential to buildings and other structures.

Expansive soils can be tentatively identified by visual observation and positively identified by appropriate laboratory testing. Although various visual methods for identification exist, only a professional soils engineer or engineering geologist should be relied on to identify this potential hazard. Jones (1976) has provided some clues to help identify naturally occurring, potentially expansive soils. These criteria for recognition are depicted in Table 2-4.

Table 2-4. Recognition Criteria for Expansive Soils

Dry conditions

- Soil hard and rock like; difficult-to-impossible to crush by hand.
- Glazed, almost shiny surface where previously cut by scraper, ditcher teeth or shovel.
- Very difficult to penetrate with hand pick or shovel.
- Ground surface displays cracks occurring in a more or less regular pattern.
- Surface irregularities, such as tire tracks, cannot be obliterated by foot pressure.

Wet soil conditions

- Soil very sticky and clinging; exposed soil will build up on shoe soles to a thickness of from 2–4 inches when walked upon for a short distance.
- Can be easily molded into a ball by hand; hand molding will leave a nearly invisible powdery residue on hands after they dry.
- A shovel will penetrate soil quite easily and the cut surface will be very smooth and will tend to be shiny.
- Freshly machine scraped or cut areas will tend to be very smooth and shiny.
- Heavy construction equipment, such as metal wheels and compacting rollers will develop a very thick soil coating that may impair their function.
- In semiarid areas having distinct wet and dry seasons expansive soils that have been undistrubed for 10–15 years or more may display a pattern of closed ridges spaced regularly on 10–50 foot centers.

Any condition

- Creep ridges (visible evidence of solifluction) on slopes; generally an indication of incipient slope instability as well as potentially expansive soil.
- Tops of fence posts tilted downhill; this may give obvious indications of downhill creep movement in situations where solifluction patterns are indistinct.
- Extensive visible cracking in walks, streets, driveways, patios and often in buildings; this does not always mean that expansive soils are present, but they often are when such symptoms are visible.

Source: Jones (1976).

Expansive soils are a nationwide problem and so common in urban areas that complete avoidance is generally unfeasible. However, the widespread distribution of swelling soils must be recognized and precautions taken to require special engineered foundations and floor system designs, as well as detailed maintenance procedures. Engineered design and soil stabilization are considered the only appropriate corrections of these adverse conditions. Thus, expansive soil damage can be minimized by a combination of five methods: (1) use of soil stabilization and moisture control techniques on building sites, (2) engineered foundation design, (3) well-planned site drainage systems, (4) landscaping to enhance drainage, and (5) careful attention to interior construction details.

Riverine Flooding

Periodic flooding of lands adjacent to streams and rivers has been an ever-present characteristic of the human geophysical environment. Caused by stream flows that exceed the capacity of the normal water course, riverine flooding involves the spillover of above-normal stream flows onto lands immediately adjacent to the normal watercourse. Those adjacent lands subject to such intermittent flow-determined flooding are referred to as the stream or river's floodplain.

Hydrologically, stream flow is measured in terms of thousands of cubic feet of water flow per second and a flood is defined as any stream flow event in which the flow exceeds the capacity of the normal watercourse. Such events are probabilistic phenomena. Assuming no change in the stream flow capacity of the normal watercourse, these events are a function of precipitation levels and water runoff volumes within the watershed of the stream or river. The recurrence interval of a flood is defined as the average time interval, in years, that may be expected to lapse between the occurrence of a flood of a specified magnitude and an equal or larger flood. For any given river or stream, a specific peak flow level is associated with each flood recurrence interval for the stream (flood return period). The return period of the event is an indicator of the magnitude of the event; that is, for a specified return interval the level of flood depth and velocity are the real measures of flood magnitude.

That portion of the stream's floodplain that will be inundated by spillover waters resulting from a stream flow level of a specified flood-return frequency is referred to by the name of that frequency interval. Thus, a flood having a two year return period will inundate all of the two year zone of the floodplain, and a flood having a 100 year return period will inundate all of the 100 year zone of the floodplain. For land use purposes, floodplains can be divided into six levels of flood classes, or six hazard zones: A through F (where A is the most hazardous, e.g., subject to most frequent flooding in any 100-year period, and F the least). These zones constitute the areas that would be inundated by the 2-5, 5-10, 10-25, 25-50, 50-100, and greater than 100 year floods. Along any section of any given river, local terrain characteristics and stream factors determine the geographic size of the floodplain, the size of any given zone within the floodplain, and the depth that floodwaters will reach over the lands comprising any given floodplain zone during a flood of specified magnitude (return period).

The damage sustained by structures during flood episodes is a function of (1) the type, strength, and elevation of the structure, (2) the depth of the floodwaters during the episode, (3) the force exerted against the structures by the moving floodwaters, (4) impacts of floating debris against the structures, and (5) the adverse effects of "wetting" on the structure and its contents as a result of the intrusion of floodwaters.

Measures for mitigation of flood hazards may be classified into three broad categories: (1) structural mitigations that involve the construction of area-protecting dams, dikes, levees, and/or the deepening or improvement of the normal water course (floodway), (2) avoidance mitigations that prevent or deter the occupancy of flood-prone lands, and (3) building mitigations such as the elevation, strengthening, and/or floodproofing of structures located with the floodplain. A possible fourth category involves reducing the rates of water runoff from land areas during storms. Obviously, as these runoff rates are increased by activities such as the paving of land areas, the "normal" flood flow is increased and the magnitude of the hazard enlarged.

Specific building mitigations suggested by Johnson (1978) for application to floodplain structures include installation of temporary or permanent closures for openings in structures, raising structures in place, constructing new structures on fill or columns, constructing small walls or levees around structures, relocating or protecting damageable property within existing structures, and the use of water-resistant materials in new or existing structures.

Storm Surge

A storm surge is an event in which the coastal water level rises above the level associated with normal tidal action. The most heavily affected areas of the United States are the Atlantic and Gulf Coasts. Increases in water level along the Gulf and Southern Atlantic coastline generally are caused by hurricanes and other less intense tropical storms. Along the northern Atlantic coastline storm surges are caused mainly by extratropical winter storms called "northeasters."

When these storms approach land, extremely large waves may be superimposed on the high waters moving toward the coast. The depth of the surge depends on interactions between the path, intensity, speed, and size of the storm, as well as the configuration of the sea bottom and coastline. Local conditions often cause a depth variability of several feet within a few miles (Friedman and Roy, 1966).

The tropical weather storms that produce storm surges usually are classified into three categories: (1) tropical depressions with maximum wind speeds less than 40 mph, (2) tropical storms with maximum wind speeds of 40 to 73 mph, and (3) hurricanes with maximum wind speeds in excess of 73 mph. The coastal storms of the northeast are extratropical in nature, with wind speeds below the hurricane level.

The most devastating storm surges are caused by hurricanes. Hurricanes originate in tropical climates where atmospheric conditions are entirely different from the temperate zone (Maldus, 1957). As a hurricane or coastal storm approaches the coast, the normal water level can be altered by five distinct processes (Harris, 1963):

1. The pressure effect–The low central core pressure of a hurricane or tropical depression causes the water level to rise. A pressure lrop of one inch of mercury will cause the water level to rise approximately 13 inches in the open ocean.
2. Onshore wind effect–The winds affecting coastal areas can be broken into two components: one normal (onshore) to the coast, and one parallel (alongshore) to the coast. Onshore winds from the normal component cause water to pile up at the shoreline. This effect, often called the "wind setup" may be particularly pronounced in shallow bays and estuaries.
3. Coriolis effect–The earth's rotation produces acceleration in a westerly direction on objects moving from the equator toward the North Pole. Thus, alongshore currents produced by alongshore winds can be forced toward the shore.
4. Wave effect–The energy possessed by surface waves moving in open waters is not significant, but as they approach coastal areas, the rapid decrease in water depth creates fast-moving waves. As they strike the shore, their momentum may cause runups with surge heights up to twice that of the wave itself. The waves may also cause a piling up of the water near the shore producing a "wave setup" analogous to the wind setup.
5. Rainfall effect–Hurricanes can dump as much as 12 or more inches of rain in 24 hours over large areas. This excessive rainfall can often cause a significant rise in coastal water levels, especially in estuaries, bayous, and other coastal wetlands.

Storm surges generally are associated with two types of damage: (1) still water effects associated with simple inundation of a structure or land area and, (2) wave action effects induced by the pounding of waves against a structure. Inundation (still water) damage is similar to that caused by riverine inundation of corresponding depths, except for the increased damage from the more corrosive salt water. Once a structure has been "salted" the hygroscopic nature of salt will forever keep the structure moist, causing continuing mildew, corrosion, and other damage to the buildings and its contents. The depth of the storm surge wave, as modified by local marine factors, determines the size of the area inundated, and the height of the inundation above mean sea level. The latter is the major factor in determining the "still water" damage that will be sustained by a structure constructed at any given elevation above mean sea level.

Storm surge mitigations include the same avoidance and building mitigations applicable to riverine flooding. In addition, the construction of seawalls protect whole areas within storm surge zones.

Tsunami

A tsunami, or seismic sea wave, is a large oceanic wave produced by an earthquake, submarine volcanic eruption, or large submarine landslide. The waves generated are formed in groups having great length from crest to crest and having long periods. In deep oceans, wave lengths may be 100 miles or more, while wave heights, from crest to trough, are only a few feet. Nevertheless, the intrinsic wave energy is impressively great.

As a tsunami enters shallow waters in the coastal areas, the wave velocity diminishes and the wave height increases. The water that runs up onto the land is measured in terms of runup elevation, the elevation above the tide level at the time of the tsunami. If the trough precedes the initial crest, the arrival of the tsunami is first noticed by a gradual recession of coastal waters; if a crest precedes, there is a rapid rise in water level. Waves following the initial wave can crest at heights of more than 100 feet and strike with devastating force. (Hwang and Divoky, 1971).

No actual measurements have been made of the forces exerted by tsunami waves when running up onto the land, and only a few estimates are available (Weigel, 1970). Damage resulting from tsunami is a result of two factors: (1) inundation and (2) the force exerted on structures by the impacting wave. Damage can be very great, and, in addition to death and building damage, extensive damage often occurs to boats that break and drag their moorings and pound against other boats in the docks, or against buildings as they are carried ashore. The forces associated with tsunamis are often so great that the only positive means of protection for most structures is to avoid areas subject to impact by a tsunami. In some areas of the Hawaiian Islands subject to tsunamis, very large and significant structures are being constructed, but designed to withstand tsunami forces. For example, the transtructural elements of the lower portions of the structures have been designed so that they are breakaway and expendable, while the structural support elements (i.e., vertical columns) are designed to withstand the force. The U.S. Government has set criteria for nuclear power plants requiring structures to be designed for and constructed to withstand the greatest possible tsunami.

In the last 50 years, more lives have been lost from tsunamis in the United States than directly from seismic activity. In 1946, a tsunami generated near the Aleutian Islands killed 173 persons in Hawaii, and in 1964 the tsunami caused by the Alaskan earthquake killed 126 people. Therefore, if man is to continue to live in the coastal zones, it is important that the risk associated with a tsunami be given consideration prior to coastal land development.

A first step in developing effective plans for mitigating the tsunami hazard is to define local impact or damage areas. Detailed studies of the shoreline, harbor areas, and ocean bottom configurations are necessary for the accurate determination of areas subject to tsunami inundation. Only when the risk areas have been determined should methods for protecting lives and property be instituted.

Tornado

The term tornado comes from the Spanish "tornada," meaning thunderstorm, and was derived from the Latin, "tornare," meaning to "make round by turning." These events generally are defined as violently rotating columns of air that are in contact with the ground. A tornado is visible as a cloud base or rotating dust cloud rising from the ground.

Tornadoes are the most violent windstorms or weather phenomena known to man. Rotating at velocities up to 500 mph, their funnel-shaped clouds can affect areas ranging from ¼ to ¾ of a mile wide and upward of 16 miles long. Extreme events have been known to travel over areas measuring up to one mile wide and 300 miles long. Tornadoes have been recorded in every state, but the Midwest and Southeast are the most vulnerable areas. While tornadoes have been known to occur throughout the year, weather conditions between April and June generally produce the greatest number. No other disaster associated with the weather strikes with the suddenness of a tornado.

Tornadoes may form in association with squall lines, with both isolated thunderstorms and thunderstorms accompanying frontal passages, and with hurricanes. As the storm system moves, one or more tornadoes may form at intervals along the path, travel a few miles, lift, and then appear farther down the track. Small vortices may travel around a main tornado axis with as many as five of these satellite vortices having been observed in a tornado at one time. The main vortex of the tornado has tangential, radial, and vertical airflow and an associated atmospheric pressure drop.

Lasting minutes or hours, tornadoes exhibit much higher wind speeds than hurricanes but affect smaller impact areas. One system for measuring the intensity of a tornado is the Fujita classification, which assigns a numeric value to a tornado based on its wind speed, broken down into ¼ mph increments. Still another, the Pearson Scale, uses an identical scheme to classify tornadoes by their path width and length. The combination of these two systems is the FPP classification (Fujita, 1970a, b and Fujita and Pearson, 1973).

For any building type, the expected damage from tornado exposure increases as a function of the Fujita rating of the occurrence. Thus, a Fujita-rated 5 tornado (F5) would be expected to cause the collapse of 65.0% of all exposed wood frame commercial and industrial structures while a Fujita-rated 1 tornado (F1) would cause the collapse of only 1% of such structures. Table 2-5 provides a general relationship between Fujita rating and damageability.

Tornadoes are so violent and destructive that they frequently are identified as the "number 1" natural disaster killer in the United States. During the past 50 years, 9000 deaths have been attributed to tornadoes (U.S. Office of Emergency Preparedness, Executive Office of the President, 1972).

Since tornadoes are localized, low-pressure storms in an overall low pressure system, the atmospheric pressure inside a building will greatly exceed the outside pressure. This differential will cause a building to tend to explode. Building

Table 2-5. General Fujita Rating-Damage Relationship

Distribution of recent tornadoes (%)	Type of damage	Fujita rating
20	light	F0
43	moderate	F1
25	considerable	F2
9	severe	F3
3	devastating	F4
–	incredible	F5

Source: Abernathy (1976).

explosions have, however, been greatly overstated; almost all damage results from the extreme winds and flying missles (Abernathy, 1976).

Mitigating the effects of tornadoes is extremely difficult. Little can be done to keep structure development and population out of the path of a tornado, since the events are random, and significant wealth is already at risk in the more tornado-prone areas. However, the vulnerability of structures exposed to tornadoes can be reduced through the use of building-strengthening technologies. This means that mitigations should concentrate on improved structural design. New buildings can be designed for greater lateral load-carrying capacity, storm cellars can be constructed, old buildings can be reinforced, mobile homes can be tied down, and windows can be boarded up.

Experience has shown that short spans of the roof or floor structures easily remain intact after an event, since short spans limit the amount of uplift caused by the winds. Framed construction also usually remains intact. Any structural system, rigidly framed together, tends to be superior to long load-bearing walls. Reinforced concrete, poured in place, will usually withstand the storm. Also rigidly connected frames will withstand tornado forces. However, it is important in both the poured-in-place concrete and steel frame construction that roof and floor systems be securely fastened to the supports. Generally, the heavier the floor or roof system, the more resistant it is to uplifting forces. Windows are no match for the extreme winds or missiles in a tornado. Also, in order to reduce the pressure differential on the building, it is proper to open windows, especially on the leeward side.

Hurricane

A hurricane is an event involving devastating wind, flood producing rain, and, sometimes lethal storm surge. However, only hurricane wind is discussed here.

Hurricanes are large cyclones of air in which the counterclockwise spiraling winds move around a relatively calm center (the core or eye) at speeds of 73 mph or more. Formed in the tropics, the typical hurricane system measures about 400 miles in diameter, moves at a forward speed of about 15 mph, and has an average life-span of nine days. The highest wind speeds in a hurricane system are measured approximately 20 to 30 miles from its center. The greatest wind speed recorded in any hurricane was 197 miles per hour (Hurricane Inez) but gusts between 73 and 120 mph can be expected to extend 40-100 miles from the center.

Hurricanes differ from tornadoes in the following respects: (1) the change in pressure associated with the passage of a hurricane over a structure is rather gradual and has very little effect on the structure, whereas the change in pressure associated with the passing of a tornado is very large, often causing structures to literally explode, (2) hurricane wind speeds are generally much smaller than those of a tornado, and (3) while both windstorms are rotating air masses, the hurricane rotating air mass covers a much greater area. The vertical wind speeds in hurricanes are very low in comparison to tornadoes. Thus, hurricanes have little capacity to hold aloft and transport large, heavy objects.

In recent years, several researchers have expended considerable effort to learn more about hurricane occurrence, including Ho et al. (1975). The most significant portion of the United States coastline at risk is along the Gulf and Atlantic Coasts. Despite the increasing population density in this area, the number of lives lost due to hurricane has been decreasing since the beginning of the twentieth century as a result of improved prediction and warning systems. However, property losses continue to rise, and there is continuing concern that the increasing population, along with ineffective building codes, may lead to a major hurricane disaster along the Gulf or Atlantic Coasts.

Although hurricane damage to buildings is predictable, the most modern building codes do not require engineers to design buildings to withstand the extreme forces of hurricane winds. Hurricane-resistant building standards should therefore be developed for application in identified hurricane hazard zones. These minimum standards should be based on the same mitigation criteria recommended for tornadoes.

Severe Wind

A severe wind is a windstorm with winds in excess of 50 mph, but not associated with a hurricane or tornado. Events of this type can occur in virtually any area of the United States; but damage can be minimized through application of a minimum building code. Since damage associated with the severe wind hazard is typical of that associated with lower velocity hurricanes and tornadoes, similar mitigations should be employed.

Mitigation Strategies

While mitigation is often taken to mean increasing structural strength, it is more useful to view it within a broader context: mitigation is a management strategy that balances current actions and expenditures with potential losses from future hazard occurrences.

Some mitigation actions may be taken by structural design teams. More generally, mitigation activities are those that eliminate or reduce the probability of occurrence of a hazard event, or those that reduce the impacts of hazard occurrence. Preparedness activities are necessary to the extent that other mitigation measures cannot prevent disasters. In the preparedness phase, governments, organizations, and individuals develop, test, and maintain plans to save lives and minimize disaster danger. They also seek to enhance disaster operations during the emergency. Response activities follow an emergency or disaster. Generally they are designed to provide emergency assistance for casualties, and reduce the probability of secondary damage. Recovery activities continue until all systems return to acceptable, normal, or better levels. Short-term recovery activities return vital life-support systems to minimum operation standards; long-term activities may continue for several years until acceptable performance levels are achieved.

Mitigation may be pursued through a variety of approaches. Unfortunately, there is a prevalent tendency to erect a hazard defense around one principal approach and not rely on a combined, often more economical, strategy. One taxonomy divides mitigation approaches into seven categories:

1. Approaches that involve measures to minimize the probability of hazard occurrence (abatement) and/or to protect areas and building sites from the hazard (area protection). Actions intended to minimize the runoff of heavy rain waters are illustrative of the former, while the construction of dams and levees are illustrative of the latter (see Table 2-6).
2. Approaches that focus on the strengthening of buildings exposed to hazards or which focus on site level systems for protecting buildings from such hazards. The use of shear panels to protect frame buildings from earthquake vibrations are illustrative of the former, while small levees around structures in floodplains are illustrative of the latter (see Table 2-7).
3. Approaches that give attention to site-development schemes for protecting structures from hazards. Suggestive of this approach is the use of more appropriate grading practices to protect against landslides, and use of soil stabilization methods to limit the shrink-swell behavior of potentially building-damaging soils (see Table 2-8).
4. Approaches, such as floodplain mapping and restrictive floodplain zoning, that involve the identification of hazard-prone sites and the application of means to prevent or restrict their development and use (see Table 2-9).
5. Loss recovery, relief, and community rehabilitation approaches. Illustrative are floodplain insurance programs and the low interest loans authorized by the Federal Disaster Relief Act (see Table 2-10).
6. Hazard warning and population evacuation systems, such as those employed to deal with approaching hurricanes or riverine floods.
7. Approaches that provide policy makers with the information and decision-assisting tools that facilitate rational and effective hazard management decision making.

Table 2-6. Methods for Minimizing the Probability of Hazard Occurrence (Abatement) and for Protecting High-Risk Areas and Building Sites (Protection)

Hazard	Specific action or example	Purpose of mitigation
Subtype one—abatement		
Riverine flood	• Reforest or reseed denuded areas • Mechanical land treatment of slopes, such as contour farming, to minimize runoff • Protect watersheds from forest and grass fires • Control or eliminate large scale clear-cutting of forest land, as in redwood park area • Manage watersheds to minimize erosion, (e.g., preventing overgrazing) • Design and construct new urban development so as to provide temporary on-site water retention	Decrease in runoff coefficient for treated area
	• Construct catchment basins for debris and sediment • Clear debris and sediment from upper streams • Create water holding areas (i.e., small lakes) • Restore or preserve natural detention areas such as sloughs, swamps, small lakes	Decrease in peak flow downstream
Storm surge	• Seed hurricanes to reduce pressure differential (techniques still in research stage)	Decrease in storm intensity
	• Maintain and preserve barrier islands and coastal wetlands • Preserve, maintain, and replace coastal sand dunes	Decrease water depth and/or reduced wave action
Earthquake	• Relieve earthstrain resulting from tectonic movement (techniques still in research stage)	Reduced potential and/or severity of earthquake occurrence

Table 2-6. (continued)

Hazard	Specific action or example	Purpose of mitigation
Subtype two—protection by structural measures—construction of engineering works to control or protect against hazard (change the basic action mechanism)		
Riverine flood	• Construct and maintain levees and floodwalls as physical barriers • Construct and maintain channel improvements to increase flow capacity • Construct dams and reservoirs to store water and release it gradually	Elimination of low return period floods (may have increased damage for rare events)
	• Emergency flood control • Minimize failures of engineering works with inspection, rehabilitation, and removal from use of substandard structures	Number of structures protected
Storm surge tsunami	• Construct and maintain breakwaters, levees, and floodwalls as physical barriers	Elimination of low return period floods
Headwater flood	• Local stream and storm drainage management with drains, street gutters, culverts, etc.	Decrease in runoff coefficient, increase in flow capacity (may increase downstream flow volume)
Expansive soils	• Cover site with blanket of nonswelling soil	Counteracts swelling pressures
	• Insure proper drainage	Reduces infiltration
Landslides	• Reforest and reseed denuded areas • Protect hillside areas from forest and grass fires • Manage hillside areas to minimize vegetation removal and erosion	Reduces, prevents or controls infiltration and erosion

Table 2-7. Methods for Structural Strengthening and/or Protection of Exposed Buildings

Hazard	Specific action or example	Purpose of mitigation
Riverine flood	• Floodproof individual structure *to minimize damage*: Anchored superstructures Anchored fuel tanks Drains in heating ducts Positive drainage for basement ceilings Separate circuits and water proofing for floodable electric lines Water resistant flooring, cabinetry, carpets Slope on gas piping Manual sewer cutoff valves Basements with break areas to admit water prior to collapse of walls Space usage consistent with flood hazard (i.e., sensitive machinery and equipment on upper floors) Water resistant doors, plywood, insulation Design lower structure to withstand flooding Design structure to withstand swift flow and uplift pressures	Decrease in depth-damage relation for modified buildings
	• Floodproof individual structures *to keep water out*: Strengthen structure to survive water pressure loads Provide for emergency operation of electricity, water, and sanitary services Use glass block for windows Close unneeded wall openings Thoroseal coating on masonry to reduce seepage Masonry cracks sealed with hydralic cement Sump pump and drain to eject seepage Steel bulkheads for doorways and other openings (install on warning)	

Table 2-7. (continued)

Hazard	Specific action or example	Purpose of mitigation
Wind, tornado, hurricane	• Anchor roof systems • Require roofs to have horizontal shear panel	Resistance of roof system to uplift pressures
	• Require lateral force bracing or shear panels • Increase window thickness • Limit window size	Lateral resistance to wind pressure
	• Shutter windows	Eliminate exposed area
Tornado	• Limit window size	Surface area reduction for missile impact
	• Construct strong (or strengthened) room in homes	Lateral resistance to tornado wind pressure and reduced missile exposure area
Earthquake	• *Structural modifications* includes basic redesign or reinforcement of: Structural frame Foundation Concrete walls and slab Floor diaphragms Precast panel connections	Increase resistance to lateral earthquake induced forces
	• *Nonstructural modifications* includes proper anchoring, bracing and structural isolation of elements: Masonry Precast panels Plumbing HVAC caps Electrical (including lights) Elevators Window systems Partitions Acoustical ceilings Miscellaneous metals	Decrease in expected damage when exposed to an earthquake event
	• *Building management*: Relocating certain essential operations into buildings or parts of buildings of more resistance	Minimize the potential for loss of life and operation

Table 2-7. (continued)

Hazard	Specific action or example	Purpose of mitigation
	Develop personnel training program that would insure that essential operations would continue to function in the event of an earthquake (e.g., train individuals to operate local power generating equipment in hospital facilities) Develop proper evacuation procedures that would insure the safety of individuals leaving a building after an earthquake (e.g., high rise buildings)	
Expansive soils	• Structural reinforcement	Keeps stresses within allowable ranges
	• Structure/foundation isolation: Footings or caissons placed below swelling zone Air gaps between foundation, slabs and the ground Place frangible material between poured concrete and soil	Decreases or eliminates damage due to swelling
	• Control moisture variability: Place impervious apron Proper drainage Protect from leaks Relocate trees	Reduces soil moisture variability
	• Eliminate swelling: Inject lime slurries Increase downward loads	Prevents or controls swelling
Landslides	• Relocate structure: Remove from hazard area Setback from head of slope	Eliminates or decreases slide potential
	• Modify foundation: Deep foundations at top of slope Shallow foundations at base of slope	Decreases critical stresses
	• Insure proper surface drainage	Controls erosion and water infiltration

Table 2-7. (continued)

Hazard	Specific action or example	Purpose of mitigation
	• Slope stabilization: Flattening of slope Pressure berms at toe Internal drains Electroosmosis Consolidation Compaction Grouting Rock bolts Piles Retaining walls	Increases factor of safety

Table 2-8. Methods for More Appropriately Preparing Building Sites

Hazard	Specific action or example	Purpose of mitigation
Tornado and hurricane	• Cleanup debris	Reduction of potential missile density of area
Wind	• Create wind breaks (e.g., trees, fences)	Reduction in wind speed due to increased surface roughness
Earthquake	• Stabilize and compact soil to reduce risk of liquefaction and differential settlement	Increased soil strength
Expansive soils	• Place blanket of nonswelling soil	Counteracts swelling pressures
	• Modify existing clay soils Inject lime Chemical stabilization	Reduces absorption and chemical reactions
	• Compaction techniques	
	Wetter than optimum	Reduces absorption
	Add lime	Lowers plasticity index
	Add granular material	Increases porosity
	Add various salts	Increases ion concentration
Landslides	• Avoid undercutting toe of slope	Maintains slope stability
	• Insure proper drainage	Controls erosion and infiltration

Table 2-8. (continued)

Hazard	Specific action or example	Purpose of mitigation
	• Analyze slope stability Soil borings Soil testing Factor of safety	Determine landslide potential
	• Construct retaining walls	Reduces stresses on slip plane
	• Slope stabilization	Increase factor of safety

Table 2-9. Methods for Identifying and Avoiding Hazard Prone Areas through Geophysical Studies and Land Use

Hazard	Specific action or example	Purpose of mitigation
Riverine flood	• Delineate floodway, floodway fringe, flash flood, or other hazard areas	Reduction in number of structures subject to flooding
Storm surge	• Restrict usage so as to prevent increased water depth or flow velocities	
Tsunami	• Restrict usage so as to minimize and control runoff	
	• Alert people to the true degree of hazard at their property	
	• Public purchase of fee-interests of flood-prone lands to restrict lands to open-space use	
	• Public acquisition of easements used to prevent or control development	
	• Tax adjustments to encourage open-space use and control development	
	• Flood-conscious governmental policies in extension of public services to flood-hazard areas	
	• Regulations to prevent obstruction of floodways: Zoning Subdivision Encroachment regulations Regulations of dikes, levees, dams, sea walls, or channel modifications	

Table 2-9. (continued)

Hazard	Specific action or example	Purpose of mitigation
	• Require sellers and brokers disclose flood hazards in real estate transaction • Relocation of existing structures	
Wind, hurricane and tornado	• Microzonation to establish severe wind areas	Reduce density of exposed structures
Earthquake	• Regional investigation of the seismic or earthquake history of the area Geologic map of regional and/or local faults Map(s) of earthquake epicenters Fault strain and/or creep map	Identification of areas of high risk and reduce density of exposed structures
	• Local investigation or the geologic conditions at or near the site that might indicate recent fault or seismic activity • Surface investigation: Geologic mapping Study of aerial photographs Review of local ground water data • Subsurface investigation: Trenching Soil borings Geophysical surveys	Identification of areas of high risk and reduce density of exposed structures
	• Land-use management: Limit types of use in high risk areas (i.e., no hospitals in an active fault zone)	Reduce density of exposed structures and population
Expansive soils	• Determine swelling potential from soil index properties: Colloid content Plasticity index Shrinkage limit Liquid limit Activity Grain size composition	Identification of high risk areas
	• Microzoning to establish swelling prone areas	Identification of high risk areas
	• Alert people as to the hazard to their property	Allows owner to take action

Table 2-9. (continued)

Hazard	Specific action or example	Purpose of mitigation
Landslides	• Regional mapping and investigations: Topography Precipitation Geology	Identify potential slide areas
	• Local investigations: Soil borings Geophysical surveys On-site inspections	Identify potential slide areas
	• Determine slide potential from soil investigations: Slope geometry Soil or rock strength Water table Densities Seepage forces Analysis	Identify factor of safety against sliding
	• Microzonation to establish slide-prone areas	Identification of high risk areas
	• Limit use in high-risk areas subject to catastrophic sliding	Reduce exposure of people and structures
	• Alert people as to the hazard at their property	Allows owner to take action

Table 2-10. Methods for Relief and Rehabilitation by Providing Public Money to Assist Victims in Recovering from Their Losses

Hazard	Specific action or example	Purpose of mitigation
Riverine flood, storm surge, tsunami, earthquake, expansive soils, landslides, tornado, hurricane, severe wind	• Insurance: Spreads the loss Provides monies for rehabilitation, another form of disaster relief (because of heavy subsidy) Required in designated flood areas for federal assistance Couple with land use controls	Reduces individual losses Redistributes economic loss
	• Loans: Reduce rate on existing federal loans Reduced rate on the reconstruction loans	Forcing mechanism to comply with local regulations

Table 2-10. (continued)

Hazard	Specific action or example	Purpose of mitigation
	Presently required flood insurance (flood hazards only) Presently requires flood damage mitigation program (flood hazards only)	
	• Grants: To individuals To local governmental agencies, presently requires local flood damage mitigation program (flood hazards only) To states for preparing plans for disaster preparedness and prevention	Incentive to include hazard mitigation in planning process
	• Feeding and sheltering of victims: Food stamps Free temporary housing Free clean up supplies Free repair and reconstruction of public structures • Disaster unemployment insurance	Eliminate disaster hardship and aid in speedup of recovery
	• Tax deductions	Assist in recovery through tax relief
	• Technical assistance and advice to states and local governments in: Performance of essential community services Warning of further risks and hazards Public information and assistance in health measures Public information and assistance in safety measures Management and control Reduction of immediate threats to public health and safety Distribution of medicine, food, and other consumables	Assist in achieving effective planning and public awareness of hazard potential

Mitigation Strategies in Developing Countries

The extensive loss of life and property caused by catastrophic events, such as earthquakes, floods, and windstorms, in other countries in the world may be significantly reduced through the application of improved siting, design, and construction practices. However, effective hazard mitigation strategies are those which are compatible with local conditions, rather than those which are blindly adopted from the United States and other developed nations. It is important first to understand the constraints that govern natural hazard mitigation policies, as well as to develop a full understanding of the cultural and socioeconomic constraints in other countries.

For example, strategies applied in developing countries cannot be based on conditions prevailing in countries with a highly developed technologic base. Also, it must be understood that within the same country local conditions may vary from region to region or even within the same region. In Iran, for example, modern buildings located in the Fars regions were designed by Tehran engineers using standards that did not take into account the geologic characteristics of the area. An earthquake that devastated the southern Fars portion of Iran resulted in over 60% of the victims perishing as a result of the collapse of improperly constructed buildings (U.S. Department of Commerce, 1974).

Criteria and standards also must be compatible with local economic capabilities, and should avoid those technologic solutions that require exorbitant construction costs. Properly formulated codes and standards should, therefore, provide sufficient flexibility to allow the adoption of feasible new technologic improvements that reduce the economic impact. For example, stabilized adobe can be produced in vast quantities at relatively low cost. Once stabilized, adobe exhibits superior structural, architectural, thermal, and acoustic properties. Bamboo can be prevented from deteriorating by applying chemical perservatives that can be manufactured easily in developing countries. Of cource, building techniques that can improve the quality of masonry construction should be furthered. New construction materials, such as jute fiber, reinforced concrete with vegetable or plastic reinforcement, stabilized adobe, and lightweight structural sandwich systems, which are now being utilized elsewhere in the world, could be incorporated in the development of safe, low-cost housing in developing countries (U.S. Department of Commerce, 1974).

In many underdeveloped countries, code enforcement has tended to be unsatisfactory. Social and economic factors affecting building construction include (1) shortage of funds, (2) heavy migration of rural population to the urban centers, (3) rapid population growth, (4) markets of insufficient size to ensure economies of scale and prefabricated construction, (5) insufficiently developed transportation and distribution systems, (6) shortage of skilled labor, and (7) generally low standards of workmanship. Further, user resistance to certain construction materials and systems must not be overlooked. In many cases, such resistance can be overcome, but only through carefully planned efforts.

The development, adoption, and implementation of improved natural hazard resistant technologies in developing countries can be effectively stimulated by appropriate action in the areas of financing, design, construction, and education. Governmental and financing institutions can contribute to improving such technologies by requiring that the projects approved and financed be designed and built in accordance with hazard mitigation requirements. Cooperative programs involving groups of multidisciplinary experts are necessary to recommend actions aimed at reducing losses from natural hazards. The United Nations Industrial Development Organization (UNIDO) in cooperation with developed countries is encouraging development, introduction, and promotion of new technologies, such as the application of plastics in the building industry. A UNIDO-sponsored working group on the "use of plastics in the building industry" formulated a set of general recommendations that might be initiated by the developing countries. The significant recommendations are summarized below.

1. The development and introduction of new building products should be *gradually* and *systematically* organized within countries where the climatic and socioeconomic conditions permit.
2. The education of architects, engineers, and building contractors in the proper selection and use of new building components is essential.
3. Appropriate specifications for the use of building components must be formulated to facilitate the ready acceptance by the local building industry.

The United Nations and other international organizations and agencies should:

1. Provide interdisciplinary teams of experts to developing countries to examine requirements and determine the possibility of producing new building materials locally with an emphasis on native resources.
2. Fellowship programs for experts from developing countries should be initiated and organized through interregional seminars.
3. Joint participation with outside professional organizations should be expedited.
4. Criteria and standards for evaluating buildings systems incorporating new technologies must be established. Information should be made available at the request of any developing country and assistance provided in evaluating the cost, performance, and adaptability of any proposed system to the local climatic and socioeconomic conditions.

There is a considerable potential for technical improvements leading to structures that are more resistant to natural hazard forces. Decisions regarding where to apply the limited resources should be a function of the specific needs of each country.

References

Abernathy, J. J. 1976. Tornado Protection—Selecting and Designing Safe Areas in Buildings. DCPA Report No. TR 83B. U.S. Govt. Printing Office, Washington, D.C.

AIA Research Corporation. 1975. Architects and Earthquakes. U.S. Govt. Printing Office, Washington, D.C.

Broms, B. 1975. Landslides. *In*: H. F. Winterkorn and H. Y. Fang (Eds.), Foundation Engineering Handbook. Van Nostrand-Reinhold, Princeton, New Jersey, pp. 375–401.

Culver, C., H. S. Lew, G. C. Hart, and C. W. Pinkham. 1975. Natural Hazard Evaluation of Existing Buildings. Building Sciences Series 61. U.S. Department of Commerce, Washington, D.C.

Flawn, P. T. 1970. Environmental Geology: Conservation, Land Use Planning and Resource Management. Harper & Row, New York.

Friedman, D. G., and T. S. Roy. 1966. Simulation of Total Flood Loss Experience of Dwellings on Inland and Coastal Flood Plains. Travelers Insurance Company, Hartford, Connecticut.

Fujita, T. T. 1970a. Estimate of Areal Probability of Tornadoes from Inflationary Reporting of Their Frequencies. SMRP Research Paper No. 89. Department of Geophysical Sciences, University of Chicago, Chicago, Illinois.

Fujita, T. T. 1970b. A Proposed Characterization of Tornadoes and Hurricanes by Area and Intensity. SMRP Research Paper No. 91. Department of Geophysical Sciences, University of Chicago, Chicago, Illinois.

Fujita, T. T., and A. D. Pearson. 1973. Results of FPP classification of 1971 and 1972 tornadoes. Presented at the Eighth Annual Conference on Severe Local Winds, Denver, Colorado, October 1973.

Harris, D. L. 1963. Characteristics of Hurricane Storm Surge. Technical Paper No. 48. U.S. Weather Bureau, Washington, D.C.

Ho, F. P., R. W. Schwerdt, and H. V. Goodyear. 1975. Some Climatological Characteristics of Hurricanes and Tropical Storms, Gulf and East Coast of the United States. Report NWS-15. National Oceanographic and Atmospheric Administration, Washington, D.C.

Hwang, L. S., and D. Divoky. 1971. Tsunamis. Underwater and Information Bull. 3(5):207–219.

Johnson, W. K. 1978. Physical and Economic Feasibility of Nonstructural Flood Plain Management Measures. U.S. Army Corps of Engineers, Institute for Water Resources, Ft. Belvoir, Virginia.

Jones, D. E., Jr. 1976. The expansive soil problem. Presented at Symposium on Expansive Soils. Geological Society of America, Denver Colorado.

Maldus, J. S. 1957. The origin of hurricanes. Scientific American. 197(2):33–39.

Richter, C. F. 1958. Elementary Seismology. Freeman Press, New York.

Terzaghi, K. 1950. Mechanism of landslides. *In*: Application of Geology to Engineering Practice. Geological Society of America, New York, pp. 84–123.

U.S. Department of Commerce, National Bureau of Standards. 1974. Design, Siting, and Construction of Low-Cost Housing and Community Buildings to Better Withstand Earthquakes and Windstorms. U.S. Govt. Printing Office, Washington, D.C.

U.S. Office of Emergency Preparedness, Executive Office of the President. 1972. Disaster Preparedness. U.S. Govt. Printing Office, Washington, D.C.

Weigel, R. L. 1970. Tsunamis. *In*: R. L. Weigel (Ed.), Earthquake Engineering. Prentice-Hall, Englewood Cliffs, New Jersey, pp. 253–306.

Whitman, R. V., S. T. Hange, and J. W. Reed. 1973. Damage Statistics for High Rise Buildings in the Vicinity of the San Fernando Earthquake. School of Engineering, Massachusetts Institute of Technology, Cambridge, Massachusetts.

3
Public Policy Approaches to Management of Natural Hazards

The Matrix of Policy Authority

Within the United States, legal responsibilities and authorities for dealing with the causes and effects of natural hazard exposures are assigned to a bewildering array of agencies and entities at local, state, and federal levels of government. In addition, important roles in mitigating the potentially adverse effects of natural hazard exposures are played by private insurance companies, lending institutions, and associations of governmental and nongovernmental officials.

Under the federal system of government that prevails within the United States, primary responsibility and authority for dealing with the problems associated with natural hazard exposures is assigned to the states, and through specific delegations of authority, to their component municipalities. Depending on the state, important responsibilities also may be vested in counties, towns, and special districts. Through their power to tax and to expend public funds, state governments may engage in a wide range of activities relevant to control of the adverse effects of natural hazard exposures, including: the construction of dams, levees, and seawalls; and investment in programs which identify and map hazardous areas, predict future hazards, and educate the public. Through exercise of the police power, each state government is legally able to engage in a wide range of regulatory activities, unless specifically prohibited from doing so by its state constitution. Among these activities are those related to the planning and control of land use, the establishment and enforcement of standards to regulate the subdivision of land and the sale or exchange of real estate, the prescription of codes and standards to govern the construction of new housing units and other selected types of structures, and the establishment of prohibitions concerning the structural integrity and use of existing structures.

In practice, most states have given local units of government the power to pursue these activities, although the pattern of delegation is not uniform. In California, for example, both cities and counties are legally able to adopt and enforce regulatory policies concerning the use and subdivision of land, structural integrity of new and existing buildings and structures, and to engage in other similar functions. On the other hand, in Texas only incorporated cities are vested with such a broad array of power and other local units of government, such as counties, are dependent on regulatory activity by state entities or by cities, which bear some limited authority to regulate activities in their unincorporated, peripheral (extraterritorial) areas.

However large their authority, the capacity of state governments to deal with the problems posed by natural hazard exposures may nonetheless be constrained by geophysical, ecologic, and sociopolitical factors. Historically, states and local units of government have relied heavily on such federal entities as the U.S. Corps of Engineers, which engages in massive construction projects necessary for the control of riverine and coastal flooding. The mapping of lands has long been viewed as a function of the U.S. Geological Survey. Responsibility for research and other related activities that can be conducted more economically at a level higher than the state have traditionally been viewed as a function of the federal government. Ecologic and geophysical factors may also limit the capacity of state and local governments to cope with their problems. In terms of flooding, for example, upstream activity that may alter traditional water runoff volumes or otherwise affect the expected annual streamflow may limit the real capacity of downstream state and local governments to deal with this problem. Similarly, economic factors may severely limit the ability of state and local governments to respond rapidly and effectively to the problems created by major catastrophic occurrences. Thus, a pattern of interdependence has developed between the several levels of government.

There are also interdependencies between the public and private sector. For example, private lending and insurance institutions may impose constraints on builders and property purchasers, thereby reducing the vulnerability of the mortgaged or insured property to natural hazard exposure. In addition, associations of private and public officials develop model standards, codes, and criteria for the guidance of governmental law-making bodies.

Further complicating the allocation of responsibility and authority for natural hazards management is the institutional pluralism associated with all three levels of government and the private sector. At no level of government is responsibility and authority for the management of natural hazard exposures assigned to a single organizational entity. At the local level, departments of city planning, public works, building and safety, and public health may individually exercise control over one or more of the activities essential to a comprehensive system of natural hazards management. Much the same situation exists at state and federal levels where major departments, regulatory commissions, and independent agencies may have individual responsibilities for specific elements of the hazard management system.

At the federal level problems associated with this situation produced the view that disaster related programs located in several agencies might be administered more effectively under one management structure. Thus, the Federal Emergency Management Agency (FEMA) was formed in 1979. Under the terms of the reorganization proposal, five existing entities were merged into one: the National Fire Prevention and Control Administration, the Federal Insurance Administration, the Defense Civil Preparedness Agency, the Federal Disaster Assistance Administration, and the Federal Preparedness Agency.

In addition, the President designated FEMA the lead federal agency for all disaster preparedness and hazard mitigation planning functions. In submitting the reorganization proposal to Congress, the President noted that his efforts had been guided by four major principles: (1) federal authorities responsible for actions related to major civil emergencies should be supervised by one official responsible to the President; (2) an effective civil defense system requires the most efficient use of all available emergency resources; (3) emergency responsibilities should, when possible, remain within the regular missions of federal agencies; (4) federal hazard mitigation activities should be closely linked with emergency preparedness and response functions.

Consistent with these principles, the President transferred to the new agency the oversight responsibility for the earthquake hazards reduction program previously assigned to the U.S. Office of Science and Technology Policy of the Executive Office of the President (1978). He also assigned FEMA responsibility for the Federal Emergency Broadcast System, the Federal Dam Safety Program, and the program that provides assistance to communities in the development of readiness plans for weather-related emergencies, including floods, hurricanes, and tornadoes. The new agency also was assigned responsibility for coordinating natural and nuclear disaster warning systems and conducting preparedness planning to reduce the consequences of major terrorist incidents. Because of the powers granted the new agency, it is possible that FEMA may become the principal staff arm of the President for policy making and program planning related to hazards and disasters. To the extent that the officials of the new agency are able to exercise this role, considerably improved coordination and integration of federal programs, policies, and activities may occur, politics permitting.

Types of Hazard-Related Public Policies

Substantial annual investments are being made each year by federal, state, and local governments to meet the following objectives: protect communities from hazard occurrences, force the use of loss-reducing technologies by builders and building owners, reduce and spread the losses sustained by individuals exposed to hazards, identify and promote the avoidance of high-hazard areas, warn exposed persons of impending hazards, and advance human understanding concerning hazard occurrences and possible mitigations. These objectives are the focus of a mix of public policies which may be ordered into the following ten categories:

1. Action-forcing policies: Adopted by higher level jurisdictions and intended to force loss-reducing activities by lower units and jurisdictions of government.
2. Attention-focusing policies: Intended to stimulate citizen, group, and governmental interest in losses produced by natural hazards and to promote voluntary state, local, and private action to reduce such losses.
3. Disaster recovery policies: Intended to assist personal, familial, neighborhood, community, and state recovery from the damages sustained as a result of a natural hazard.
4. Technology development policies: Focused on development of new knowledge and technology to support hazard mitigating policies.
5. Technology transfer policies: Focused on transfer of knowledge to consumers, governments, and others, and the use of that knowledge in the long term (as in hazard analysis programs) and the short term (disaster warnings).
6. Regulatory policies: Regulate the decisions and behaviors of private parties and governmental entities to reduce losses associated with exposure to natural hazards. Such policies may force the use of avoidance, building strengthening, site preparation, and other mitigation methods.
7. Investment and cost allocation policies: Specify conditions governing acquisition and allocation of resources to sustain the activities described above and below. Such policies determine how much will be spent, when, for what purpose, where, and at whose expense.
8. System management policies: Intended to fix responsibilities, specify the means used, and define the restrictions to be met by hazard mitigation programs.
9. System optimization policies: Intended to ensure that other policies are effective, compatible with system goals, and internally consistent.
10. Direct action policies: Authorize direct governmental action to implement a policy, such as physical construction or removal of structures.

This taxonomy of public policies was used to guide the following analysis of current federal, state, and local legislation. It was constructed on the assumption that complex problems might well require policy actions across the range of posited policy types. Development of the taxonomy was guided by the simpler classic model which holds that there are four broad types of public policies: *allocative* or *distributive*; *regulatory*; *structural*; and *redistributive* (Lowi, 1964; Anderson, 1975; Jones, 1977).

U.S. Federal Natural Hazard Policies and Programs

Numerous federal programs provide assistance to parties engaged in natural hazards planning and mitigation efforts, particularly with regard to the flood hazard. Figure 3-1 identifies the principal agencies and the general activities conducted by each. Presumably, the Presidential directive which established FEMA, and others concerned with specific hazards, such as the flood related Executive Order

11988, (May 24, 1977) will provide for improved future central coordination of this complex mix of federal programs and services and for a more cost-effective approach to hazard planning and mitigation.

FEDERAL AGENCY OR ACTIVITY	PROGRAM	EARTHQUAKES	HURRICANE WINDS	TORNADOES	TSUNAMIS	LANDSLIDES	RIVERINE FLOODS	STORM SURGE	SEVERE WINDS	EXPANSIVE SOIL
Farmers Home Administration (DOA)	Emergency Loans	□	□	□	□	□		□	□	□
	Watershed Protection and Flood Prevention Loans						■			
Agricultural Stabilization and Conservation Service (DOA)	Emergency Conservation Measures	□	□				□		□	
Soil Conservation Service (DOA)	Resource Conservation and Development						■			
	Water Shed Protection and Flood Prevention						■			
Department of the Army Office of the Chief of Engineers (DOD)	Flood Control Works (Rehabilitation)		□		□		□	□	□	
	Flood Fighting and Rescue Operations						□	□		
	Flood Plain Management Services						■	■		
	Emergency Bank Protection						□	□		
	Small Flood Control Projects						■	■		
	Snagging and Clearing						■	■		
	Planning Assistance to States						■	■		
	Dam Inspection						■			
Public Health Service	Emergency Medical Service Planning	□	□	□	□	□	□	□	□	
Housing Protection and Mortgage Credit/FHA (HUD)	Mortgage Insurance: Disaster Victims	□	□	□	□	□	□	□	□	
Disaster Response and Recovery (FEMA)	Disaster Assistance	□	□	□	□	□	□	□	□	
Federal Insurance Administration (FEMA)	National Flood Insurance Program					■*	■	■		
Office of Plans and Preparedness (FEMA)	Energy Management Assistance	■ □	■ □	■ □	■ □	■ □	■ □	■ □	■ □	
	State Disaster Preparedness Grants	■	■	■	■	■	■	■	■	
	Earthquake and Hurricane Loss Study and Contingency Planning	■	■							
Office of Disaster Response and Recovery (FEMA)	Disaster Assistance	□	□	□	□	□	□	□	□	□
Defense Electric Power Administration (DOI)	Electric Power Planning for Emergencies	□	□				□		□	
Bureau of Reclamation (DOI)	Reclamation Projects						■			
National Science Foundation	Basic Research	R	R	R	R	R	R	R	R	R
Small Business Administration	Loans to Natural Disaster Victims	□	□	□	□	□	□	□	□	
Tennessee Valley Authority	Water Resources Development and Flood Control						■			
Federal Crop Insurance Corp. (DOA)	Crop Insurance		□				□		□	
U.S. Geological Survey (DOI)	Prediction	■			■	■	■			
National Weather Services (NOAA)	Hazard Warnings		■	■	■		■	■	■	
National Meterological Center (NOAA)	Hazard Warnings			■						
National Science Storm Forecast Center (NOAA)	Hazard Warnings			■						
Radar Report and Warning Coordination Circuit (NOAA)	Hazard Warnings		■	■						
GOES - Satelite System (NOAA)	Hazard Warnings		■	■	■					
National Hurricane Center (NOAA)	Hazard Warnings		■							
National Tsunami Warning Center (DOI)	Hazard Warnings				■					
Office of Coastal Zone Management (NOAA)	Coastal Hazard Mitigation	■	■		■	■	■	■	■	

Legend:

*mudslides only

- ■ Programs with pre-disaster functions (disaster mitigation)
- □ Programs with post-disaster functions (disaster relief)
- R Programs with natural hazard research functions

Figure 3-1. Key federal agencies and programs related to natural hazards.

LEGISLATION OR PROGRAM	TYPE OF POLICY									
	ACTION-FORCING	ATTENTION-FOCUSING	DISASTER RECOVERY	TECHNOLOGY DEVELOPMENT	TECHNOLOGY TRANSFER	REGULATORY	INVESTMENT AND COST ALLOCATION	SYSTEM MANAGEMENT	SYSTEM OPTIMIZATION	DIRECT ACTION
NATIONAL FLOOD INSURANCE PROGRAM	●	●	●		●	●	●	●		
NATIONAL DAM INSPECTION PROGRAM		●			●	●	●			
DISASTER RELIEF ACT OF 1974		●	●	●	●		●	●		
EARTHQUAKE HAZARDS REDUCTION ACT OF 1977		●	●	●	●			●	●	
CORPS OF ENGINEERS FLOOD CONTROL WORKS										●
COASTAL ZONE MANAGEMENT ACT OF 1976	●	●		●	●	●		●	●	
MOBILE HOME CONSTRUCTION & SAFETY STANDARDS ACT OF 1974	●	●		●	●	●				●

Figure 3-2. Major disaster-related legislation by policy type.

Federal legislative responses to natural hazards have ranged from authorization for massive investments in area flood control facilities, through the operation of disaster warning systems, to extension of postdisaster assistance.

As shown in Figure 3-2, major natural hazard legislation by the federal government encompasses all ten policy types listed under the taxonomy of mitigation policies. Most frequently represented in federal legislation are technology transfer and attention-focusing policies, and only under the Corps of Engineers flood control program is the federal government significantly engaged in direct action without operating through other levels of government. Coastal Zone Management, National Flood Insurance Programs, and the Mobile Home Construction and Safety Standards are the only federal activities that involve *action forcing* policy. Following is a discussion of the most significant federal policies.

Flood Hazard Policies

Of the nine natural hazards examined, riverine floods have received the greatest attention from the federal government. Legislation dealing with riverine floods has included: (1) The Flood Control Act of 1936, et seq.; (2) The National Flood Insurance Act of 1968; (3) The Flood Disaster Protection Act of 1973; and (4) The National Dam Inspection Act of 1972 and Executive Orders 11988 and 11990.

The first of these acts signaled the beginning of a massive federally funded effort to reduce flood hazards through construction of dams, levees, floodway improvements, and other similar measures. The flood insurance and flood disaster protection acts brought about a shift in policy and an awareness that floodplain avoidance measures should be given a priority equal to or exceeding the earlier commitment to structural approaches.

1. Flood Control Act of 1936, et seq.–Administered by the United States Army Corps of Engineers. The Flood Control Act of 1936 designated the Corps of Engineers as the primary agency for the construction of flood control works. Later legislation expanded this program to include a range of water resources development functions. Under this broadened program, the Corps expended a total of $1.8 billion in fiscal 1974 and $2.1 billion in fiscal 1975 on rivers, harbors, and flood control projects. In this two year period the Corps undertook 24 major new flood control projects. The Corps estimates cumulative damages prevented by its flood control projects during the period 1936–1975 to be $59.6 billion, compared to project costs of $10.2 billion.

In addition to direct construction activities the Corps provides information, technical and planning assistance, and guidance to communities in identifying the magnitude and extent of the flood hazard and in planning wise use of floodplains. The agency issues Floodplain Information Reports containing flood area maps, tabulations, hydraulic data, and descriptions including some flood history and estimates of the frequency of future floods. Nonstructural alternatives also are part of the activities of the Corps, such as the acquisition of wetlands for natural storage areas and the development of recreation facilities in floodprone lands (U.S. Army Corps of Engineers, 1974–1975).

2. National Flood Insurance Act of 1968 (P.L. 90-448)–Administered by the Flood Insurance Administration. Although the possibility of insurance against flood hazard was considered after virtually every major flood, no federal action to establish such a program was taken until 1968. Hurricanes and floods in California and the northeast in 1955 led to the enactment of the Federal Flood Insurance Act of 1956 [P.L. 1016, 84th Congress, 70 Stat. 1078], but because of disagreement with the insurance industry and doubts over the effectiveness of the proposals, funds were not appropriated. The Alaska earthquake of 1964 and Hurricane Betsy in 1965 provided the final impetus for federally subsidized flood insurance (U.S. Congress, 1966a). With rising concern over the losses from natural disasters, the increasing volume of federal funds demanded for relief of victims, and the limited success of local communities in managing their floodplains (Platt, 1976a), the National Flood Insurance Act of 1968 (P.L. 90-488, Title XIII) was adopted. Through it, the federal government became involved with the land use planning process on nonpublic lands, in what the Administrator of the National Flood Insurance Program (NFIP) called the "first constructive land use bill in the Nation" (U.S. Congress, 1975).

The land use ramification of flood insurance programs was recognized in the report by HUD to the President prior to passage of the 1968 act. The report espoused the view that the best long-run solution to losses in the nation's floodplains would be one that encouraged a shift in land use from residential to industrial and recreational. The report stated that zoning, building permits, extension of public services, and other public actions could provide guides to private investment which could work toward the same end. In HUD's view, the management of flood-prone areas involves more than the extension of flood insurance to properties within flood areas, but involves the use of insurance as a facilitating force toward development and implementation of long-range flood-mitigating land use policies (U.S. Congress, 1966b).

Approximately 16,000 local governmental units exercise jurisdiction over 7% of the United States land area that is subject to flooding (Platt, 1976a). A U.S. Geological Survey study of 26 U.S. cities has revealed that over half of the floodplain areas of these cities has been developed at urban densities and that floodplains comprise a large fraction of their urban areas (Schneider and Goddard, 1974). In the cities examined, over 16% of their total urban area was found to be located within a floodplain. In Chicago, about 10%, or 132 square miles, is subject to flooding, and that area includes some of the most densely developed portions of the city. Much of the increase in occupancy of urban floodplains is attributable to the pressures of general economic growth and encroachment on flood zone lands which were avoided when other land was available for development. Along with this growth and development, property and life losses from flooding have also increased (Cheatham, 1977) (see Table 3-1). If current trends continue, the annual expected flood caused property losses will be in excess of $3 billion dollars in the year 2000.

The Flood Insurance Act of 1968 was adopted, in part, to deal with the problems posed by continued development of the nation's floodplains. Section 1305 of the Act states that "flood insurance will be made available in only those states or areas which have evidenced a positive interest in securing flood insurance coverage under the program." Section 1315 states: "After June 30, 1970, no new flood insurance coverage shall be provided under this title in any area . . . unless an appropriate public body shall have adopted permanent land use and control measures . . . consistent with the comprehensive criteria for land management and use under section 1361." The latter section of the statute establishes criteria for land management and use, supporting development of *local* measures for land use, flood control, flood zoning, and flood damage protection. Procedures include (a) adopting measures to constrict the development of land exposed to flood damage; (b) guiding the development of proposed construction away from locations threatened by flood hazards; (c) assisting in reducing damage caused by floods; and (d) otherwise improving the long-range management and use of flood-prone areas.

A National Flood Insurance Fund was created with borrowing authority for the Secretary of HUD of up to $250 million. The ceiling on outstanding insurance was placed at $2.5 billion. The Act also made flood insurance available for the first time to owners of flood-prone one to four family dwellings and to small businesses.

The insurance program was to be managed by the National Flood Insurers Association (NFIA), a pool of insurance companies that sold and serviced flood insurance and distributed profits and liabilities to its members. The federal government made premium equalization payments to the insurers' pool to make up the difference between below-cost premiums received and the actuarial cost of the insurance. The federal government supported private firms by guaranteeing to pay claims in excess of the financial capacity of the privately financed pool.

Federally subsidized rates totalled about 10% of actuarial rates, and were

Table 3-1. Flood Losses of Life and Property in the United States, 1941–1974

Property losses in thousands of dollars

Year	Property losses	Life losses
1941	39,524	47
1942	98,507	68
1943	199,732	107
1944	101,079	33
1945	165,798	91
1946	70,813	28
1947	272,328	55
1948	229,959	82
1949	93,931	48
1950	176,050	93
1951	1,028,741	51
1952	254,064	54
1953	122,204	40
1954	106,842	55
1955	995,491	302
1956	64,688	42
1957	360,303	82
1958	218,255	47
1959	141,255	25
1960	92,976	32
1961	154,033	52
1962	75,237	19
1963	177,946	39
1964	651,642	100
1965	788,046	119
1966	117,004	31
1967	375,218	34
1968	339,399	31
1969	902,654	297
1970	225,453	135
1971	287,525	74
1972	4,465,135	554
1973	1,894,493	148
1974	576,203	121

Source: Comptroller General of the United States (1980).

available only for existing structures in the floodplain; new structures could only be insured at the actuarial rates, which are prohibitive. Thus, the scheme involved the hope that floodplains would eventually be cleared of structures subject to flood damage. In order for a property owner to purchase federally subsidized flood insurance, the entire community was required to become eligible for inclusion in the program by adopting adequate floodplain management measures.

However, two weaknesses in the 1968 Act soon became apparent: (1) The program was voluntary. Although it may have been in the long-term interest of the federal taxpayer to be partially relieved of the disaster relief burden, many communities apparently were little interested in placing a short-term hardship on their residents living in exposed areas. (2) Detailed studies of individual communities were required before rates could be determined. The Secretary was to identify flood zones, and within five years after enactment, establish a set of actuarial flood insurance premiums based on the flood-zone statistics. As a result of these limiting factors only four of the nation's 16,000 flood-prone communities became eligible under the program during its first year and only 20 policies were purchased (Platt, 1976a).

In 1969 Congress modified the original statute to permit communities to take part in the program with reduced amounts of coverage, while the lengthy rate-making studies were completed. Under the regular program, coverage is double what it is under the emergency program, and premiums are based on actuarially determined rates. More stringent floodplain management procedures also were required. By May 1973, however, there were only 2200 eligible communities in the program (U.S. Congress, 1975). In order for the intent of the program to be carried out, federal spokesmen asserted that all flood-prone communities would have to participate and this participation was made mandatory by the Flood Disaster Protection Act of 1973 (P.L. 93-234).

3. Flood Disaster Protection Act, 1973 (P.L. 93-234)–Administered by the Flood Insurance Administration. Under the provisions of the 1973 Act, no federally related financial assistance can be provided for acquisition or development of any identified flood-prone property unless (a) the community in which it is located has entered the National Insurance Program, and (b) the applicant for such financing has purchased a flood insurance policy. Federally related financing includes direct federal assistance of any kind and also private loans insured or regulated by federal agencies such as the Federal Deposit Insurance Corporation (Platt, 1976a).

Title I contains an expanded the National Flood Insurance Program by providing for coverage of losses due to erosion and the undermining of shorelines. The bill more than doubled the flood damage coverage for homeowners and businesses, authorizing insurance for single family dwellings in amounts up to $35,000, and for their contents in amounts up to $10,000. Other residential buildings were authorized to be insured for up to $100,000, and their contents up to $10,000. Subsidized low-cost flood insurance was made available at 25 cents per $100 of coverage in most areas, with the federal government paying from 70 to 90% of the cost.

Title II of the 1973 Act deals with disaster mitigation requirements by requiring federal identification of flood-prone communities and notifying them of the designation. Upon notification the community is required to apply for participation in the flood insurance program or prove that it is not flood-prone. The identification of flood-prone areas and criteria for land use management are to be established in consultation with elected local officials. A community is designated as flood-prone if it has a 1% chance of being vulnerable to a serious flood in any year. Once it is so designated, it has one year to enter the flood insurance program before penalties are imposed.

The bill retained a provision of the 1968 National Flood Insurance Act requiring communities, as a condition of their participation in the insurance program, to adopt land use and control measures to restrain construction projects in areas exposed to possible flood damage.

In signing the bill, President Nixon said he expected it would both help to reduce losses from floods, and to provide faster and fairer assistance to victims than was achieved under previous disaster relief loan programs. However, the mandatory penalties for noncompliance stirred a great deal of controversy and opposition. Subsequent amendments modified certain elements of this bill. Exemptions from the ban on mortgage lending were provided in P.L. 94-375 (passed on July 20, 1976) under conditions where loans were (a) used to purchase a residential dwelling occupied before March 1, 1976; (b) for up to $5000 to improve existing residences; (c) used to finance the purchase of a building occupied by a small business before January 1, 1976; and (d) used to finance improvements for agricultural purposes on a farm.

The bill was amended in May 1977 to remove the provision prohibiting federally insured lending institutions from making loans to property owners in federal government designated flood-prone areas that were not participants in the flood insurance program. The amendment instead required lending institutions to notify recipients that they would not be eligible for federal disaster relief and that the federal government could not assist individuals who undertook development in flood-prone areas without adequate floodproofing measures. Despite the controversy, more than 85% of designated flood-prone communities chose to participate in the program. The federal government has encouraged participation because the insurance and preventive land-use control measures are assessed as less costly than general disaster relief after the fact. The Federal Water Resources Council estimated that the government could save about $1 billion annually if all flood-prone communities engaged in floodplain management and all of their residents were insured. Table 3-2 compares the insurance indemnifications for Hurricane Agnes, which occurred before flood insurance was made mandatory, and Hurricane Eloise, which occurred after the passage of the Flood Disaster Act of 1973. Although the total flood damage to all structures and contents was 2½ times greater in Hurricane Agnes, the insurance indemnification for flood damage was over eleven times greater for Hurricane Eloise.

The federal government believes the flood insurance program can realize substantial savings both to itself and the private sector by reducing the costs of disaster relief well below existing levels. The mechanisms for reducing these costs are primarily land use and building code regulations that incorporate floodproofing requirements.

Table 3-2. Agnes (1972) versus Eloise (1975)[a]

State	Agnes	Eloise	Agnes	Eloise
	Flood damage ($)[b]		Number of flood insurance policies in force	
Pennsylvania	410,060,000	112,800,000	683	40,180
Virginia	69,650,000	6,000,000	667	9,224
New York	41,386,000	27,420,000	2,046	25,121
Maryland	15,000,000	8,000,000	693	5,610
Florida	11,207,000	20,000,000	21,208	128,051
New Jersey	8,112,000	1,500,000	9,259	41,069
Connecticut	3,000,000	2,000,000	115	4,875
District of Columbia	1,000,000	400,000	c	c
All states	569,741,000	228,780,000	61,228	258,744
	Amount of insurance protection in force ($)		Number of flood insurance claims filed	
Pennsylvania	8,654,000	869,576,000	146	8,000
Virginia	13,606,000	302,217,000	209	450
New York	43,657,000	704,264,000	7	1,650
Maryland	14,395,000	176,780,000	99	600
Flordia	356,756,000	3,871,549,000	1,019	500
New Jersey	193,065,000	1,206,133,000	107	375
Connecticut	1,997,000	135,383,000	8	150
District of Columbia	c	c	c	c
All states	1,064,140,000	7,433,230,000	1,614	12,375
	Insurance indemnification for flood damage ($)			
Pennsylvania	772,260	37,500,000		
Virginia	1,601,354	4,200,000		
New York	35,000	2,438,000		
Maryland	783,933	2,900,000		
Florida	1,778,659	2,000,000		
New Jersey	90,511	562,000		
Connecticut	40,000	1,200,000		
District of Columbia	c	c		
All states	5,146,717	58,000,000		

Source: National Flood Insurers Association insurance experience and the American Insurance Association property claim services. Reprinted in U.S. Congress, Oversight on Federal Flood Insurance Programs, 1975, p. 270.

[a]Comparison of the actual insurance indemnification for flood damage under the national flood insurance program in which the insurance purchase requirement was optional with the insurance indemnification under the national flood insurance program as amended by the limited insurance purchase requirement in the Flood Disaster Protection Act of 1973.

[b]Flood damage to structures and contents.

[c]District of Columbia entered Flood Program on October 31, 1975, no insurance in force during Agnes and Eloise.

Since adoption of the 1973 Act there has been steady growth in the program. Almost 1.8 million policies were in effect in 16,732 communities at the end of 1979 with a total policy face value coverage of about $73,000,000,000. Much of the coverage was stimulated through setting the requirement that in order for victims of floods to obtain federally funded relief or disaster loans they would be required to have purchased flood insurance. The FIA believes the program can still expand and projects a market for 7 million policies. It should be noted that a few communities have chosen to drop out of the program rather than comply with the floodplain regulation and floodproofing building requirements. However there have been numerous appeals about specific map delineations or flood elevations. Under the terms of the Housing and Community Development Act of 1977 (P.L. 95-128) some of the mandatory requirements in previous legislation were weakened in respect to issuance of mortgages and loans by banks and federally related institutions in identified flood hazard areas in nonparticipating communities (National Science Foundation, 1980).

In its current form, the Flood Insurance Program is projected both to reduce total flood caused economic losses and to alter the distribution of economic loss. As shown in Table 3-3, the program is projected to save significant amounts of both tax revenue and funds for disaster victims themselves. By the year 2015, the flood insurance program is scheduled to become self-sufficient, although this possibility may be doubtful.

Although the National Flood Insurance Program directly affects land use, the Federal Insurance Administration has suggested that it is not intended to be a general land use program. In testimony before the Senate Committee on Banking, Housing and Urban Affairs, then Acting Director Robert Hunter emphasized that land use controls are state powers delegated to local communities. The National Flood Insurance Program only requires the local adoption of floodplain management regulations, addressing new construction within flood-prone areas only for the purpose of reducing future flood losses. According

Table 3-3. Distribution of Economic Losses (in Millions)

Year	From insurance funds premiums less expenses	From tax revenues ($)	From disaster victims ("self-insurance") ($)
		No flood insurance program	
1975	0	188	283
1980	0	290	436
1990	0	629	943
2000	0	1,284	1,925
		Flood insurance program	
1975	269	173	29
1980	384	158	29
1990	678	194	31
2000	1,129	176	31

Source: U.S. Congress, Oversight on Federal Flood Insurance Programs, 1975.

to Hunter the floodplain management standards are performance standards, focusing on reducing or eliminating future flood damage to new construction—ends that the Congress cited as a primary objective of both the 1968 and 1973 acts (U.S. Congress, 1975).

In summary, the National Flood Insurance Program is intended to reduce economic and other losses due to floods, and was implemented as a result of rising costs to the federal government in providing relief to flood victims. Although it was not intended to be a federally monitored land use program, the mandatory nature of the program after 1973 has put the federal government in the position of establishing performance standards for land uses and building construction requirements in floodplains. The 1977 amendments, however, reduce some of the action-forcing authority of the federal government.

4. National Dam Inspection Program (P.L. 92-367)–Administered by the Secretary of the Army. This act authorized the Secretary of the Army to carry out a national inspection program aimed at more than 28,000 nonfederal dams in the United States. It does not apply to dams under the jurisdiction of the Bureau of Reclamation, the Tennessee Valley Authority, the International Boundary and Water Commission, or the Federal Power Commission; dams that were inspected in the 12 months preceding adoption of the legislation; or dams that pose no threat to human life or property. The Secretary was directed to submit a report to Congress by July 1, 1974, which would include an inventory of all dams, a review of inspection findings, recommendations for corrective action, and a suggested national program for dam safety regulation.

A report by the House Public Works Committee (H. Rep. 92-1232) supporting the passage of the act noted that state programs for licensing and inspection of nonfederal dams varied greatly in scope and effectiveness. It cited the dam failures at Buffalo Creek, West Virginia, and Rapid City, South Dakota, as examples of disasters that might be prevented by passage of the bill. The report said the bill would "provide an accurate assessment of the scope of the problems and an appropriate sharing of responsibility between federal, state and local governments and public and private interests." The estimated cost of the program was $90 million. However, the implementation of this bill was delayed after its passage, due to the assignment of a relatively low priority to this subject by the executive branch.

A major disaster due to dam failure occurred with the collapse of the Teton Dam in Idaho on June 5, 1976. Eleven people died, 1 million acres were destroyed, 16,000 head of livestock killed, and $1 billion lost in property damage. Payment of claims arising from the disaster cost the United States $549 million. This was a Bureau of Reclamation project and thereby exempt from the dam inspection legislation. It was, however, the first dam failure in the Bureau's 75 year history. The investigation panel of the House Government Operations Committee issued a report (H. Rep. 94-1667) on September 23, 1976, with the findings that:

(a) There exists within the bureau of Reclamation a "bureaucratic momentum" to build dams, and that once construction is begun the decision to halt construction is no longer an option. Safety problems are generally met with unquestioned reliance on the Bureau's abilities to "engineer" workable solutions.

(b) The Bureau was deficient in its examination of the geologic site of the Teton Dam.

(c) The Bureau was deficient in responding to warnings that the site might be dangerous.

(d) The United States Geological Survey was deficient in withholding information about geologic hazards at the Teton Dam site from the Bureau for six months.

Although this report and the disaster did not result in any legislative action, President Carter reactivated the 1972 Dam Inspection Program on November 28, 1977 (Los Angeles Times, 1977). Earlier that year, Congress had taken the initiative and voted \$15 million for federal inspection of private dams, but it was the November 5 collapse of the never inspected dam near Toccoa, Georgia, that led President Carter to implement this program. The President indicated he was committed to a multiyear program so that all high-hazard dams could be inspected within four years. The administration further sought to encourage states to share inspection responsibility. California's Dam Safety Program was cited as exemplary. It involves expenditure of about \$2 million annually for inspection and regulation of more than 1000 private, municipal, and state-owned dams.

5. Executive Orders 11988 and 11990. In addition to the above cited statutory measures, two comparatively recent executive orders have focused on the flood hazard. Executive Order 11988, was issued May 24, 1977, in furtherance of the National Flood Insurance Act of 1968, the Flood Disaster Act of 1973, and the National Environmental Policy Act of 1969. The Executive Order was intended to bring unity to the flood programs by requiring that each federal agency, through its normal activities, provide leadership and take action to reduce the risk of flood loss; minimize the impact of floods on human safety, health, and welfare; and restore and preserve the natural and beneficial values served by floodplains. The Executive Order set forth specific procedures for federal agencies to observe in carrying out this policy. These are depicted in Figure 3-3.

Executive Order 11990, of the same date, requires similar consideration for wetlands and was issued in furtherance of the National Environmental Policy Act of 1969. Under its terms federal agencies are prohibited from taking any action in a wetland unless the importance of the site clearly outweighs the requirements of the Executive Order, and are directed to (a) avoid the destruction or modification of wetlands; (b) avoid direct or indirect support of new construction in wetlands; (c) minimize the destruction, loss, or degradation of wetlands; and (d) preserve and enhance the natural and beneficial values of wetlands.

Thus, the two Executive Orders emphasized the need for sound floodplain management to guide future development as well as attempts to restore floodplains to enhance their natural values. However, progress under the Executive Orders has been limited. General information collected as of January 1980

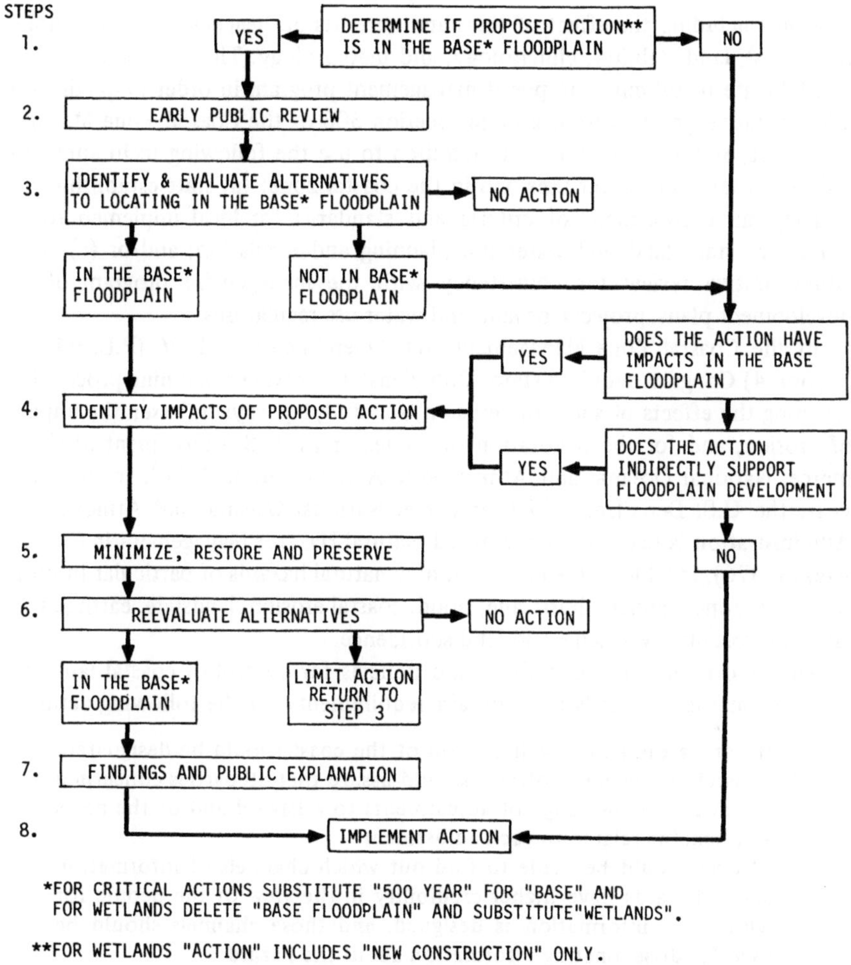

Figure 3-3. Decision-making process for E.O. 11988. (From Federal Register, 1980.)

indicated that of the 13 cabinet level agencies, seven had published proposed regulations, and three had developed informal drafts. For the 18 independent agencies, only eight had published final or interim procedures, and of the 49 subagency units, only 14 had published proposed or interim procedures (Thomas, 1980).

Federal Policies for Mitigation of Coastal Hazards

Although not emphasized in the statutory language, mitigation of the effects of coastal natural hazards is one of several objectives of the Coastal Zone Management Act of 1972 (P.L. 92-583). Among the subjects required to be addressed in the coastal plans required by the Act are floods and flood damage prevention, erosion (including the effect of tides and currents on beaches and other shoreline areas), land stability, climatology, and meteorology. The Act requires each coastal state to submit a proposed management program in order to qualify for administrative grant assistance under Section 306 of the Coastal Zone Management Act, and such programs are required to use the following techniques for control of land and water uses within the coastal zone: (a) state establishment, review, and enforcement of criteria and standards for local implementation; (b) direct state land and water use planning and regulation; and/or (c) state administrative review for consistency with the management program of all development plans, projects, or land and water use regulations.

In the Coastal Zone Management Act Amendments of 1976 (P.L. 94-370 Section 4) Congress required state "306 plans" to provide a planning process for assessing the effects of shoreline erosion, for studying ways to lessen the impact of erosion, and for the restoration of eroded areas (U.S. Department of Commerce, National Oceanic and Atmospheric Administration, 1976). In the same year, the U.S. Department of Commerce, National Oceanic and Atmospheric Administration issued a report, entitled Natural Hazards Management in Coastal Areas (1976), in which were identified nine natural hazards of particular interest to coastal zone planners: hurricane, flood, coastal erosion, landslide, earthquake, tsunami, volcano, avalanche, and land subsidence.

The report recommended that state policies for control of coastal lands be targeted on these hazards and contain requirements for the following actions:

1. Hazard areas along each section of the coast should be designated.
2. Coastal management offices should assure that all parties concerned are aware of the range of adjustments to a hazard and of the costs and benefits related to each adjustment.
3. Efforts should be made to find out which channels of information about hazards have higher credibility in the view of the people for whom the information is designed, and those channels should be used for disseminating information about the hazard.
4. Descriptions of the proposed change in adjustments to hazards also should discuss the existence or creation of the necessary powers to promote the new work within state or local agencies. Specific consideration should be given to ways in which planning for natural hazards in coastal areas can be linked with emergency planning for disasters under Section 201 of the Disaster Preparedness Act of 1974.

Federal Policies for Disaster Relief

Under the terms of the Disaster Relief Act of 1974 (P.L. 93-288) the authority to direct and supervise most federal disaster assistance activity was assigned to the office of the President. Under Executive Order 11795 (July 11, 1974) the Secretary of HUD was delegated the responsibility to coordinate the government's diffused disaster response programs. Executive Order 12148 transfered most of these functions to the Federal Emergency Management Agency.

In its current form the Disaster Relief Act of 1974 supplements other legislation and authorizes federal assistance to both public and private parties who experience losses in disasters. It further provides for the development and implementation of long-range recovery programs for major disaster areas. The Act applies to major catastrophes caused by the following phenomena: hurricane, tornado, storm, flood, high water, wind-driven water, tidal wave, tsunami, earthquake, volcanic eruption, landslide, mudslide, snowstorm, drought, fire, explosion, or other catastrophe that is so determined by the President. Federal assistance to disaster areas is initiated upon declaration by the President that a major disaster or emergency exists in a specified place. Although primarily concerned with providing postdisaster relief, this bill does contain some disaster mitigation measures.

For example, the Act provides for development of an effective disaster warning program, utilizing Civil Defense or other communications systems, and contains provisions aimed at achieving better coordination and delivery of previously authorized disaster services. Also the Act encourages states and local governments to apply for federal grants which provide up to $250,000 for the development of a *comprehensive program for disaster preparedness and prevention*, including mitigation, warning, emergency operations, rehabilitation, recovery, application of science and technology, and research. The preparation of such plans was first authorized by legislation enacted in 1969 and all but one state was utilizing federal funds for plan development purposes by September 1976.

Also, states, local governments, and individuals are encouraged by the Act to protect themselves by obtaining *insurance coverage* to supplement or replace government assistance. Owners of property that has been repaired or restored with federal disaster relief funds are required to obtain disaster insurance as a condition for receiving future federal disaster assistance. This requirement does not apply to individual homeowners, but only to state and local governments, nonprofit institutions, and public works projects.

The possible role of insurance in disaster relief and recovery operations was examined in a Senate committee report on the implementation of the 1969 Disaster Relief Act following Hurricane Camille. In the report, the special subcommittee on Disaster Relief, of the Senate Public Works Committee recommended that the use of some form of all-risk insurance be stimulated by the federal government. Testimony presented to the committee indicated that existing insurance coverage was inadequate protection against disasters like Camille, and that expanded insurance programs might have to be supervised and financed by the federal government. Property insurance coverage for extreme natural events also was recommended in President Nixon's message to Congress on disaster relief and Hurricane Camille (April 22, 1970). More recently (July, 1980) Congressman Patterson from California conducted hearings on the issue of availability and affordability of earthquake insurance.

The use of loss reducing *hazard mitigation measures* by state and local governments and other entities is also recommended by the Act. It requires that selected hazard evaluation and mitigation activities be undertaken in areas in which federal financing assistance has been provided for the repair or replacement of selected types of properties, including those owned by federal, state, and local governments, as well as those owned by nonprofit entities. Under Section 406 of the Act, FEMA is to provide technical assistance regarding hazard evaluation methods and hazard mitigation codes, standards, and specifications for the repair or reconstruction of public and private structures.

Other major provisions of the Act require or authorize appointment of a federal coordinating officer to operate in the affected area; mobilization of federal personnel as emergency support teams; assistance of federal agencies in distributing food, supplies, and medicine, removal of debris, etc.; use of local firms for assistance and recovery work; nondiscrimination in disaster assistance; priority for public housing assistance; repair and restoration of damaged public facilities; temporary housing assistance; unemployment assistance and increased benefits; relocation assistance; family and individual counseling for mental stress; and loans and grants to local governments for economic recovery.

Amendments and supplemental appropriations are occasionally added to deal with special situations. For example, one set of amendments (P.L. 92-209) authorizes federal financial assistance to private nonprofit medical care facilities damaged or destroyed by major disasters occurring after January 1, 1971.The adoption of this amendment which made available to these facilities the same type of grants offered to publicly owned facilities was prompted by the damage sustained by several hospitals in the Los Angeles area as a result of the earthquake in February 1971.

Another form of federal assistance to private owners of disaster damaged property was provided by a 1972 enactment which authorizes tax deductions for disaster losses (P.L. 92-418). Persons who suffer property losses in the first six months of a taxable year due to an event in a Presidentially declared disaster area are authorized to file an amended federal income tax return for the previous year, claiming uninsured losses as a deduction. It should be noted that this form of tax transfer has important implications to the equity of public policies designed to provide relief to disaster victims. A recent Comptroller General report (1980) states that less well-to-do residential property owners are in lower tax brackets and thus bear a higher percentage of disaster losses than those in the higher tax brackets. The resulting inequities are further reinforced because lower-income groups tend to constitute a higher proportion of floodplain residents than higher income groups. Figure 3-4 shows the affect of the 48% and 21% tax brackets on the percent of costs born by disaster victims as a result of interest rates on government loans, maturity periods, and effective income tax rates. Direct federal payments for disaster assistance during the period 1953 to 1973 under this and preceding acts are shown in Table 3-4.

Federal Earthquake Policies

The basis for federal policy related to the earthquake hazard is defined in the Earthquake Hazards Reduction Act of 1977 (P.L. 96-124). This Act recognizes earthquake as a major natural hazard and deals with prevention and mitigation

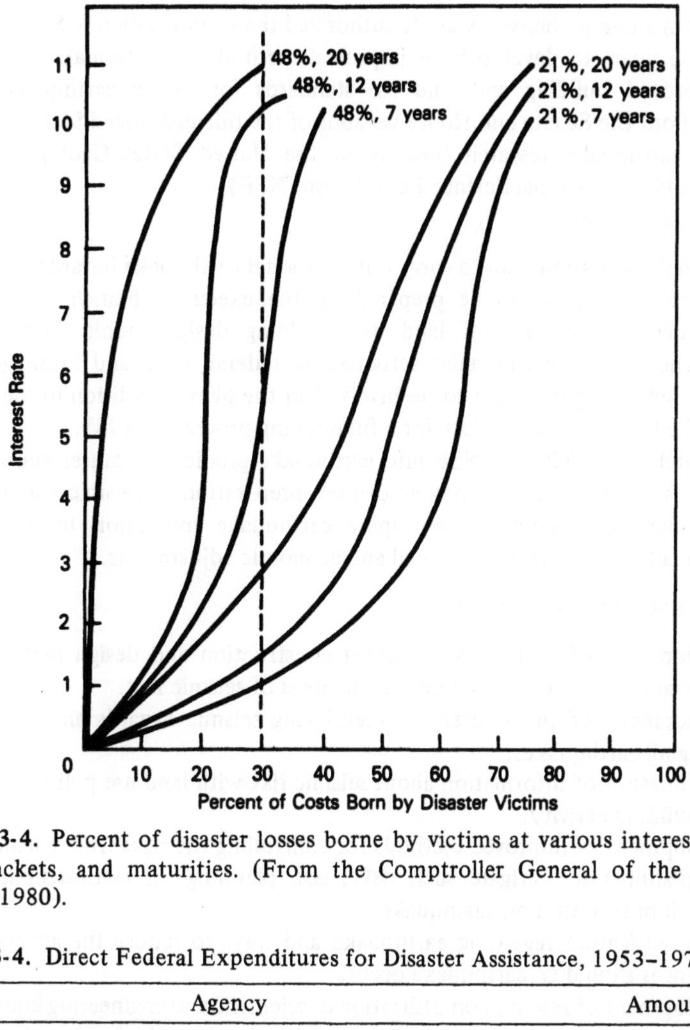

Figure 3-4. Percent of disaster losses borne by victims at various interest rates, tax brackets, and maturities. (From the Comptroller General of the United States, 1980).

Table 3-4. Direct Federal Expenditures for Disaster Assistance, 1953–1973

Agency	Amount ($)
1. Federal Disaster Assistance Administration (FDAA), formerly Office of Emergency Planning and Office of Emergency Preparedness (OEP)	1,844,827,290
2. Small Business Administration	809,254,922
3. Farmers Home Administration	448,180,766
4. Department of Agriculture	18,415,159
5. Federal Highway Administration, formerly Bureau of Public Roads	484,637,000
6. U.S. Army Corps of Engineers	299,341,940
7. Veterans' Administration	2,000,000
8. Office of Education	102,330,691
9. Federal Insurance Administration	46,774,000
Total	4,051,761,768

Source: Cochrane, 1975.

methods in a comprehensive way. It authorized the appropriation of $250 million over three years to develop both improved methods of earthquake prediction and improved building and land use standards for use in earthquake-prone regions. Both the Senate and House versions of the bill authorized $102.5 million for the earthquake research budgets of the United States Geologic Survey (USGS) and the National Science Foundation (NSF).

The Act requires:

1. An implementation plan to carry out a national earthquake hazards reduction program. The plan, to be prepared by the executive branch is to ensure informed coordination of land use, building design, public information, insurance, warning and relief activities by federal, state and local agencies. Eight federal agencies are to be involved in the plan in addition to the USGS and the NSF. The target date for a functioning program is 1985.
2. Research by USGS and NSF into earthquake prediction, causes and mechanisms of earthquakes, zoning guidelines, preparation of seismic risk analysis for emergency planning, developing earthquake mitigation techniques in man-made structures and in social and economic adjustments.

The main objectives of the Act are:

1. Development of earthquake resistant construction and design methods for public and high occupancy buildings in areas of seismic risk.
2. Development of procedures for identifying seismic hazards and predicting damaging earthquakes.
3. Coordination of information about seismic risk with land use policy decisions and building activity.
4. Development of improved methods for controlling the risks from earthquakes and planning to mitigate such risks; also planning for reconstruction and redevelopment after an earthquake.
5. Public education regarding earthquake and ways to reduce the adverse consequences should an earthquake occur.
6. Development of research on utilization of scientific and engineering knowledge to mitigate earthquake hazards. The social and economic, legal, and political consequences of earthquake prediction, and ways to assure the availability of earthquake insurance or some functional substitute.
7. Development of research applied to control or alteration of seismic phenomena.

In a report to the U.S. House of Representatives (proposing adoption of the bill), the Committee on Science and Technology (May 11, 1977) stated:

> Of all the potential mechanisms to avoid earthquake hazards, the simplest and most direct would be zoning. Although cities cannot be relocated and undeveloped high-risk areas may be potentially very valuable, several courses of zoning action may be feasible:
>
> 1. Risk zoning of critical parts of already developed areas to turn them into park land or other nonhazardous use as opportunity arises.
> 2. Risk zoning of high risk undeveloped areas to prevent future hazardous development.

> 3. Development of systematic techniques for collection and evaluation of data for use in microzoning (zoning of comparatively small areas) and the establishment of criteria for microzone levels of risk.
> 4. Adoption of building codes which require higher levels of earthquake resistance in higher risk areas.
>
> Since many areas of high seismic risk are already heavily developed and populated, the promotion of earthquake resistant building practices is receiving priority consideration as an effective approach to minimize earthquake damages from ground movement, which causes over 90 percent of the damages. Even if an earthquake is anticipated and residents have evacuated the area, there is still a need to reduce the damage to buildings and other facilities. If there has been no advance warning and no evacuation, the need for safer building is obviously critical.
>
> Since the greatest earthquake hazards result from inadequacies in building construction, this is obviously an area in need of close attention. The current state of the art in defining seismic design criteria and in earthquake resistant construction techniques leaves much to be desired. Therefore, additional research to advance the state of the art is of critical importance. Buildings not constructed in accord with adequate design provisions should be evaluated and, if found to be hazardous, should be strengthened or replaced. So that new construction should not add to the earthquake hazards, seismic design criteria providing appropriate resistance should be incorporated in building regulations and enforced.
>
> The costs of adopting building standards designed to reduce earthquake hazards vary considerably. Witnesses estimated that the implementation of seismic standards increase new building costs one to six percent.
>
> The costs can run much higher for special buildings such as hospitals. The earlier earthquake resistance is introduced into the construction process, the less the cost increase.

Advocates of the legislation contended it was needed because more and more Americans live in high-risk areas on both coasts, scientific advances make accurate prediction a real possibility, and seismological data are needed to evaluate site selection for nuclear power plants and other critical facilities.

Other Federal Hazard-Related Policies

In addition to the major policies discussed above, the federal government administers numerous other hazard-related programs. Included are predisaster services (emergency preparedness to mitigate the effects of the disaster), postdisaster services (emergency preparedness for disaster relief), and natural disaster warning activities (see Figure 3-1).

For example, the Farmers Home Administration (FmHA) provides Watershed Protection and Flood Prevention Loans. As described in the Catalog of Federal Domestic Assistance,

> Loan funds may be used to help local sponsors provide the local share of the cost of watershed works of improvement for *flood prevention*,

irrigation, drainage, water quality management, sedimentation control, fish and wildlife development, public water based recreation and water storage and related uses. (Program Number 10.419)

Other agencies with programs in flood hazard mitigation include the Bureau of Reclamation and the Soil Conservation Service, as well as regional agencies such as the Tennessee Valley Authority. Most other federal programs appear to be directed toward disaster relief with the exception of those administered by the National Science Foundation and the U.S. Geological Survey, which sponsor basic and applied research. Disaster relief programs include loans to victims, insurance for crops and buildings, emergency physical rehabilitation programs, and planning for disaster preparation.

Through the National Oceanic and Atmospheric Administration, the federal government also issues warnings for five of the hazards studied here (floods, hurricanes, tornadoes, tsunami, and severe wind). Although no earthquake warning system has been developed, the U.S. Geological Survey is performing significant research in this field and currently provides data relative to earthquake and landslide-prone areas. Storm surge warnings are included as hurricane warnings. No warning is provided for expansive soil, since it poses no sudden and unexpected threat.

State Land Use Policies

Although state level policies that specifically address natural hazards are limited in number, all states have adopted land use policies that may serve as a framework for natural hazard mitigation. Kusler (1976) has summarized and analyzed state statutes authorizing local governments and state agencies to adopt zoning regulations, subdivision controls, building codes, and special flood hazard regulations, with emphasis on land use control for the regulation of flood-prone areas. *The Seventh Annual Report of the Council on Environmental Quality* summarized state land use and planning authority in categories complementary to those devised by Kusler (Council on Environmental Quality, Executive Office of the President, 1976). The CEQ noted that, by mid-1976, most states had an active interest in land use controls—something that was rare only five years earlier when *The Quiet Revolution in Land Use Control* was published by CEQ as a review of "pioneering efforts of a few states and regions to restructure government and to fashion policies for improving land development decisions."

Developed from the Kusler and CEQ reports, Figure 3-5 summarizes the current statutory authority of the 50 states with respect to land use policy, with special reference to flood mitigation authority. In general, the following situation prevails.

1. General Authorization. All states have authorized the adoption of zoning regulations, but only twenty have adopted requirements for subdivision regulations at the state level. Typically, these statutes require or allow the adoption

State	General Authorization: Subdivision	General Authorization: Building Code	Home Rule Powers: General	Home Rule Powers: Special Reference to Land Use	Extraterritorial Controls: Subdivision	Extraterritorial Controls: Zoning	Extraterritorial Controls: Building Code	Prior Planning Requirements: Comprehensive Plan for Zoning	Prior Planning Requirements: Comprehensive Plan for Subdivision	Prior Planning Requirements: Flood or Drainage Language	Enabling Authority for Adoption of Interim Regulations: Zoning	Enabling Authority for Adoption of Interim Regulations: Subdivision	Sensitive Land Uses Included in Regulations — Subdivision: Public Utilities	Sensitive Land Uses — Subdivision: Agriculture	Sensitive Land Uses — Subdivision: Large Lots	Sensitive Land Uses — Zoning: Agriculture	Sensitive Land Uses — Zoning: Non-Conforming Uses Regulated	Sensitive Land Uses: Wetlands	Sensitive Land Uses: Power Plant Siting	Sensitive Land Uses: Designation of Critical Areas	Comprehensive Permit System	Coordinated Incremental Planning	Mandatory Local Planning	Differential Assessment Laws	Statutes Authorizing Adoption of Flood Plain Regulations by State Agencies	Local Flood Hazard Regulations Specifically Authorized: Zoning	Local Flood Hazard Regulations Specifically Authorized: Subdivision	Local Flood Hazard Regulations Specifically Authorized: Building Code	State
AL		•	•	•	•			•	•	•				•	•	•			•						•	•	•	•	AL
AK								•					•	•	•	•	•		•			•		•			•		AK
AZ	•		•		•	•				•			•						•			•		•	•	•	•	•	AZ
AR					•	•				•			•	•	•	•			•					•	•	•	•	•	AR
CA	•	•	•		•			•		•	•		•			•	•		•			•		•	•	•	•		CA
CO	•		•		•			•	•	•	•		•	•	•	•			•	•				•	•	•	•	•	CO
CT		•	•					•	•	•			•	•	•	•		•	•			•		•	•	•	•		CT
DE			•					•					•			•		•				•		•		•	•		DE
FL	•	•	•			•		•	•				•	•		•	•	•	•	•	•	•	•	•		•	•		FL
GA	•	•	•	•	•	•		•	•	•			•	•			•	•				•				•			GA
HI	•							•					•	•	•	•			•	•	•	•		•	•	•	•	•	HI
ID			•		•			•			•	•	•				•						•	•					ID
IL			•		•	•				•									•					•	•	•	•	•	IL
IN		•			•	•			•	•			•	•	•	•	•					•		•	•	•	•	•	IN
IA	•	•	•	•	•	•		•					•	•	•		•							•	•	•			IA
KS	•		•		•	•		•	•				•	•	•										•	•	•		KS
KY					•	•		•	•	•	•	•							•					•	•	•	•		KY
LA			•	•				•		•			•		•	•	•	•								•	•	•	LA
ME	•	•	•					•	•				•	•		•	•	•	•	•	•	•	•	•	•	•			ME
MD		•	•	•	•			•	•	•			•		•		•	•	•	•		•		•	•		•		MD
MA		•	•										•	•	•	•		•	•					•	•	•	•		MA
MI	•		•	•				•	•	•	•		•	•		•								•	•	•	•		MI
MN	•	•	•		•	•		•			•		•	•	•	•		•	•	•		•		•	•	•	•		MN
MS								•					•		•		•	•							•				MS
MO			•		•	•		•	•									•	•					•	•				MO
MT	•	•			•	•	•	•		•	•		•						•	•		•	•	•	•	•	•		MT
NE			•		•	•	•	•	•				•	•		•			•				•	•	•	•			NE
NV	•	•	•					•		•			•			•	•		•	•		•	•	•		•	•		NV
NH								•			•		•	•	•	•		•	•					•			•		NH
NJ			•		•	•		•		•			•			•		•	•					•	•	•	•		NJ
NM	•				•	•	•	•		•			•	•	•	•	•		•			•		•		•	•		NM
NY	•	•	•					•		•			•	•	•	•	•	•	•	•	•	•		•	•	•	•	•	NY
NC		•			•	•	•	•					•	•			•	•				•		•	•	•			NC
ND	•				•	•		•	•				•	•	•				•					•		•			ND
OH		•	•		•			•	•				•	•					•					•					OH
OK			•					•	•	•			•		•									•	•				OK
OR	•	•	•		•	•		•			•		•	•		•			•	•		•	•	•			•		OR
PA			•					•		•			•			•	•	•	•	•				•	•	•	•		PA
RI		•	•					•					•	•	•	•		•	•			•		•		•			RI
SC		•			•			•	•	•			•	•		•	•		•					•		•	•		SC
SD					•	•		•	•	•	•		•	•	•	•	•			•				•					SD
TN		•	•		•	•			•				•	•	•				•							•	•		TN
TX			•	•	•			•					•		•	•	•	•						•		•	•	•	TX
UT			•					•	•	•	•		•		•	•	•					•		•		•	•	•	UT
VT	•							•		•	•	•	•	•	•	•	•	•	•		•	•		•	•	•	•	•	VT
VA		•			•					•			•	•	•	•		•					•	•		•	•		VA
WA	•	•	•		•			•			•		•	•		•	•	•	•			•		•	•	•	•		WA
WV			•	•					•	•			•	•	•										•	•			WV
WI	•	•	•		•	•		•		•	•		•	•	•	•		•	•	•		•			•	•			WI
WY					•			•	•				•	•	•	•			•			•	•	•					WY

Figure 3-5. Land use and building code authority in the United States.

of subdivision regulations by local government units under certain statewide guidelines. Thirteen of the states refer to the Interstate Land Sales Act (Arizona, California, Colorado, Florida, Georgia, Hawaii, Iowa, Kansas, Minnesota, New York, North Dakota, Oregon, and Washington). To the extent that natural hazard policy is to be established at the state level, many important provisions can be inserted into existing statutes within the current regulatory framework, without having to prepare and adopt a basic legislative package.

2. Home Rule Powers. Thirty-four states have allowed home rule for at least some local government units, and eight of these states make special reference to land use regulation in their home rule statutes. Some other states grant special powers to their largest cities. Home rule can be important for communities in adopting flexible and sensitive natural hazard mitigation policies appropriate for their jurisdictions.

3. Extraterritorial Controls. Of the 50 states, 31 grant extraterritorial subdivision controls to their municipalities, generally extending three to five miles beyond corporate limits. Twenty-one states grant extraterritorial zoning controls to their municipalities. In a few states, there is concurrent jurisdiction between municipalities and counties, and in others the county government has primary responsibility for subdivision and zoning. Four states that authorize both extraterritorial zoning and subdivision controls also authorize extraterritorial building code authority.

4. Prior Planning Requirements. In 32 states, zoning is required to be in accordance with a comprehensive plan, and 10 states require even more explicit planning prior to the adoption of zoning regulations. Twenty-two states require a comprehensive plan, master plan, or at least a street or transportation plan prior to the adoption of subdivision regulations. Twenty-seven states include specific flood or drainage language in their comprehensive plan enabling authority. In practice, however, a comprehensive planning requirement is nearly universal in the United States, since it has been an integral part of the HUD 701 planning process.

The Model State Zoning Enabling Act, published by the U.S. Department of Commerce in 1926, directs that zoning shall be "in accordance with a comprehensive plan," and municipal zoning ordinances typically have incorporated this terminology. A comprehensive plan for a municipality is important for several reasons beyond that of generally promoting "orderly development." Kusler (1976) points out that the extent of encroachment allowed in a floodplain depends partly on the availability of land elsewhere for similar purposes. If land elsewhere is available for, say, industrial uses, such should be excluded from floodplains even if the buildings could be floodproofed. Second, land uses in areas surrounding a critical area for natural hazards may help to define the boundaries of high-risk areas. Again, it is a matter of relating the level of restrictions to the potential for loss resulting from a given occurrence of a natural event.

In addition, a comprehensive plan may specify density of development and, to a certain extent, building types, to support development less susceptible to damage during a given type of natural disaster. Also, the circulation element in a comprehensive plan can help ensure that the street network will not easily be disrupted in an emergency.

5. Interim Regulations. Several states permit local jurisdictions to adopt interim regulations. Ordinances that suspend all development for a specified period of time are the most common. Such ordinances, sometimes called "Holding Zones," establish that no new development may take place in a given area for a period ranging from six months to two years, until a comprehensive plan or acceptable compromise development plan can be adopted. Fourteen states authorize some type of interim zoning regulations, and three states have similar authority for subdivisions. Florida allows the adoption of interim regulations with respect to building codes. Interim regulations can be particularly helpful in a period of transition, when a municipality or county is attempting to prepare equitable regulations in the face of increasing development pressures or citizen controversy with respect to a hazard-prone area. Even without specific interim regulation authority, however, the zoning power generally allows for the designation of low-density special permit zoning districts, which can serve a similar purpose if adopted prior to protracted controversy surrounding a site development plan.

6. Sensitive Land Uses. In addition to the general procedural and substantive regulations discussed above, some states have enacted regulations specifically addressed to hazard mitigation subjects such as the siting of power plants in hurricane, earthquake, and other hazard zones. However, numerous states have enacted statutes which exempt certain types of property from the regulation standards contained in land use or subdivision regulations. For example, four states exclude certain public utilities from these standards. These exemptions are significant because, as Kusler (1976) points out, public utility facilities such as roads, bridges, and levees are major offenders in blocking flood flows. They can also pose threats in an earthquake and other extreme natural events. Seventeen states exempt agricultural uses from subdivision regulation, and the same number (although not the same states) exempt them from zoning regulations. Agricultural fills, dikes, fences, and buildings also may block floodways and are subject to flood damage; and they can be hazards should an earthquake or tornado occur.

The subdivision of large lots, often over 2½ acres, is subject to regulation in only 29 states. If large lots are exempt, fewer regulatory measures can be applied to sensitive areas.

Nonconforming land uses, defined by the ASPO Model Zoning Ordinance (American Society of Planning Officials, 1966) as lots, structures, or uses lawful before the zoning ordinance was passed but which would be prohibited or restricted under the ordinance, are exempt from zoning regulations in 38 states. Such uses can pose particular hazards in an extreme natural event because the uses are generally unspecified; they tend to be older, unharmonious uses conflicting in intensity and scale with permitted uses in the zoning district. They may exist essentially unregulated until a natural disaster removes them, but meanwhile they affect development patterns and influence risk levels in surrounding areas. Although supporters justify the exemptions of these uses on the basis of a variety of arguments it is undeniable that such exemptions have the capacity to reduce the effectiveness of natural hazard mitigation regulations.

Other sensitive uses regulated by some states include, first, inland wetlands. Twenty-two states either have established uniform permit procedures for these areas, or uniform regulations. All eligible states (a total of 30) are participating in the federally funded coastal zone management program authorized by the Coastal Zone Management Act of 1972. Second, 34 states have the authority to determine the siting of power plants and related facilities. This is an important land use tool particularly with respect to earthquake mitigation. Third, 13 states have established regulations for the identification and designation of areas of critical state concern, such as environmentally fragile areas.

In addition to the above policies, another important land use regulatory measure that can affect natural hazard mitigation is the requirement that permits be secured prior to the initiation of certain types of development. Five states have adopted legislation containing such requirements, and 24 states have established mechanisms to coordinate land use planning activities. Nine states require their local governments, not merely authorize them, to establish a mechanism for land use planning through zoning, a comprehensive plan, and a planning commission.

Related to the direct regulation of land use are the financial policies that support certain land use configurations. For example, the continuing conversion of agricultural land to other uses at the fringe of urban areas, resulting partly from increased tax assessments of peripheral agricultural land, can have serious consequences for a comprehensive planning effort. At this time, 42 states have developed tax measures designed to give property tax relief to owners of agricultural or open space lands. A similar policy could help to regulate lands subject to hazards such as landslides, or along a seismic fault, subject to appropriate development by the property owners.

Both Kusler and the CEQ have identified those states which have authorized the adoption of floodplain regulations by state agencies. They indicate that 30 states have adopted regulations which may offer models for more general hazard mitigation measures. Finally, Kusler discusses those states that have included specific flood hazard regulations in various land use measures. A total of 39 states have zoning enabling legislation, and 34 states have enacted similar authorizations for subdivision control. The most common regulatory measure is the power to secure "safety from flood."

State Coastal Zone Regulations

In addition to statewide authority for natural hazard mitigation, several states have developed regulations relating specifically to the coastal zones adjoining the oceans and the Great Lakes (see Figure 3-6). These activities have been supported by the Coastal Zone Management Act of 1972.

As noted by Platt (U.S. Department of Commerce, NOAA, 1976), California, Washington, and Rhode Island have pioneered the concept of state coastal zone management. While there are major differences between the three states' coastal zone management programs, several common features can be identified. First, relevant planning areas consist of entire shorelines, not simply discrete landforms or problem areas. Second, administration is largely a state function with specific responsibilities delegated to certain local and regional entities in California and

	COMPREHENSIVE COASTAL OR LAND USE	SHORELAND ZONING	STATE FLOOD-PLAIN REGULA-TIONS	CRITICAL AREAS	COASTAL WETLANDS	SETBACK OF ENCROACHMENT LINES	BEACH AND SHORE PRESERVATION
ALABAMA	a						
ALASKA							
CALIFORNIA	b		x				
CONNECTICUT			x		x	x	
DELAWARE	c				x		x
FLORIDA	d			x		x	x
GEORGIA					x		
HAWAII	e	x	x			x	
ILLINOIS							
INDIANA							
LOUISIANA					x		
MAINE		x	x	x	x		
MARYLAND	f		x	x	x		x
MASS.					x		
MICHIGAN		x	x				
MINNESOTA		x	x	x	x		
MISS.					x		
N.H.					x		
NEW JERSEY	g		x		x		
NEW YORK			x		x		
N.C.	h		x	x	x		
OHIO							
OREGON	i			x			
PA.							
R.I.	j				x		
S.C.							
TEXAS					x		
VIRGINIA					x		x
WASHINGTON	k		x		x		
WISCONSIN		x	x				

[a] ALA. COASTAL ZONE DEVELOPMENT ACT OF 1973.

[b] CALIF. COASTAL ZONE CONSERVATION ACT OF 1972 (FINAL PLAN ADOPTED AUGUST, 1976). SAN FRANCISCO BAY CONSERVATION AND DEVELOPMENT COMMISSION ACT OF 1969.

[c] DEL. COASTAL ZONE ACT OF 1971.

[d] FLORIDA LAND AND WATER MANAGEMENT ACT OF 1976.

[e] HAWAII STATE LAND USE ZONING ACT.

[f] MD. STATE LAND USE ACT OF 1974.

[g] N.J. COASTAL AREA FACILITIES REVIEW ACT OF 1973.

[h] N.C. COASTAL AREA MANAGEMENT ACT OF 1974.

[i] ORE. LAND CONSERVATION AND DEVELOPMENT ACT OF 1973.

[j] R.I. COASTAL MANAGEMENT ACT OF 1971.

[k] WASH. SHORELINE MANAGEMENT ACT OF 1971.

Figure 3-6. State laws relating to coastal hazard mitigation. (From Platt, 1976b).

Washington. Third, state coastal authority extends inland to embrace activities and physical features associated with the coastline.

According to Platt, these programs may be readily adapted to incorporate new perceptions of natural hazards. For example, the Rhode Island Coastal Management Council denies permits for development on "undeveloped" beach areas and seeks to limit construction on dunes or beaches. As summarized in Figure 3-6, selected provisions of state policies are as follows.

1. Shoreline Zoning. Shoreline zoning is practiced in Hawaii, Maine, Michigan, Minnesota, and Wisconsin. Local governments must adopt land use regulations for their river and lake shoreline areas or such regulations will be adopted on their behalf by the state. In either case, administration of regulations remains with the local government.

2. State Floodplain Regulations. New York has adopted mandatory floodplain zoning measures for all communities with recognized flood hazard areas. Eleven other coastal states have adopted other forms of state-level floodplain management legislation.

3. Critical Areas Programs. The Model Land Development Code proposed in 1975 by the American Law Institute suggested that states assume particular responsibility for "critical aieas," as defined by physical, cultural, economic, or aesthetic criteria. As applied to coastal zone management, the critical areas approach is more restrictive geographically than the techniques described above, according to Platt. Critical areas programs have been adopted by Maine, Maryland, Florida, North Carolina, and Oregon.

4. Coastal Wetland Programs. Several states rely on their coastal wetlands permit programs for coastal zone management purposes, according to Platt (U.S. Department of Commerce, NOAA, 1976). With the recent attention given to the importance of coastal wetlands in the ecologic food chain, and the extent of their loss due to development along the Atlantic and Gulf Coasts, many states have implemented coastal wetland programs.

5. Mandatory Setbacks. A mandatory setback or "encroachment line" may be legally imposed to restrain all further development of fill within a specified distance of a body of water. In Connecticut, encroachment lines have been established to protect the 100 year floodplain portions of the Connecticut River and certain other streams.

Great Lakes states are turning to mandatory setbacks as a response to severe erosion occuring due to high lake levels. Michigan is proposing a statewide setback for lakeside development equivalent to 30 years of erosion (the average term of a mortgage). Illinois is considering a 100 year erosion setback, a distance of possibly 200 feet along its "North Shore" bluffs.

6. Development Moratoria in Coastal Areas. Platt notes that most measures described above require extensive research and planning studies as a prerequisite to final implementation. Accordingly, some states (Washington, California, New York, and Florida) have adopted an interim period of control through state legislation to be superseded by a final plan.

Building Code Policy and Authority

Traditionally, the enactment and enforcement of building codes have been a local government concern. In his survey of flood-related statutory authority, Kusler found that only 23 states had adopted one building code to be used by all governments, or otherwise regulated building code provisions over the entire state. Twelve of these states specifically authorized flood hazard regulations and four states extended building code authority of municipalities to extra-territorial zones. Field and Rivkin (1975) note that over 15,000 localities issue building permits, and approximately 8000 have their own building code, either based on a national code or developed locally.

Four national model building codes have been prepared by various entities, thus providing states and local governments with the option of enacting all or part of the recommended code provisions. The four code associations are influential in different parts of the country, although there is some overlap (Figure 3-7). Field and Rivkin (1975) note:

> The International Conference of Building Officials (ICBO) claim is in the West, Building Officials and Code Administrators (BOCA) in the Northeast, and Southern Building Codes Congress (SBCC) in the South. In the North Central region, BOCA and ICBO actively compete against each other for city members. American Insurance Association (AInA), creator of the National Building Code, is active in three regions, but has no penetration in the West. AInA, unlike the other three code associations, is not an association of building officials. Established in 1905, AInA created the National Building Code as part of their underwriting procedures for ascertaining risk of insurance losses due to fires. Overall, model codes are most evident in the South and West.

In 1970 Field and Ventre (as noted in Field and Rivkin, 1975) surveyed 919 cities in the United States with respect to building codes. At that time 73% of the cities in the survey based their code on one of the four national model codes. State based codes, themselves often based on one of the four model codes, were used by 1.3% of the cities, and locally drafted codes by 10.8%. Only 2.2% of the cities surveyed (all of them small) reported having no building code. The major hazard-related provisions of the several model codes follow.

Wind

Although building code policy can be important for the mitigation of several natural hazards, their widest application has been in preventing wind damage, including hurricane winds. The four model codes specify minimum wind loads

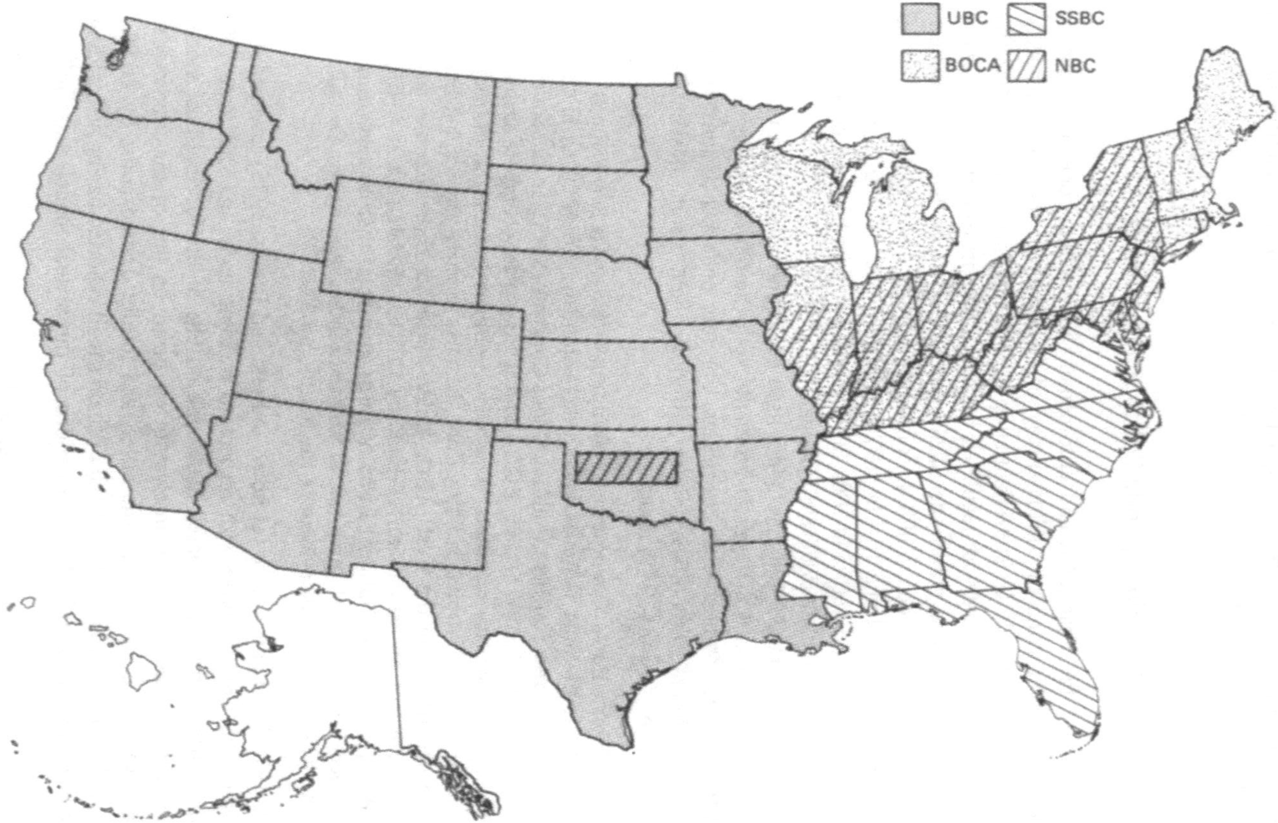

Figure 3-7. General building code application, by state.

for design, in terms of minimum wind pressure for various heights above ground level. None of these codes provides specifically for tornadoes.

The methodology and design wind pressures vary among the four model building codes. The BOCA code wind provisions specify a single set of wind pressures that vary with height. The other three codes set forth several sets of wind pressures with height; the set that is applicable for any given location depends on the expected magnitude of extreme winds for a given level of risk (such as once in 100 years).

The National Building Code has patterned its wind provisions after the recommendations of the American National Standards Institute (ANSI) A58-1-1972 Building Code (1972). First, the basic wind speed must be identified for a municipality; this is accomplished by means of a map contained in the code developed by Thom (1968), which has superimposed upon it isotachs, or contours, of the basic wind speed for a return period of 100 years.

The Southern Standard Building Code (SSBC) of the Southern Building Code Congress also utilizes Thom's 100 year wind speed map to determine the wind speed for a given locality. Wind load pressures are specified in the SSBC code for various values of the basic wind speed. The pressures specified in this code are, however, different than those specified by the NBC.

As with the National Building Code and the Southern Standard Building Code, the Uniform Building Code also utilizes a map; however, the map defines minimum allowable wind pressures at a reference height of 30 feet. From this, the wind pressures for different height zones are determined from a table in the UBC. The 1976 UBC wind provisions have changed little over the years and are essentially the same as found in its 1961 edition (Buoh and Bihr, 1975).

The method by which the wind pressures are used to obtain design forces is the same for each building code with the exception of the NBC. The pressures obtained by the BOCA, SSBC, and UBC may be used directly for design while the pressures found by the NBC must be modified by external pressure coefficients before design. For purposes of this analysis, the UBC has been chosen as the reference code.

To obtain an approximate equivalency in design wind loads among the model codes, the wind pressures specified by the BOCA code must be multiplied by the factors shown in Table 3-5, to arrive at the same wind forces that would be calculated by the method used by UBC. Tables 3-6 and 3-7 are the multiplication factors that need to be applied to the wind pressures of the NBC method to

Table 3-5. Multiplication Factors to BOCA Wind Pressures for Approximate Equivalency to UBC Design Wind Loads for All Heights

UBC wind pressure zone (psf)						
20	25	30	35	40	45	50
1.25	1.50	1.90	2.10	2.35	2.70	3.00

Table 3-6. Multiplication Factors to NBC Wind Pressures for Approximate Equivalency to UBC Design Wind Loads for Heights Less than 100 Feet

NBC basic wind speed (mph)	UBC wind pressure zone (psf)						
	20	25	30	35	40	45	50
50	1.90	2.20	2.80	3.00	3.50	3.90	4.40
60	1.40	1.60	2.00	2.20	2.50	2.90	3.30
70	1.00	1.15	1.44	1.60	1.80	2.05	2.35
80	0.75	0.85	1.07	1.15	1.35	1.50	1.70
90	0.55	0.65	0.85	0.90	1.05	1.20	1.30
100	0.45	0.52	0.65	0.72	0.85	0.95	1.05
110	0.40	0.45	0.55	0.60	0.70	0.77	0.85
120	0.30	0.36	0.45	0.50	0.57	0.65	0.72
130	0.26	0.30	0.36	0.42	0.47	0.55	0.60

obtain the equivalent design wind forces by the UBC. Tables 3-8, 3-9, and 3-10 show the multiplication factors to obtain equivalency of the SSBC to the UBC design wind forces (Hart, 1976).

Of the four model codes, the BOCA code specifies just a single set of wind pressures. These pressures are generally significantly less than that which would be required by any of the three other codes. BOCA does not seem to recognize that higher design pressures may be required. Section 716.0 of the Code does state that "for building and structures located . . . in geographical regions subject to higher wind loads than herein specified, the design wind load shall be determined by the prevailing conditions." However, no performance standards or guidance are provided as to what the wind loads should be or what regions may be subject to higher wind loads.

The UBC, although allowing for geographical variation in the minimum

Table 3-7. Multiplication Factors to NBC Wind Pressures for Approximate Equivalency to UBC Design Wind Loads for Heights Greater than 100 Feet

NBC basic wind speed (mph)	UBC wind pressure zone (psf)						
	20	25	30	35	40	45	50
50	1.60	2.10	2.50	3.00	3.30	3.70	4.00
60	1.20	1.60	1.90	2.20	2.40	2.80	3.00
70	0.90	1.10	1.30	1.50	1.70	1.90	2.10
80	0.65	0.85	1.00	1.10	1.30	1.45	1.60
90	0.55	0.65	0.80	0.90	1.00	1.15	1.25
100	0.40	0.50	0.65	0.70	0.80	0.90	1.00
110	0.35	0.43	0.50	0.60	0.65	0.75	0.80
120	0.30	0.35	0.42	0.50	0.57	0.65	0.70
130	0.25	0.30	0.35	0.40	0.47	0.55	0.60

Table 3-8. Multiplication Factors to SSBC Wind Pressures for Approximate Equivalency to UBC Design Wind Loads for Heights Less than 100 Feet

SSBC basic wind speed (mph)	UBC wind pressure zone (psf)						
	20	25	30	35	40	45	50
70	1.25	1.45	1.80	2.00	2.30	2.60	2.90
80	0.95	1.12	1.40	1.55	1.75	2.00	2.25
90	0.75	0.90	1.13	1.25	1.42	1.60	1.80
100	0.60	0.72	0.90	0.98	1.15	1.30	1.45
110	0.50	0.60	0.75	0.82	0.95	1.07	1.20
120	0.43	0.50	0.62	0.68	0.78	0.90	1.00
130	0.36	0.42	0.53	0.58	0.67	0.76	0.85

allowable wind pressures, is based on wind information that was collected through 1951. This suggests that a reevaluation of the UBC wind requirements may be in order so that more current wind data are included.

In addition to the wind provisions of the four model codes, local communities subject to severe wind levels have adopted building requirements to meet their local needs. For example, the Corpus Christi, Texas, building code specifications for wind standards were developed from their experience with previous hurricanes.

Figure 3-8 provides an illustration of the relationship between the wind pressure criteria of the standard codes. The precise relationship between code design level forces is somewhat more complex because the relative level of wind loads presented in the various codes is a function of building height, wind speed, and experience.

In addition to the wind provisions of the standard building codes, certain hazard-prone states have developed extensive building code standards for natural hazard mitigations. For example, the Texas Coastal and Marine Council (1976, 1977) and Lesso (1976) analyzed the natural hazard threat along the Texas

Table 3-9. Multiplication Factors to SSBC Wind Pressures for Approximate Equivalency to UBC Design Wind Loads for Heights between 100 and 500 feet

SSBC basic wind speed (mph)	UBC wind pressure zone (psf)						
	20	25	30	35	40	45	50
70	1.10	1.40	1.70	2.00	2.20	2.50	2.70
80	0.85	1.10	1.30	1.50	1.65	1.95	2.10
90	0.67	0.85	1.00	1.20	1.30	1.50	1.65
100	0.55	0.70	0.80	0.95	1.08	1.25	1.35
110	0.45	0.57	0.68	0.80	0.87	1.00	1.10
120	0.38	0.47	0.56	0.67	0.74	0.85	0.95
130	0.32	0.40	0.50	0.57	0.63	0.73	0.80

Table 3-10. Multiplication Factors to SSBC Wind Pressures for Approximate Equivalency to UBC Design Wind Loads for Heights Greater than 500 Feet

SSBC basic wind speed (mph)	UBC wind pressure zone (psf)						
	20	25	30	35	40	45	50
70	1.00	1.25	1.45	1.70	1.90	2.20	2.40
80	0.75	0.95	1.10	1.30	1.45	1.65	1.80
90	0.57	0.75	0.87	1.00	1.15	1.30	1.45
100	0.47	0.60	0.70	0.85	0.95	1.05	1.15
110	0.38	0.50	0.60	0.70	0.75	0.87	0.96
120	0.33	0.42	0.50	0.57	0.65	0.75	0.82
130	0.28	0.36	0.42	0.48	0.55	0.62	0.70

Gulf Coast and established procedures for determining the degree of exposure to "reasonable probable" hurricane conditions. The council also developed a model minimum building standard designed to reduce hurricane damage if implemented as an adjunct to a standard building code.

Earthquake

The repeated occurrence of earthquakes has prompted governments in earthquake-prone areas to incorporate lateral force criteria in building regulations.

According to Slosson and Krohn (1977), after the 1933 Long Beach earthquake the Uniform Building Code was amended in 1935 to include earthquake-resistant design requirements. In addition, the 1933 Long Beach earthquake led to the passage by the California legislature of the Riley Act and the School House Safety Act (Field Act).

The Riley Act mandated application of earthquake-resistant design requirements to commercial and industrial buildings. The Field Act required that school buildings be designed to sufficiently withstand earthquakes so that no harm would come to the occupants. The Working Group on Earthquake Hazard

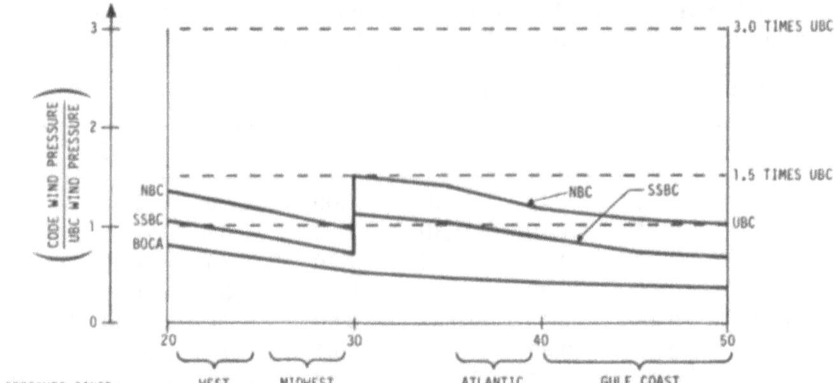

Figure 3-8. Uniform building code wind pressure (psf).

Reduction (U.S. Office of Science and Technology Policy, Executive Office of the President, 1978) state that:

> The Act's principal provisions require that all construction plans be prepared by qualified persons and that the designs be checked by an independent State agency. The independent review is generally considered the most important part of the Field Act, aimed at catching design errors or omissions, or other inadequacies that might not provide adequate earthquake resistance, before construction contracts are let.
>
> Another important aspect of the Act requires construction to be continuously inspected by a qualified person retained by the school board to see that all of the requirements of the plans are carried out. Moreover, all parties, including the architect, engineer, inspector, and contractor, must submit verified reports stating that the approved plans and specifications were complied with in construction.
>
> Knowledgeable observers consider the Field Act eminently successful in assuring reasonable compliance with acceptable levels of earthquake resistance. Almost all Field Act schools have performed well in all earthquakes since the law's passage. While some experts anticipate that some Field Act buildings will be severely damaged in future great earthquakes, there is agreement that injury or life loss will, nevertheless, be greatly reduced because of the Act's requirements.

Following the 1971 San Fernando Earthquake in California, a series of legislative initiatives were adopted by the California legislature. One of these initiatives focused on hospital design and construction, resulting in the Hospital Act of 1972. The Field Act was used as a guide in drafting this new piece of legislation according to the Working Group on Earthquake Hazard Reduction (U.S. Office of Science and Technology Policy, 1978). The law requires that design work for new hospitals, or substantial additions to or alterations of existing hospitals, be done by qualified specialists. The design must be thoroughly reviewed for safety by the Office of the State Architect.

The 1971 earthquake also precipitated amendments to the Uniform Building Code and to the code utilized by the City of Los Angeles. Specifically these amendments require new residential buildings to be equipped with diaphragm sheathing and veneer ties; and further, contain standards for framing, reinforcing of concrete in masonry chimneys, anchorage of water heaters, and cutting and notching of walls and studs. It is anticipated that these changes will greatly improve the safety and stability of residential structures with a cost increase of less than 1%. Also, there were code changes for multistory structures of 760 feet or higher. These changes require site analysis and structural dynamic analysis.

Expansive Soil

Of the four model building codes, only the Uniform Building Code specifically mentions expansive soils. Section 2904 specifies a standard test to determine a soil expansion index. In areas with expansive soil, the code only specifies that there shall be a "special design consideration."

Landslide

Many landslides are man induced and result either from the lack of adequate slope grading codes or from nonenforcement of such codes. In Contra Costa County, California, approximately 80% of the landslides are man related (Nilsen and Turner, 1975).

The effectiveness of adequate grading codes as a deterrent against man-induced landslides is exemplifed in the City of Los Angeles. In that community property losses and deaths resulting from landslides and mudflows during the 1950s and 1960s prompted the enactment of local codes and land use regulations. The heavy rains of January 1952 caused approximately $7.5 million damage within the City of Los Angeles. During the heavy rains of 1952 and 1955, loss or near loss of approximately 150 homes was induced by the Portuguese Bend landslide in Palos Verdes Hills, Los Angeles County. Damages sustained at Portuguese Bend, as well as in other areas of Los Angeles, brought about the first grading codes in 1952.

The grading codes were subsequently updated and improved by the City of Los Angeles one year after the heavy rains in 1962 which caused flood, landslides, and mudflows. This final code change in 1963 greatly reduced the loss or risk factor from man-induced landslides. The new grading code were given a test by the severe rains of 1969. An analysis of data collected by the Department of Building and Safety of the City of Los Angeles after the 1969 storm strongly suggests that landslide damage can be essentially eliminated by the proper use of scientific and engineering analysis of building sites in conjunction with realistic codes properly enforced.

International Differences in Public Policies toward Natural Hazards

Foster (1980) has observed that natural hazard-induced catastrophes most frequently affect the developing world and that such problems are relatively less significant in industrial nations. For example, he notes that "the losses from geophysical hazards such as tropical cyclones, earthquakes, floods, and droughts each year in the developing world cause an average of 250,000 deaths and $15 billion in damage prevention and litigation costs, whereas only 1000 annual fatalities and property damage equivalent to 1% of the gross national product is produced in the industrialized nations as the result of the same phenomena." These differences may be due to a combination of two factors: (1) the more serious attention given to the threat of natural hazards in the more affluent industrialized nations, and (2) the higher level of structural quality achieved in the wealthier nations as a simple function of market economics and consumer preferences, whether influenced by public policies or not.

Nonetheless, substantial differences have been observed between the hazard-mitigation policies of the several industrialized nations. For example, Wiggins and

Moran (1971) have reviewed the earthquake-mitigation provisions of building codes in 14 nations, including the United States and have found substantial differences in the safety of structures required by such codes. Because of the highly varying effectiveness of approaches to the mitigation of effects of major hazards, including floods, the United Nations has inaugurated efforts to develop model national approaches to several hazards. Typical is the 1976 report that attempted to establish guidelines for flood loss prevention and management in the developing countries. Similarly, a great many international conferences have been conducted to identify approaches to mitigation of seismic risk, and similar attention is being devoted to such hazards as tsunamis, coastal flooding, and heavy winds.

Conclusion

A variety of policy approaches have been taken by governmental entities to reduce the risks associated with the exposures of people and property to natural hazards. Typically, these approaches have required multiple levels of government to conduct specific hazard mitigation activities, have provided mechanisms for coordination of the total effort, and have made use of a mix of regulatory, capital investment, disaster relief, research and hazard zone management strategies. Each of these strategies has been examined in this chapter, which also provides the context in which the problem of natural hazard management will be examined in Part II and new public policies considered in Part III.

References

American National Standards Institute. 1972. American National Standards Buildings Code Requirements for Minimum Design Loads in Buildings and Other Structures. NASI Publication, A58-1-1972. ANSI, New York.

American Society of Planning Officials. 1966. The Text of a Model Zoning Ordinance. 3rd edition. ASPO, Washington, D.C.

Anderson, J. E. 1975. Public Policy Making. Praeger Publishers, New York.

Buoh, V. R., and J. E. Bihr. 1975. Uniform building code for wind loading. Meeting Preprint 7621. ASCE National Convention, Denver, Colorado, November, 1975.

Cheatham, L. R. 1977. An Analysis of the Effectiveness of Land Use Regulations Required for Flood Insurance Eligibility. Water Resources Research Institute, Mississippi State University, Mississippi State.

Cochrane, H. C. 1975. National Hazards and Their Distributive Effects. Program on Technology, Environment and Man, Monograph No. NSF-RA-E-75-003. Institute of Behavioral Science, University of Colorado, Boulder.

Comptroller General of the United States. 1980. Federal Disaster Assistance: What Should the Policy Be? (PAD-80-39). U.S. General Accounting Office, Washington, D.C.

Council on Environmental Quality, Executive Office of the President. 1975. Seventh Annual Report. U.S. Govt. Printing Office, Washington, D.C.

Fed. Regist. 1980. 45, No. 176.

Field, C. G., and S. R. Rivkin. 1975. The Building Code Burden. D. C. Heath, Lexington, Massachusetts.

Foster, H. D. 1980. Disaster Mitigation for Planners; The Preservation of Life and Property. Springer-Verlag, New York.

Hart, G. C. 1976. Natural Hazards: Tornado, Hurricane, Severe Wind Loss Models. Prepared for the NSF, NTIS No. PB 294594/AS. J. H. Wiggins Company, Redondo Beach, California.

Jones, C. O. 1977. An Introduction to the Study of Public Policy. Duxbury Press, North Scituate, Massachusetts.

Kusler, J. A. 1976. Statutory Land Use Control Enabling Authority for the Fifty States, with special reference to Flood Hazard Regulatory Authority. Housing and Urban Development, Federal Insurance Administration, Washington, D.C.

Lesso, W. G. 1976. Potential Wind Damage Reduction Through Use of Wind-Resistant Building Standards. Texas Coastal and Marine Council, Austin.

Los Angeles Times. 1977. Carter Orders Check of High-Hazard Private Dams. Part II. November 29: p. 8. Los Angeles, California.

Lowi, T. J. 1964. American business, public policy, case studies and political theory. World Politics 16:677–715.

Mileti, D. S., T. E. Drabek, and E. J. Haas. 1975. Human Systems in Extreme Environments: A Sociological Perspective. Institute of Behavioral Science, University of Colorado, Boulder.

National Science Foundation. 1980. A Report on Flood Hazard Mitigation, NSF, Washington, D.C.

Nilsen, T. H. and B. L. Turner. 1975. Influence of Rainfall and Ancient Landslide Deposits on Recent Landslides (1950–1971) in Urban Areas of Contra Costa County, California. U.S. Geol. Surv. Bull. 1388.

Platt, R. H. 1976a. The national flood insurance program, some midstream perspectives. J. Am. Inst. Planners 42:303–313.

Platt, R. H. 1976b. Legal aspects of natural hazards regulations in the coastal zone. *In*: U.S. Department of Commerce, National Oceanographic and Atmospheric Administration (1976, Appendix B).

Ripley, R. B., and G. A. Franklin. 1976. Congress, the Bureaucracy and Public Policy. Dorsey Press, Homewood, Illinois.

Schneider, W. J., and J. E. Goddard. 1974. Extent and Development of Urban Flood Plains. Geol. Surv. Circular 601-J. U.S. Geological Survey, Washington, D.C.

Slosson, J. E., and J. P. Krohn. 1977. Effective building codes. Calif. Geol. pp. 136–139.

Texas Coastal and Marine Council, 1976. Model Minimum Hurrican Resistant Building Standards for the Texas Gulf Coast. TCMC, Austin.

Texas Coastal and Marine Council. 1977. Hurricane Resultant Building Standards and Natural Hazards. TCMC, Austin.

Thom, H. C. S. 1968. New distribution of extreme winds in the United States. J. Struct. Div., Am. Soc. Civ. Eng. 87 (No. ST 7):1787–1801.

Thomas, F. N., Chairman, Flood Plain Management Task Force, 1980. Progress Report, January 1, 1980. Implementations of Executive Order 11988. U.S. Water Resources Council, Washington, D.C.

U.S. Army Corps of Engineers. 1974–1975. Final Report of the Chief of Engineers on Civil Work Activities. USACE, Washington, D.C.

U.S. Congress. 1966a. Insurance and Other Programs for Financial Assistance to Flood Victims. 89th Congress, Print No. 43. Committee on Public Works, U.S. Congress, Washington, D.C.

U.S. Congress. 1966b. Senate Committee on Banking and Currency. Insurance and Other Programs for Assistance to Flood Victims. Department of Housing and Urban Development Report. U.S. Congress, Washington, D.C.

U.S. Congress, 1975. Senate Committee on Banking, Housing and Urban Affairs. Oversight on Federal Flood Insurance Programs. 94th Congress, 1st Session. U.S. Congress, Washington, D.C.

U.S. Department of Commerce, National Oceanic and Atmospheric Administration. 1976. Natural Hazard Management in Coastal Areas. U.S. Dept. of Commerce, Washington, D.C.

U.S. Office of Science and Technology Policy, Executive Office of the President. 1978. Earthquake Hazards Reduction: Issues for an Implementation Plan. Working Group on Earthquake Hazards Reduction. U.S. Govt. Printing Office, Washington, D.C.

Wiggins, J. H., and D. Moran. 1971. Earthquake Safety in the City of Long Beach Based on the Concept of Balanced Risk. J. H. Wiggins Company, Redondo Beach, California.

PART II

NATURAL HAZARDS RISK ASSESSMENT AND MITIGATION ANALYSIS

Rational approaches to identification and evaluation of the range of effects produced by exposures of people and property to hazardous natural occurrences require the use of a variety of analytic methods. These methods are aimed at identifying the natural hazards present in a particular geographic area, the frequency of occurence across the range of magnitudes associated with the events, the types and numbers of structures and people exposed, and the alternative mitigation strategies available to reduce the risk arising from the exposure of structural and human populations.

In this part of the book we outline the methods employed to assess the risks within the United States which result from the periodic occurrence of nine natural hazards; landslide, earthquake, expansive soils, hurricane winds, tornado, severe winds, riverine flooding, tsunami, and storm surge. In each case, the methods resulted in identification of the areas within the United States in which these hazards are experienced, the numbers of people and specialized types of property exposed to those hazards, the vulnerability of specific classes of exposed objects to each hazard, and the measure of risk appropriate for each resulting class of exposure, involving some specified hazard and some exposed population.

4
Methodologies for Hazard Analysis

Introduction

A natural event concerns us when it results in negative consequences for people and/or their property. Such consequences are a function of the characteristics of the physical hazard (e.g., flooding, area inundated, depth of water, velocity, debris content, etc.), the nature of the area impacted (population, land use, and investment in buildings), and the vulnerability of objects exposed to the hazard. Both the physical hazard and the land use near the hazard may change significantly over time. Thus, it is necessary to understand both existing and future socioeconomic and geophysical conditions in a hazard area in order to determine the types and magnitudes of damages that may be expected now and in the future.

Appropriate evaluation of a hazard requires a determination of the probability of the occurrence of a natural event at its various intensity levels. Intensity in this context relates to such parameters as wind velocities, water depth, ground shaking, or ground movement. Since natural events are always occurring, they generally concern us only at intensity levels causing significant damage to individuals or a community. Thus, hazard intensity always must be related to the quality of exposed structures or buildings. Similarly, structures or buildings vary in their degree of vulnerability to the intensity of a hazard as a function of their design, materials, and quality of construction and maintenance. Accordingly, in this study it was necessary to classify structures by type and use.

Since the intensity of any hazard event and the vulnerability of structures to the event are important only where structures are exposed to the event, the study required an assessment of exposures in specific land areas. The general procedures used to integrate findings resulting from these several assessment steps are depicted in Figure 4-1.

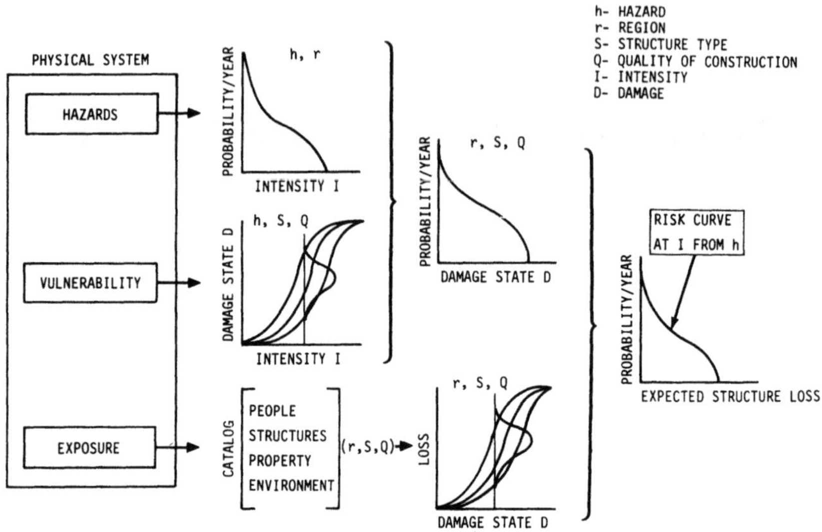

Figure 4-1. Generalized natural hazard risk determination procedure.

Risk Analysis of Natural Hazards

To appropriately assess the levels of risk associated with natural hazard exposures in the United States, and to accomplish the range of purposes that guided the development of the normative model presented in Chapter 1, we structured the analysis so as to develop information necessary to accomplish the following:

1. Identification and description of the characteristics, geographic distribution, and potential effects of nine hazardous natural events within the United States (hazard analysis).
2. Assessment of the vulnerability of several classes of buildings, and their occupants, to each hazard (vulnerability analysis).
3. Identification and measurement of the major primary, secondary, and high order effects expected to be associated with the exposure, by major geographic area, of buildings and their occupants to these hazardous natural events (loss analysis).
4. Identification and explication of the major candidate public problems associated with these effects (problem analysis).
5. Identification of the costs and characteristics of the major types of technologies appropriate for mitigating the effects induced by exposure of buildings and their occupants to each of the nine natural hazards (technology analysis).
6. Identification and description of the major types of public policies that may induce the application of hazard-mitigating technologies (public policy analysis).
7. Estimation of the economic costs and other effects associated with the use of selected technologies to mitigate the effects of selected natural hazards (cost analysis).

8. Identification of the major effects and candidate public problems that might be generated by the use of selected technologies to mitigate the effects of the nine natural hazards (problem analysis).
9. Identification and evaluation of the major problem-solving strategies and public policies relevant to the problems identified in Items 4 and 8, above (public policy analysis).

As depicted in Figure 4-2, the examination was structured into five major and three minor elements. The major elements include risk analysis, technology analysis, cost analysis, problem analysis, and public policy analysis. The minor elements include hazard analysis, vulnerability analysis, and loss analysis. The objectives and general methodologies of each of these major elements are discussed below.

Risk Analysis

The central element in the examination was the risk analysis, whose major components are shown in Figure 4-3. The performance of this analysis required the development of computer-based hazard, exposure, and vulnerability models, which were then used with appropriate damage algorithms and risk equations in a series of separate studies, each concerned with a specific natural hazard. For eight of the nine natural hazards, a computer model was developed to depict, by county and state within the United States, the probability of hazard occurrence by hazard intensity. Where appropriate, the hazard models included both regional

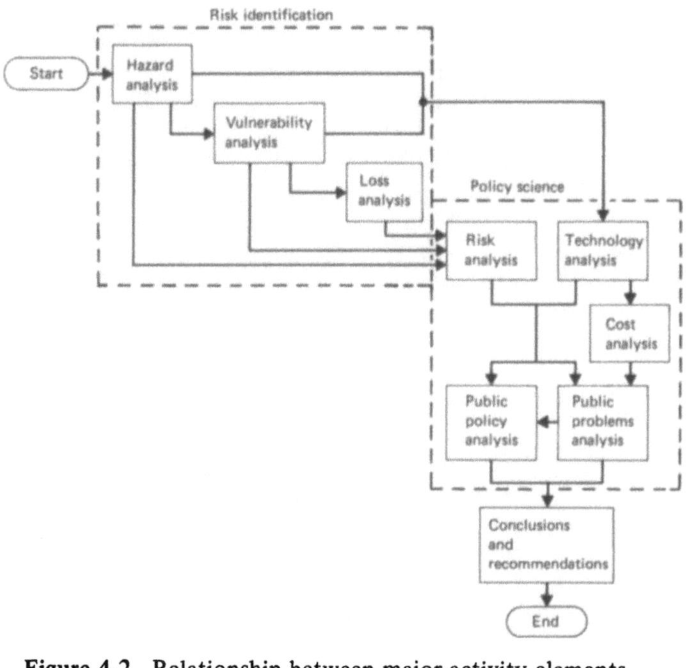

Figure 4-2. Relationship between major activity elements.

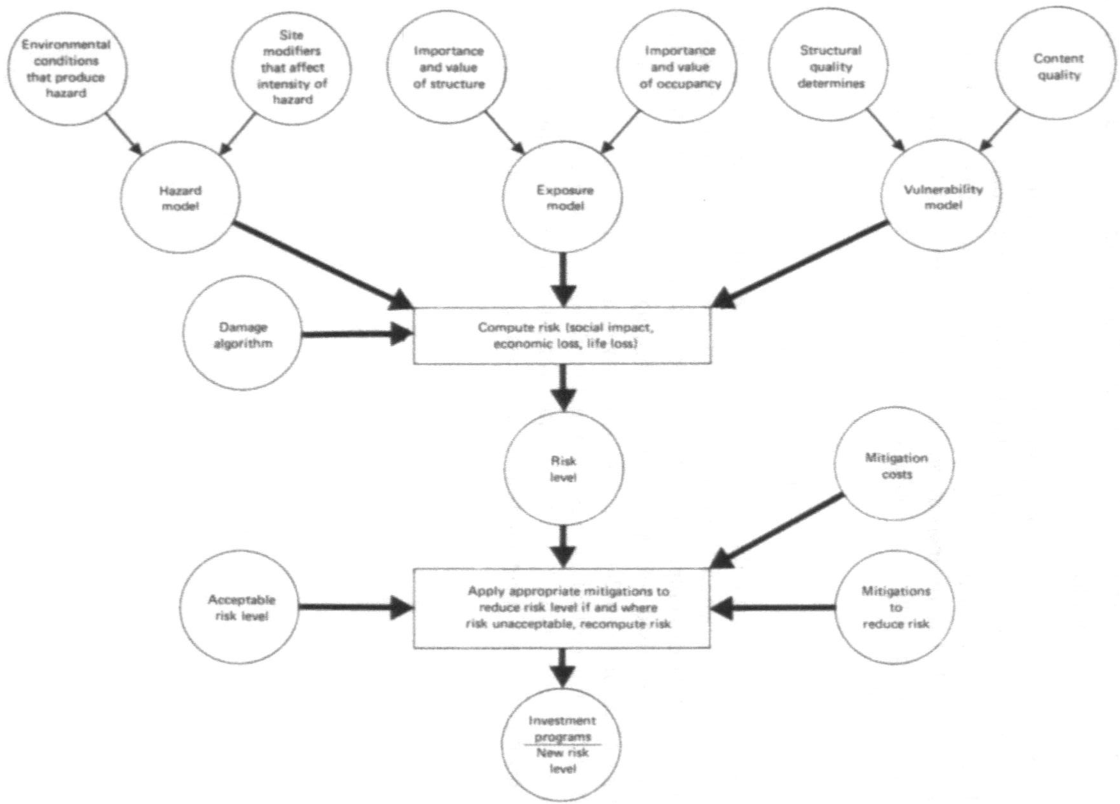

Figure 4-3. Decision process in natural hazard risk evaluation.

and site-modifier data, such as regional weather patterns and local ground elevation conditions. In contrast, the risk assessment for riverine flooding was targeted only on multistate regional areas. Based on a model originally developed by Friedman and Roy (1966), our model was applicable only to a regional combination of states. This portion of the total risk analysis cannot therefore be extended to any specific county in the United States.

The vulnerability element of the risk analysis involved the development of a computer-based exposure model for each hazard and appropriate damage algorithms related to various types of buildings. The exposure model included an inventory of the value of buildings and building contents, and the human population at risk in individual counties, subcounty hazard zones (where appropriate), and states, as well as data concerning the type and vulnerability of exposed assets within those areas. The damage algorithms that were developed stated the expected relationship between various hazard intensities and the degree and/or value of damage to structures of specific types. Typically, estimates of other types of losses such as death, building content loss, unemployment, and homelessness were related to the expected levels of damage to buildings.

The final risk equations utilized the damage algorithms and hazard, exposure, and vulnerability models to generate probabilistic annual estimates of dollar losses expected to arise from the exposure of buildings and building contents to natural hazard in the years 1970 to 2000. Except for riverine flooding, these data were generated, or are retrieavable, at the scale of individual counties, states, and the nation. The hazard-induced effects that were quantified by the risk analysis include (1) economic losses to buildings (building losses); (2) economic losses to the contents of buildings (contents losses); (3) economic losses resulting from the lack of availability of goods and services to occupants of hazard-impacted areas measured in the form of transportation costs (supplier losses); (4) economic losses in the form of lost take-home pay from employment (income loss); (5) years of unemployment induced by hazard-caused damage to buildings and their contents (unemployment); (6) years of homelessness induced by damage to residences (homelessness); and (7) loss of life resulting from the occurrence of hazardous natural events (life loss).

Technology Analysis

For each of the hazards studied, a preliminary list was compiled of the physical and management technologies of possible utility in mitigating the unwanted effects of building exposure to the hazard. These preliminary lists were grouped, for heuristic purposes, into seven major classes:

1. Structural and other technical methods for minimizing the probabilities of hazard occurrences and for protection of hazard-prone areas.
2. Structural methods for strengthening or protecting buildings exposed to hazardous natural events.
3. Methods for preparing building sites to limit the hazard exposure and structure vulnerability.

4. Methods appropriate to the identification or avoidance of hazard-prone areas.
5. Hazard forecasting, warning, and population evacuation systems.
6. Loss recovery, relief, and rehabilitation systems.
7. Information processing and decision-assisting methodologies useful in the selection and use of potential hazard-mitigating techniques.

Using this classification scheme, a second examination then was made of each hazard, and relevant technologies were identified in each class for each hazard, where appropriate.

Cost Analysis

The cost analysis focused on two information requirements: (1) the direct dollar costs associated with application of specific hazard mitigation technologies to specific types of buildings; and (2) the county-level cost/loss relationships which would be associated with mandatory community-wide use of specified hazard-mitigating technologies on specific types of new buildings.

Accordingly, the cost analysis was conducted in three steps: (1) building cost estimates, (2) county cost/loss ratio assessments, and (3) cost feasibility tests of selected mitigations.

1. Building Cost Estimates. For selected types of structures, literature reviews and surveys of expert opinion were conducted to identify the proportionate or absolute increases in costs that would be experienced were selected hazard-mitigating technologies applied to new buildings to reduce expected losses from earthquake, wind, and flooding. These estimates were developed in a form capable of yielding crude estimates of the absolute or proportionate cost increases associated with structural design levels equal to given proportionate increases in the levels specified in the 1973 Uniform Building Code. However, other estimates also were developed or utilized, such as those relating to the cost of floodproofing, soil stabilization, etc.

2. County Cost/Loss Ratios. Using these building cost estimates, a computer model was developed to generate estimates of total dollar costs that would be incurred were several structure-strengthening technologies to be applied to new buildings constructed between 1980 and 2000. The model also was designed to generate an expected cost/loss reduction ratio for mitigation of wind, storm surge, and earthquake hazard effects in the year 2000. This permitted the identification of counties that, for these hazards, would experience ratios of costs to loss reduction of 0.75 and higher, were the designated mitigation technologies to be employed.

The cost/loss reduction ratios were calculated for each of the following mitigation technologies: (1) imposition of earthquake zone 3 building code standards in each county, (2) four foot elevation of structures above floodplain levels, (3) two foot elevation of structures above floodplain levels, (4) floodproofing of structures to two foot elevations above level of floodplain, (5) escalation of building code standards to a level equaling 1.5 times UBC requirements for purposes of reducing wind losses, and (6) escalation of building code standards to a level equaling 3.0 times UBC requirements for purposes of reducing wind losses.

3. Cost Feasibility Tests of Selected Mitigations. Computer outputs and other data were used to determine the annual amortized costs of various specific mitigations under a variety of interest rates. The results of these calculations were used to identify sets of cost-feasible hazard mitigations and the net annual savings (loss reductions) achievable through the application of alternative hazard management programs and policies.

Public Problems Analysis

This phase focused on the identification of those states of affairs identified as constituting *candidate public programs*. The data generated by the risk analysis and cost analyses were utilized in this effort, but were converted into forms more amenable to problem-oriented treatment. Thus, aggregated loss data by county and state were converted into per capita loss values and other similar forms. Additional data concerning aggregated and per capita costs of the other U.S. phenomena generally viewed as constituting public problems also were collected to serve as risk referents. Further, an effort was made to identify specific loss situations that might be valued differently by various sets of public policy makers and various stakeholders in situations where natural hazard-induced consequences are expected.

Public Policy Analysis

This phase focused on (a) identification and review of major past and current hazard management policies; (b) identification of the sociopolitical and legal constraints that influence the making of hazard management public policy; and (c) identification and explication of alternative public policies for dealing with each candidate public problem.

Principal among the technology-applying public policy actions addressed in this inquiry were (1) building codes, (2) land use standards and regulations, (3) other selected regulatory policies, (4) land classification policies, (5) public capital investment policies, (6) public distributive and redistributive policies, and (7) public research, information dissemination, and education policies. Possible constraints influencing the capacity of each of the several jurisdictions of government to engage in policy making in each class also were examined.

Hazard-Exposure-Vulnerability Estimation

Methodologically, the extent of damage to buildings, their contents, and associated socioeconomic impacts due to various natural events are derived from the dollar value of exposed buildings and contents multiplied by the probability of various loss levels from a given type of occurrence. Once the value of the items subject to a level of risk is known, it is possible to derive the expected annual loss by multiplying the amount at risk by the probability of loss of some portion of that amount, summed over all contingencies. This series of probabilities is illustrated in the form of an event tree, shown in Figure 4-4.

The probabilities placed on the branches of the event tree represent the future expected frequencies of a certain natural event, by specific intensity level. This means that there is a set of contingencies and intensities of a hazard, each of which is a branch of the event tree and has an annual probability of occurrence.

The sum of the elements of the right-hand column represents the annual expected dollar loss for a specific location from a natural hazard. Again, it should be noted that this expected value does not represent loss from any specific event, but is the mean value of loss over an extended period based on the probabilities assigned to the branches of the event tree.

Many of the probabilities of annual states of nature are available from historical records; however, the determination of the dollar value of items at risk is a more subtle economic problem. Care must be taken to account for the relevant dimensions of value. At the same time, double counting must be avoided. Finally, once the dimensions of the factors of risk have been catalogued, they must be given an accurate dollar valuation. Such valuations are obvious in some instances and elusive in others.

As noted, the intensity of the hazard will result in certain losses in buildings and their contents, losses in income, and an increase in supplier transportation costs. These four types of losses are referred to as economic losses. Social

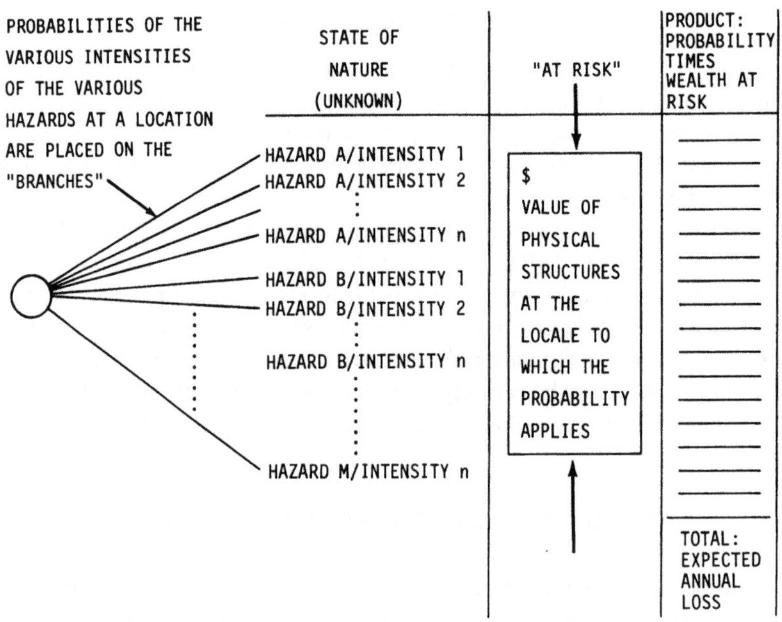

Figure 4-4. Event tree for deriving expected loss from probabilities of extreme natural events.

losses are identified as life loss, homelessness, and unemployment. Both sets of losses are derived from the expected losses in building wealth. Specifically, the determination of the impacts progress from the physical phenomena to aggregate economic and social costs. The order of computation for assessing the hazards effects or impacts is presented in Figure 4-5.

Building wealth losses are determined as a function of *damage rates*, or the proportion of replacement value needed for repair. The damage rates associated with a hazard intensity are applied to buildings of various types and to building contents within sections of the local economy.

Expected income losses are estimated through the use of a regional multi-sector model of income determination where eleven sectors are considered (e.g., farming, construction, manufacturing), and the regions are either counties or states. These estimates are computed by including the relationship between damage rate and the amount of downtime or productive capacity within those buildings. This allows estimation of loss of production due to the inactivity of productive capital.

A third expected dollar loss investigated is the loss experienced by a purchaser who, even though not sustaining direct damages, nonetheless experiences increased transportation costs for a product no longer available from a supplier suffering damage in the hazard exposure area.

Some of the social losses are described in the form of noneconomic relation-

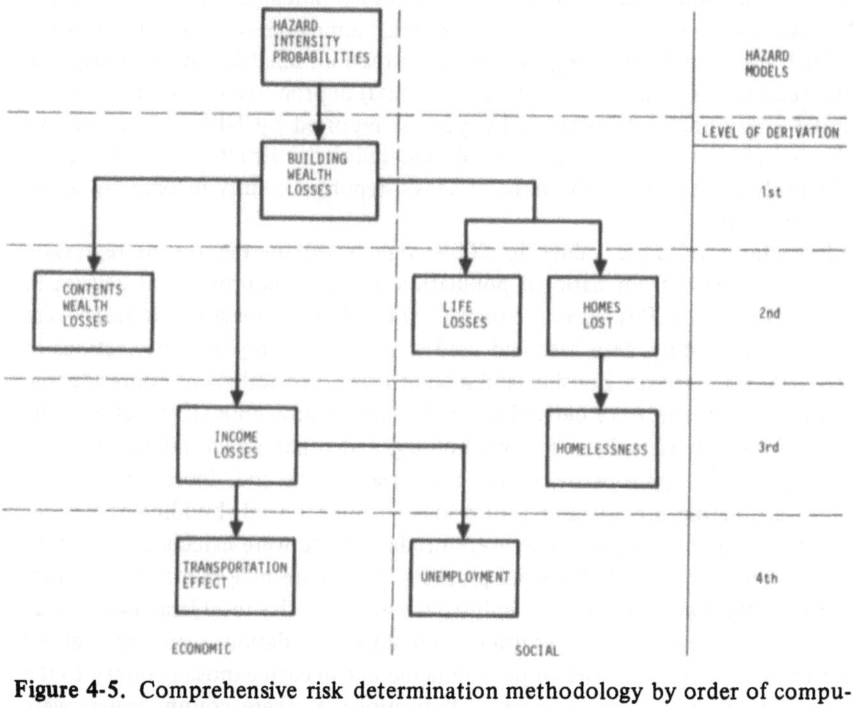

Figure 4-5. Comprehensive risk determination methodology by order of computation. (From Hirschberg et al., 1978.)

ships. Specifically, the level of homelessness is estimated in terms of estimated person days of homelessness resulting from a specific hazard. Another such aspect is the projected unemployment associated with an event. This is quantifiable in economic terms but is not included in the economic loss calculations since production loss computations already account for this part of wealth loss. In fact, adding expected unemployment to the loss computations would be double counting. Nevertheless, this dimension of loss is identified separately. In addition to the specific social and economic losses identified, expected levels of life loss are predicted.

Determining Present and Future Exposure to Hazards

Critical to the performance of a natural hazards risk analysis is the definition of the size and characteristics of the human and structural populations that are, or will be, exposed to a specific hazard at a specific intensity level. For all hazards except tsunami and riverine flooding, the quantitative identification and categorization of populations at risk of exposure to natural hazards in 1970, 2000, and intervening years were calculated exclusively from a county aggregated data base. The components of the data base are shown in Table 4-1.

Census data provided the basis for determining 1970 counts, by county, of population, residential structure (by type), and value of residential property. National 1970 values for publicly owned, commercial, and industrial buildings were derived from data published by the National Bureau of Economic Analysis (NBEA), and were allocated to the several commercial-industrial categories utilized in the study (farming, mining, construction, manufacturing, transportation, finance, insurance, real estate, and services) in proportion to their corporate capital consumption allowances by year as reported by NBEA. The national value of publicly owned, commercial, and industrial structures was allocated to individual counties on the basis of census-reported county income, by sector of the economy.

Projections of these values to 2000 were based on the use of regression equations derived from national population and income projections developed by the U.S. Water Resources Council (1974). These regression equations were input to the county data base and used to develop county-level projections of 1970-2000 population, housing, and income levels. Projections of nonresidential building values then were derived from the income projections, by sector of the economy. Projected values of both housing and other structural wealth were calculated using appropriate building survival curves and depreciation rates.

For estimating present and future losses from tsunami, building values per square mile of port facility also were needed. These were calculated by determining average values of industrial property per square mile in major U.S. cities (Manvel, 1968) and the ratio of industrial acreage to the total land area in such places. The national values resulting from these calculations were regionalized on the basis of relative level of per capita income in each exposed county to the national average per capita income. Projections of these county values were based on the use of national income and growth rates.

Table 4-1. County Data Base and Sources

Type of data	Data aggregated to level of	Date data taken	Number of data points	Source
General county data				
County population	County	1970	3,132	(U.S. Bureau of Census, 1973)
County area	County	1970	3,132	"
County name	County	1970	3,132	(National Bureau of Standards 1974)
County latitude and longitude	County	1970	3,132	(Dept. of Interior, 1970)
County coastal flag (whether on an ocean coast or not)	County	1970	3,132	"
Name of SMSA county is part of	County	1970	3,132	(U.S. Water Resources Council, 1974)
Housing characteristics data				
Percent of housing units with basements	State NSMSA, SMSA	1970	298[a]	(U.S. Bureau of Census, 1972b)
Percent of housing units with cement slab foundation	State NSMSA, SMSA	1970	298	"
Percent of housing units built before 1940	State NSMSA, SMSA	1970	298	"
Percent of housing units that are mobile homes	State NSMSA, SMSA	1970	298	"

Table 4-1. (continued)

Type of data	Data aggregated to level of	Date data taken	Number of data points	Source
Percent of housing units in structures with 4 or more stories	State NSMSA, SMSA	1970	298	"
Percent of housing units in structures with 5 or more stories	State NSMSA, SMSA	1970	298	"
Average number of persons per housing units	State NSMSA, SMSA	1970	298	"
Economic data				
Locally assessed real property values	County	1971	3,132	(U.S. Bureau of Census, 1972b)
Assessment–sales price ratios	State	1971	2,002	"
Total personnel income percapita	NSMSA, WRC, and SMSA	(1980)[b]	452	(U.S. Water Resources Council, 1974)
Total agricultural income percapita	NSMSA, WRC, and SMSA	(1980)	452	"
Total mining income percapita	NSMSA, WRC, and SMSA	(1980)	452	"
Total construction income percapita	NSMSA, WRC, and SMSA	(1980)	452	"
Total manufacturing income percapita	NSMSA, WRC, and SMSA	(1980)	452	"

Total wholesale–retail income percapita	NSMSA, WRC, and SMSA	(1980)	452	"
Total financial insurance, real estate income percapita	NSMSA, WRC, and SMSA	(1980)	452	"
Total service industry income percapita	NSMSA, WRC, and SMSA	(1980)	452	"
Total federal government income percapita	NSMSA, WRC, and SMSA	(1980)	452	"
Total state and local government income percapita	NSMSA, WRC, and SMSA	(1980)	452	"
Percent of national average of total local income percapita	NSMSA, WRC, and SMSA	(1980)	452	"
Growth projections				
Intercept of population projection[c]	NSMSA, WRC, and SMSA	(1970–2000)	452	(U.S. Water Resources Council, 1974)
Slope of population projection[c]	NSMSA, WRC, and SMSA	(1970–2000)	452	"
Intercept of income/ per capita projection[c]	NSMSA, WRC, and SMSA	(1970–2000)	452	"
Slope of income/ per capita projection[c]	NSMSA, WRC, and SMSA	(1970–2000)	452	"

[a] 247 SMSA where used in the 1970 census and 51 states, District of Columbia the 51st.
[b] Projected data.
[c] Computed linear least square fit of projected data from 1970–2020.

All study estimates of population at risk of exposure to riverine flooding were derived from a national-scope computer model developed for this analysis. Among the primary inputs to the model were the 1967 findings by the U.S. Army Corps of Engineers that 5539 cities of various sizes were subject to periodic flooding (see Table 4-2) and that 2037 U.S. communities had been equipped with area flood protection structures and facilities (Lee et al., 1976). Based on U.S. regional population loadings and city sizes, 5539 communities in the size ranges identified by the Corps therefore were assigned to the computer-modeled 10 U.S. regions. These regions were identified on the basis of physiographic, climatologic, and demographic parameters.

Utilizing data developed by others (Friedman and Roy, 1966; Schneider and Goddard, 1974), plots were constructed to show the percentage of dwelling units estimated to be located in various flood zones of the modeled cities (Figures 4-6 and 4-7). These data then were converted into tabular form to show the fraction of dwelling units in various flood zones within urban communities. As shown in Table 4-3, the number of dwelling units in Zone F (all areas outside the 100 year floodplain) is equal to the sum of the dwellings in Zone A through E (zones within the 100 year floodplain). This apportionment was not intentional, but was what the data yielded. It results in greater exposure of the national population to a type F flood than that depicted in previous models. Following application of these procedures, census reported population levels per dwelling unit in each of the U.S. regions identified in the study then were determined. These were used to build population estimates for study cities used in the riverine flood model.

National data on the numbers and value of commercial/industrial and governmental buildings were allocated to the ten study regions of the United States, and were further apportioned to cities within each region. Once the number of commercial/industrial and governmental properties in each city was established, they were apportioned to the several flood zones of the city in the same proportion as residential structures. Data from the U.S. Bureau of the Census were used

Table 4-2. Number of Urban Places with Flood Problems

City size	Number of places with flood problems (incorporated municipalities)
Greater than 100,000	140
50,000 - 100,000	204
25,000 - 50,000	390
10,000 - 25,000	903
5,000 - 10,000	992
2,500 - 5,000	1,098
1,000 - 2,500	1,812
Total	5,539

Source: U.S. Army Corps of Engineers, 1967.

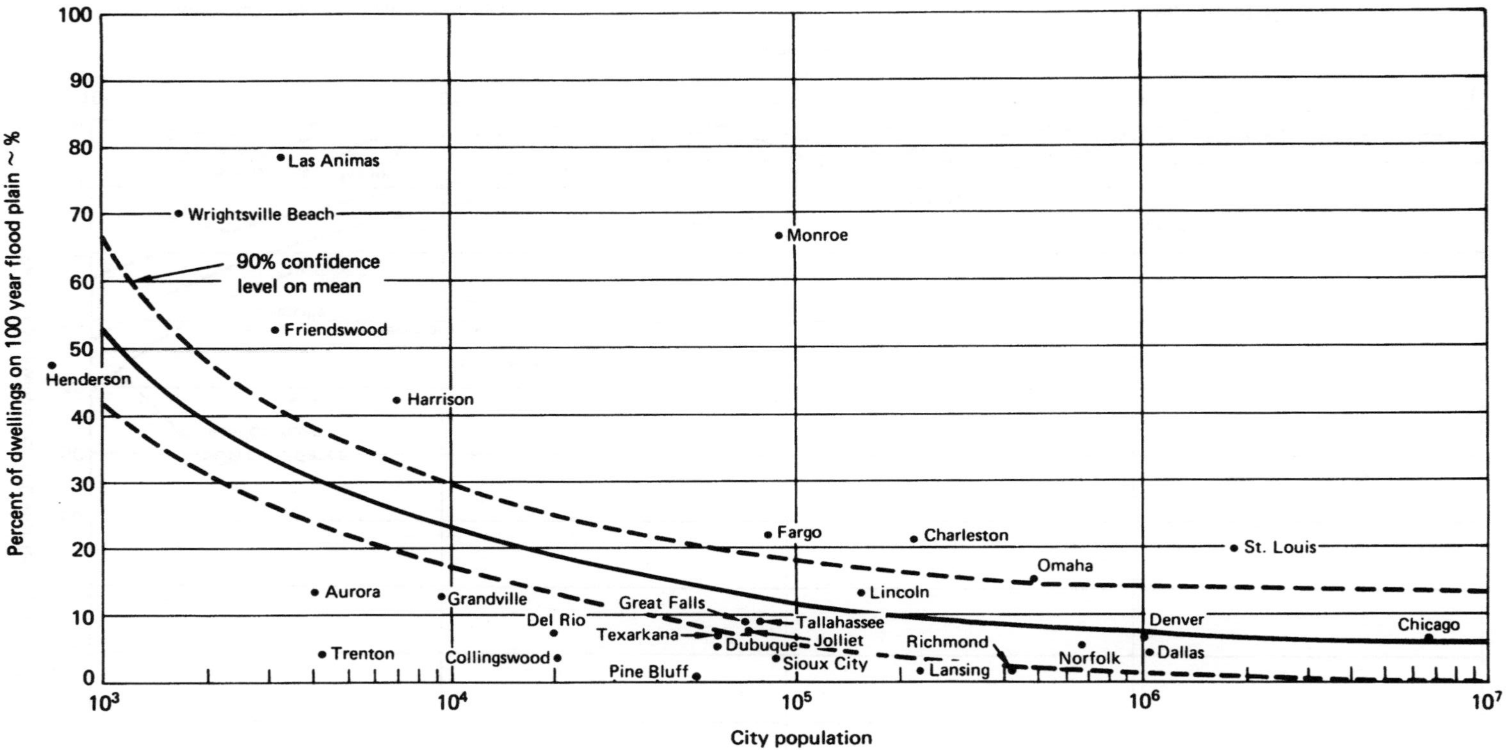

Figure 4-6. Percent of dwellings on 100 year floodplain—central and coastal plains. (From Lee et al., 1978.)

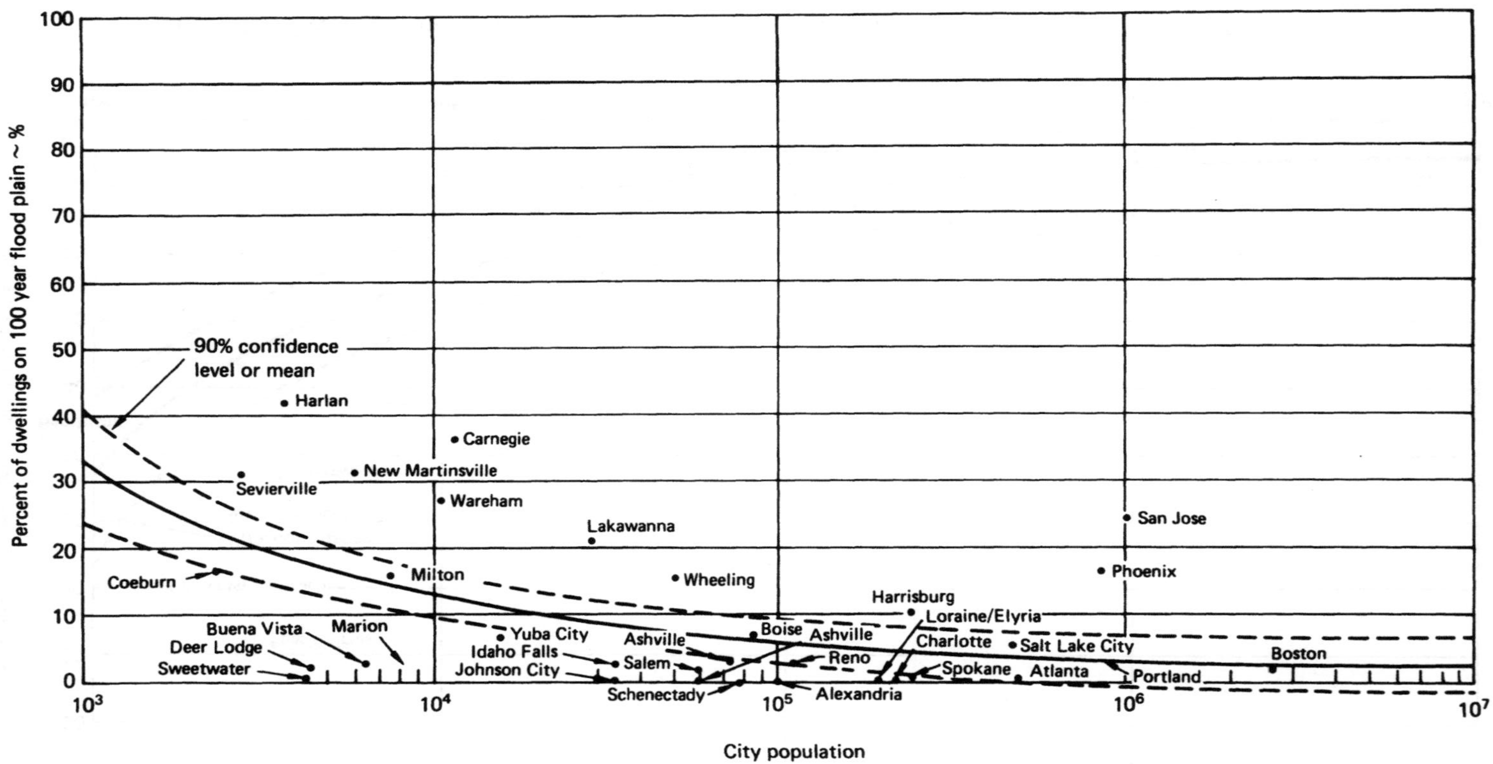

Figure 4-7. Percent of dwellings on 100 year floodplain—western and eastern highlands. (From Lee et al., 1978.)

Table 4-3. Distribution of Dwelling Units by Flood Hazard Zone

Hazard zone	Return period of flood (years)	Fraction of dwellings on 100 year floodplain[a]	
A	2–5	.135	$\Sigma = 1.0$ (A–E)
B	5–10	.150	
C	10–25	.200	
D	25–50	.245	
E	50–100	.270	
F	more than 100	1.000	

[a]There are as many dwelling in Zone F as in Zone A to E inclusive.

to determine the median value of owner-occupied housing units by size of city. Using data developed by Musgrave (1974), these values were modified to remove the value of land. To account for varying values of property by flood hazard zone, a scale factor derived from the Friedman model (Friedman and Bocaccino, 1972) was applied to the data. The result of these procedures was assignment of a value of $184,000 to each commercial/industrial and governmental property within flood modeled cities.

Estimates of Building Contents Value

Estimates of the value of building contents were developed for each type of building categorized in the study. The value of the contents of residential buildings was derived from Shavell (1971). The value of the contents of farm, manufacturing, wholesale trade, and retail trade buildings was derived from national estimates developed by Loftus (1972). The value of the contents of all other types of buildings examined in the study was derived from national estimates by Musgrave (1974). For the latter, appropriate allocations to the various classes of buildings used in this study were made on the basis of a five year moving average estimate of the capital consumption allowances for each economic sector. In all cases, projections of building contents value were calculated using appropriate content service lives, survival curves, and depreciation rates. Estimates of building contents value were made from the Institutional Investors Study Report of the Securities and Exchange Commission (Goldsmith, 1971).

The value of building contents is forecasted to increase at a faster rate than the value of buildings. As a result, by the year 2000, total U.S. building contents value is estimated to be 6.3 trillion dollars; and the total U.S. structure value, 4.9 trillion dollars. Thus, by the year 2000, the value of what a building contains will exceed the value of the structure in which it is housed. This circumstance results, of course, from the high projected increases in per capita income, and from the linkage between per capita income and consumer-business investment in building contents and consumer goods.

Estimating Earthquake Losses

Although many analysts have examined the general subject of earthquake hazards in the United States, most have only viewed the subject in terms of the probability of earthquake occurrence or magnitude. In contrast, this analysis also considers related subjects as the vulnerability of buildings to ground movements induced by earthquakes; the value of property exposed to this hazard; annual losses expected to arise from such exposures; and methods that might mitigate the consequences of earthquake exposures. The procedures used are summarized in Figure 4-8.

The MMI scale is defined in terms of *damage potential*, a reminder that actual damage may occur only when structures are exposed to the hazard. That is, where there are no structures, there can be no structural damage, regardless of the intensity of the earthquake hazard. Just as one may consider the frequency of occurence of earthquakes within a region as a function of Richter magnitude, so may one consider the frequency of occurrence of earthquake ground shaking as a function of intensity. The frequency of ground shaking, by site, as a function of intensity, constitutes the basic hazard model for earthquakes. Algermissen and Perkins (1976) have developed a ground shaking acceleration map for the United States (Figure 4-9) which has been the primary map used in developing the Uniform Building Code aseismic construction standards.

However, in order to estimate building losses as a function of intensity it was necessary to utilize an MMI map of the United States developed by Wiggins (1975) (Figures 4-10, 4-11, and 4-12). Ground shaking, in terms of MMI, was then estimated on a half-degree longitude by half-degree latitude grid system overlaid on the map.

Annual average estimates of structural damage caused by such ground shaking can be estimated from information showing the damage that will be sustained by the various types of structures exposed to the earthquake intensity range predicted as relevant to a particular geographic locale. The MMI scale, which describes intensity in terms of local effects of earthquakes, is well understood, and is most closely related to damage.

The lower limit of the intensity scale that may produce measurable damage was selected to be MMI=6. Damage at levels below an MMI=6 were considered slight and difficult to define. The upper limit of the intensity scale was set at MMI=12. At this severe intensity, damage to buildings is expected to be total.

Since force increases exponentially with intensity, it is reasonable to expect that damage will also increase exponentially with intensity. The damage algorithm is then a direct relationship between the amount of damage to a structure type as a percentage of market value and the MMI in the region.

Damage algorithms were derived for two types of construction: single family residential structures, and industrial-commercial construction. These damage algorithms are given for four different relative strengths of construction as defined by quality factors (Q) (Wiggins et al., 1974). Estimates developed by Whitman et al. (1973) permit the establishment of basic structure quality factors.

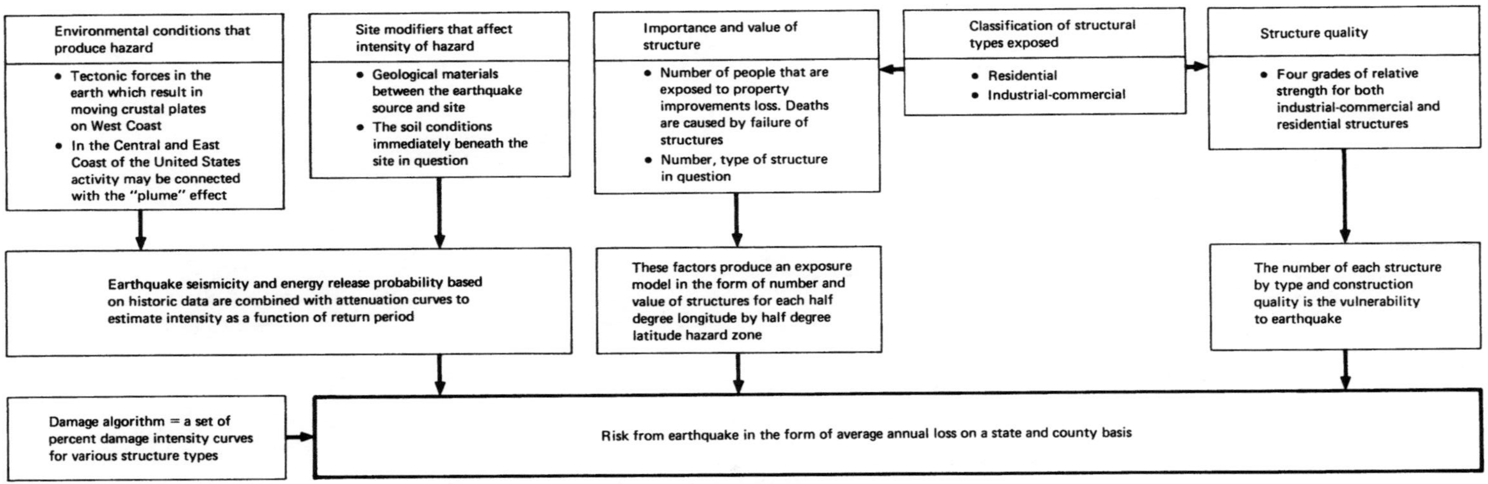

Figure 4-8. Earthquake risk determination procedure.

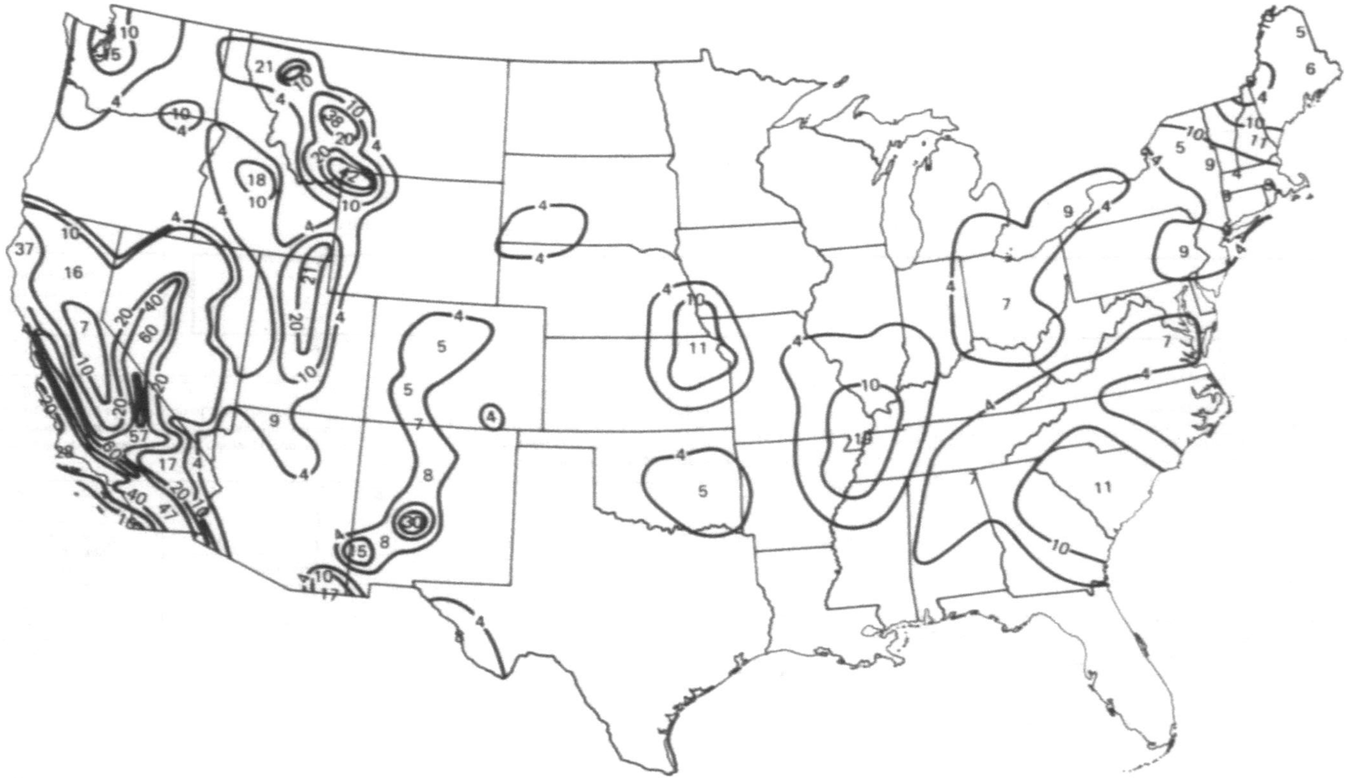

Figure 4-9. Seismic risk zone map developed by Algermissen and Perkins (1976) (also U.S. Geological Survey, 1976.) Preliminary map of horizontal acceleration (expressed as % of g) in rock with 90% probability of not being exceeded in 50 years. The maximum acceleration within the 60% contour along the San Andreas and Garlock faults in California is 80% of g.

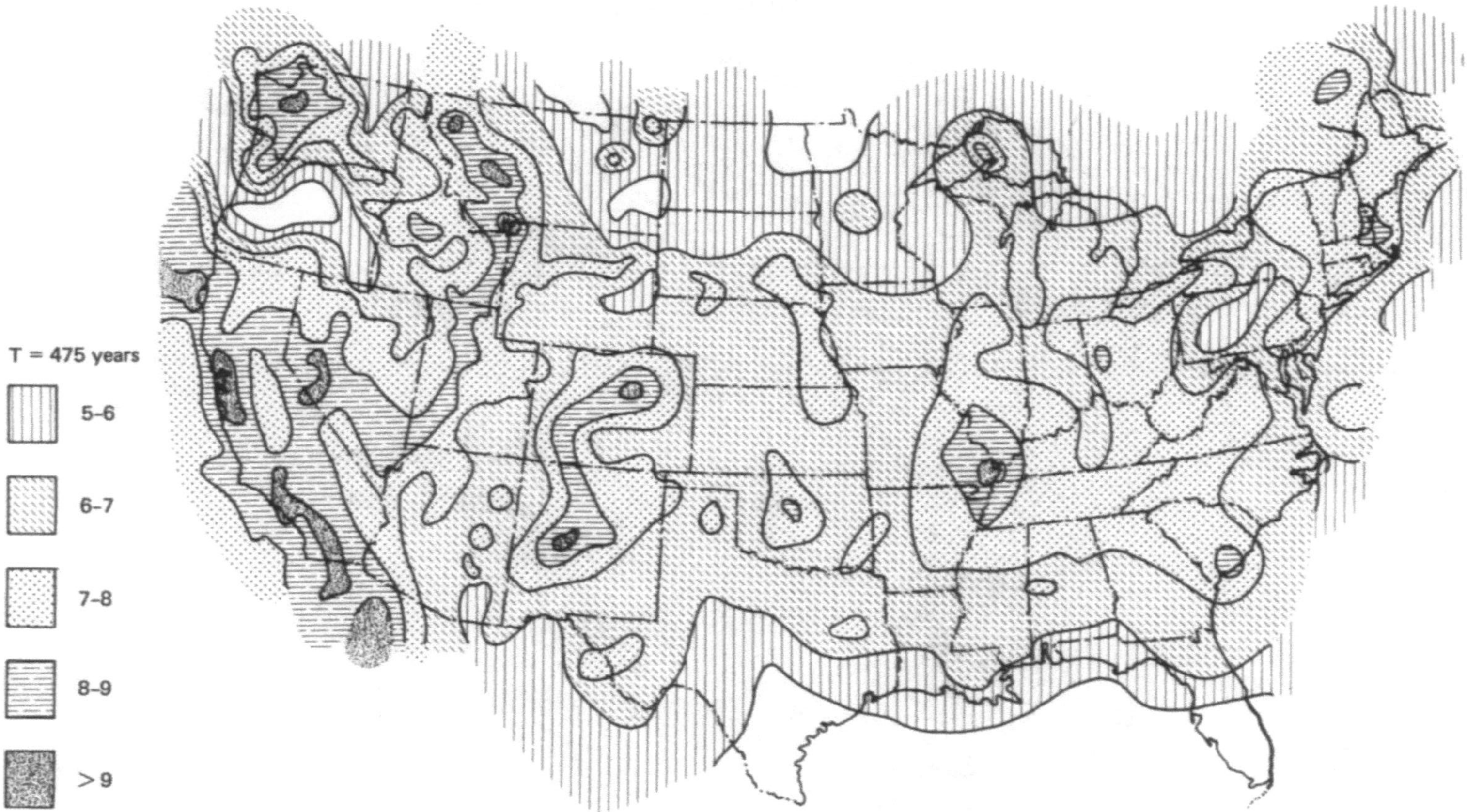

Figure 4-10. MMI for the continental United States. (From Wiggins, 1975.)

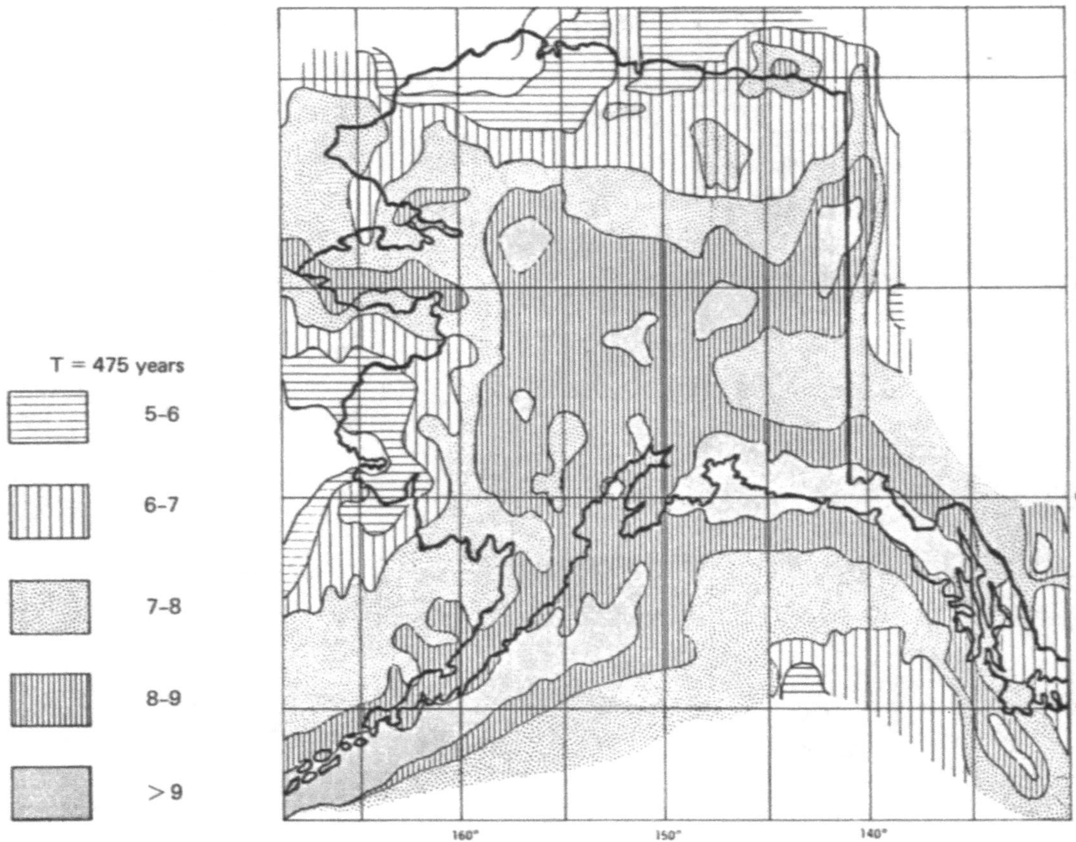

Figure 4-11. MMI for Alaska. (From Wiggins, 1975.)

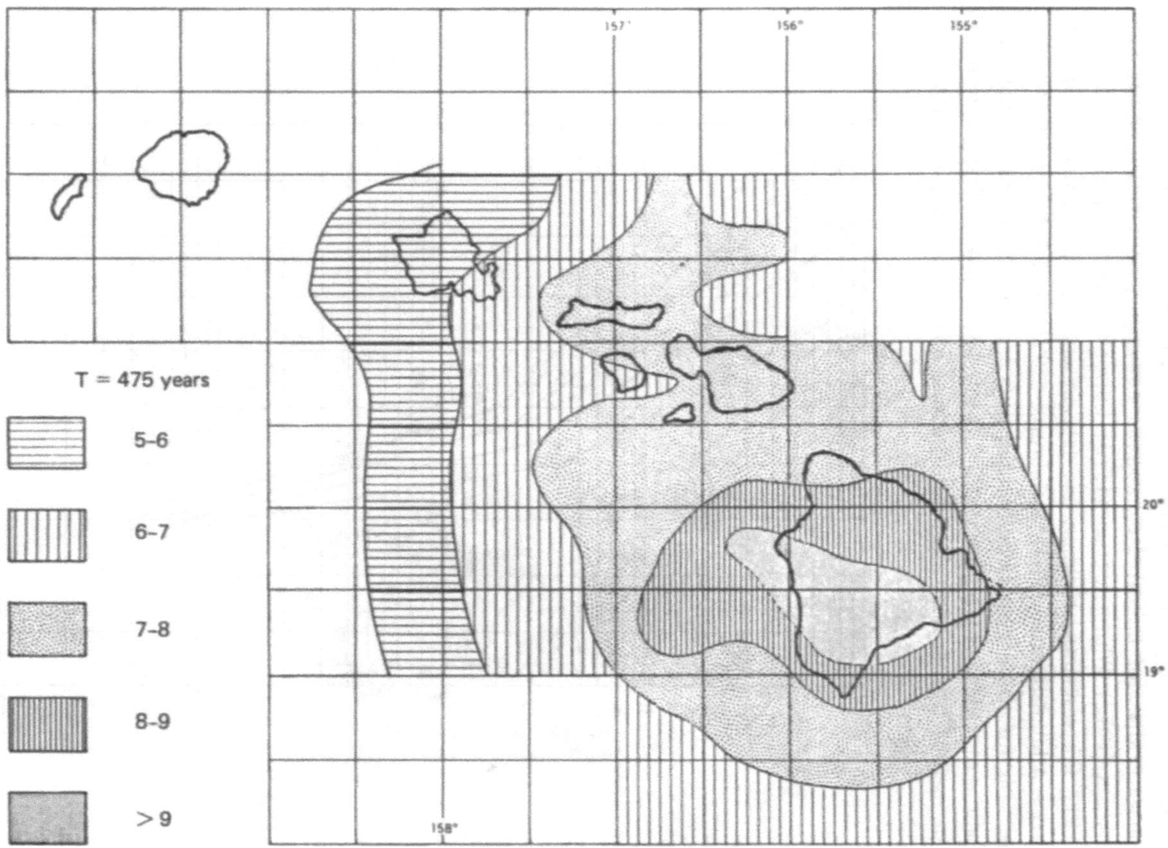

Figure 4-12. MMI for Hawaii. (From Wiggins, 1975.)

On this basis, structures in California built prior to 1933 were assigned Q=1 and structures built in California after 1933 were assigned Q=3.

For each type and strength level of building, estimates were developed showing the relationship between MMI and the percent damage sustained by structures. Earthquake intensity damage relationships were developed from the earthquake damage studies for industrial-commercial buildings presented by Barnes and Pinkham (1972), Whitman et al. (1973), Culver et al. (1975), and Steinbrugge (1971). For residential buildings the data presented by Environmental Science Service Administration (1969), Steinbrugge (1971), and McClure (1973) were the basis for the damage relationships. The damage algorithms for the four qualities of structures are presented in Figures 4-13 and 4-14. These algorithms were used both to develop estimates of annual losses for earthquakes as well as

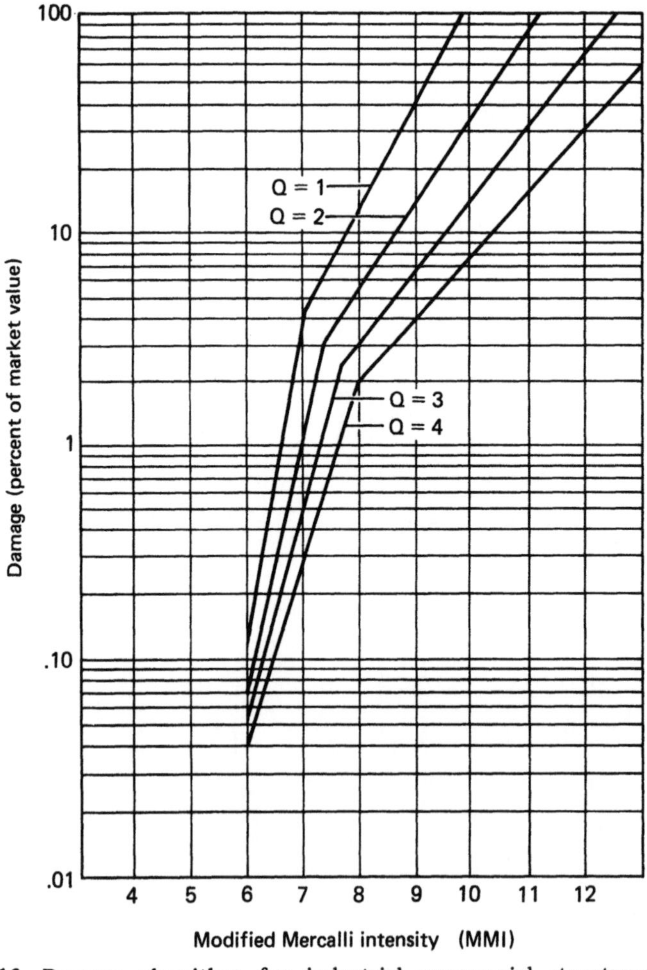

Figure 4-13. Damage algorithm for industrial-commercial structures. (From Wiggins et al., 1978.)

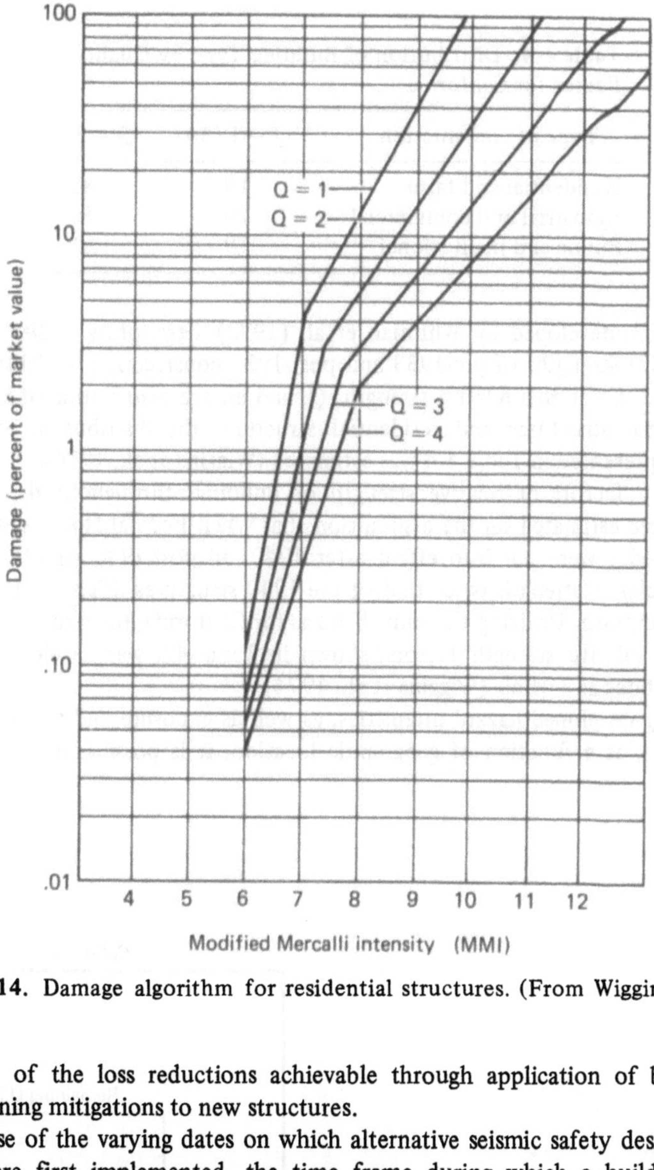

Figure 4-14. Damage algorithm for residential structures. (From Wiggins et al., 1978.)

estimates of the loss reductions achievable through application of building-strengthening mitigations to new structures.

Because of the varying dates on which alternative seismic safety design standards were first implemented, the time frame during which a building was constructed has a significant relationship to the relative strength of the structure. Accordingly, a distribution of structure type by region throughout the United States was developed for two age categories and a separate distribution was developed for California (Wiggins et al., 1974; Hirschberg et al., 1978).

The natural force behind modification of the Uniform Building Code to include increased lateral force design parameters for engineered structures was the 1933 Long Beach earthquake. Due to adoption of these changes in building code specifications in California after 1933 the ability of new construction to withstand lateral forces was increased. Using the average relative strength change

Table 4-4. Distribution of Building Type by Quality Factor for California

Type of construction	$Q = 1$ (%)	$Q = 3$ (%)
Residential and farm	18	82
Industrial and commercial	20	80
Public and institutional	19	81

after 1933, developed by Whitman et al. (1973), descriptors of Q=1 and Q=3 for relative strengths of pre-1933 and post-1933 construction in California was developed. Using this relative strength, Q, and an age distribution of industrial-commercial structures and residential structures the distribution by quality factor represented in Table 4-4 was estimated (Wiggins et al., 1974).

Quality factors or relative strength of buildings throughout the other 49 states were estimated on the assumption that when Federal Housing Authority (FHA) codes were put into effect after 1940, all post-1940 structures would have a relative strength twice that of pre-1940 structures. Figure 4-15 presents this assumption. Utilizing the same basis as for California, the structure distribution, by relative strength factors, shown in Table 4-5 was developed for the United States as a whole (Wiggins et al., 1974).

Having developed hazard intensities, as well as exposure and vulnerability of structures, as a function of geographic location, it is possible to estimate risk

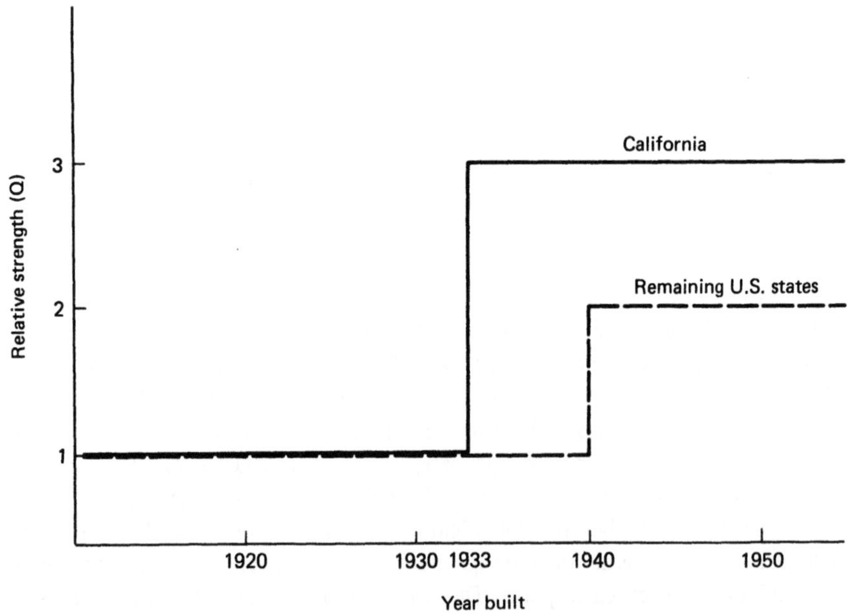

Figure 4-15. Relative structural strength by year for California and the remaining United States. (From Wiggins et al., 1978.)

Table 4-5. Post-1940 Structure Relative Strength Factors for Regions of the United States

Type of construction	Northeast		Central		South		West	
	$Q = 1$ (%)	$Q = 2$ (%)	$Q = 1$ (%)	$Q = 2$ (%)	$Q = 1$ (%)	$Q = 2$ (%)	$Q = 1$ (%)	$Q = 2$ (%)
Residential and farm	55.4	44.6	49.1	50.4	29.5	40.5	31.0	69.0
Industrial and commercial	56.1	43.9	39.9	40.1	25.8	74.2	17.6	82.4
Public and institutional	55.3	44.7	44.5	55.5	27.7	72.3	24.3	75.7

of loss for particular regions of the country. That is, by combining intensity expectancy with the damage algorithms for the exposed structures, a region's damage rate can be computed. The damage rate in this context is defined as the average annual percent loss expected to occur to exposed buildings.

Estimating Expansive Soil Losses

Four major operations were required to develop estimates of expansive soils losses in the United States: (1) the mapping of U.S. expansive soils areas; (2) the assignment of whole counties or fractions of counties to the mapped areas; (3) determination of the size of the population at risk of exposure to expansive soils in each area, and (4) the development and use of a damage algorithm (Figure 4-16).

In the mapping phase, all portions of the continential United States were assigned to one of three expansive soils zones. Criteria used to establish the low, moderate, and high zones depicted in the map correspond to shrink-swell guidelines set forth by the U.S. Soil Conservation Service. The quantitative method used by the Service (U.S. Department of Agriculture, 1971) for determining shrink-swell behavior of soil is referred to as COLE (coefficient of linear extensibility). COLE represents an estimate of the vertical component of swelling in a soil clod or ped. The following somewhat modified Soil Conservation Service definitions were used to establish the expansive soils zones depicted on the map:

High: Generally includes soils high in clay that are made up of a large percentage of montmorillonitic minerals. These soils have a COLE value usually greater than 6%.

Moderate: Generally includes soils containing moderate amounts of clay that also contain some montmorillonitic minerals. COLE values for these soils vary between 3% and 6%.

Low: Generally includes soil containing some clay; however, the clay consists mainly of kaolinite and/or other low shrink-swell clay minerals. These soild have COLE values generally lower than 3%.

Although the whole of any given local area may be assigned on the map to one of the three expansive soil zones, smaller portions of the included local area may have ratings both higher and/or lower than the rating shown for the whole of the zone to which the area is assigned (Figure 4-17).

In developing the map, all available state soils maps were used, but were extensively supplemented by comments from the various state soils scientists and the professional staff of the U.S. Soil Conservation Service. The mapped areas of moderate and high soil expansivity include essentially those areas chiefly underlaid by montmorillonitic soils. There are over 12,000 soils series throughout the entire country, of which approximately 10% contain montmorillonite mineralogy. Of these, about 250 soils series are extensive enough to be shown as mappable units on intermediate scale generalized state soil maps.

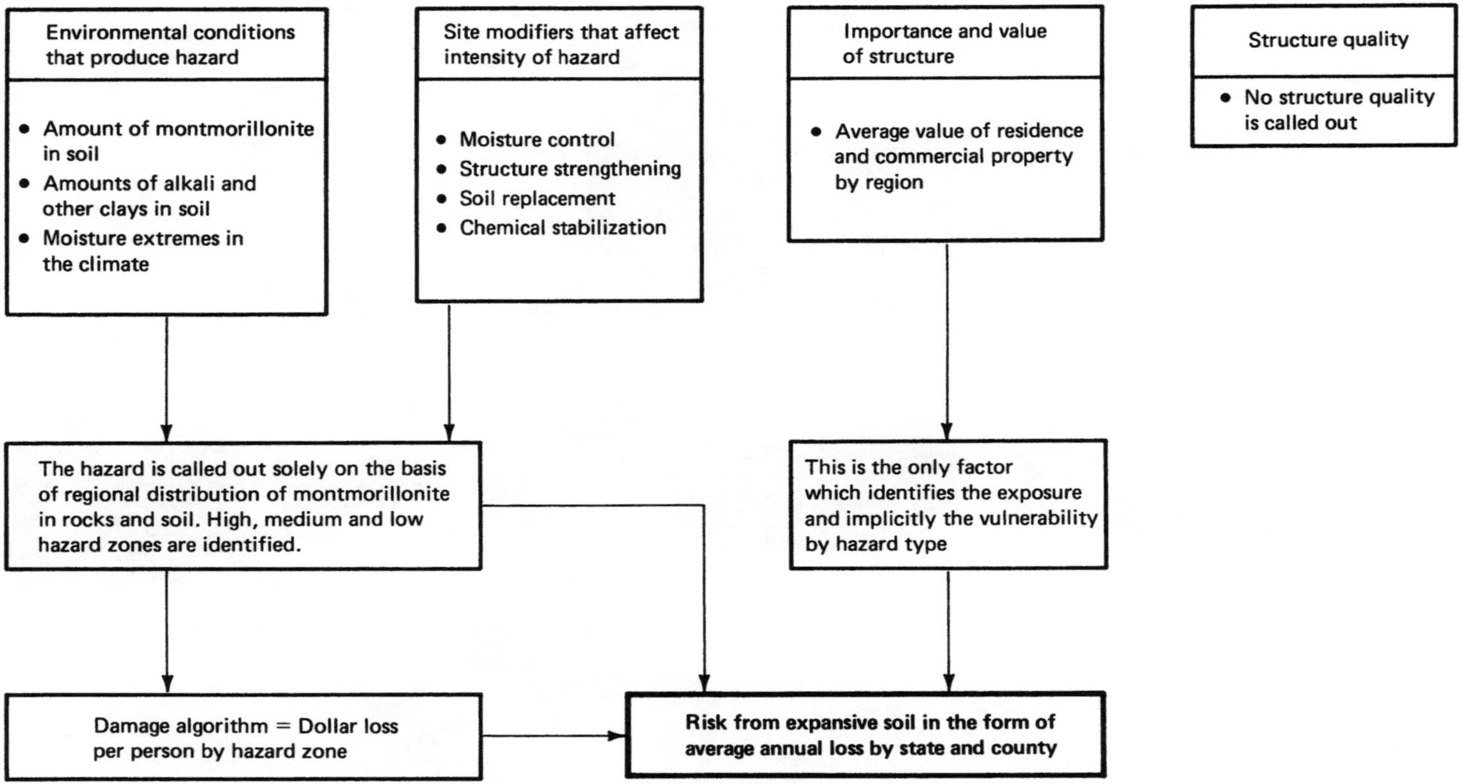

Figure 4-16. Expansive soil risk determination procedure.

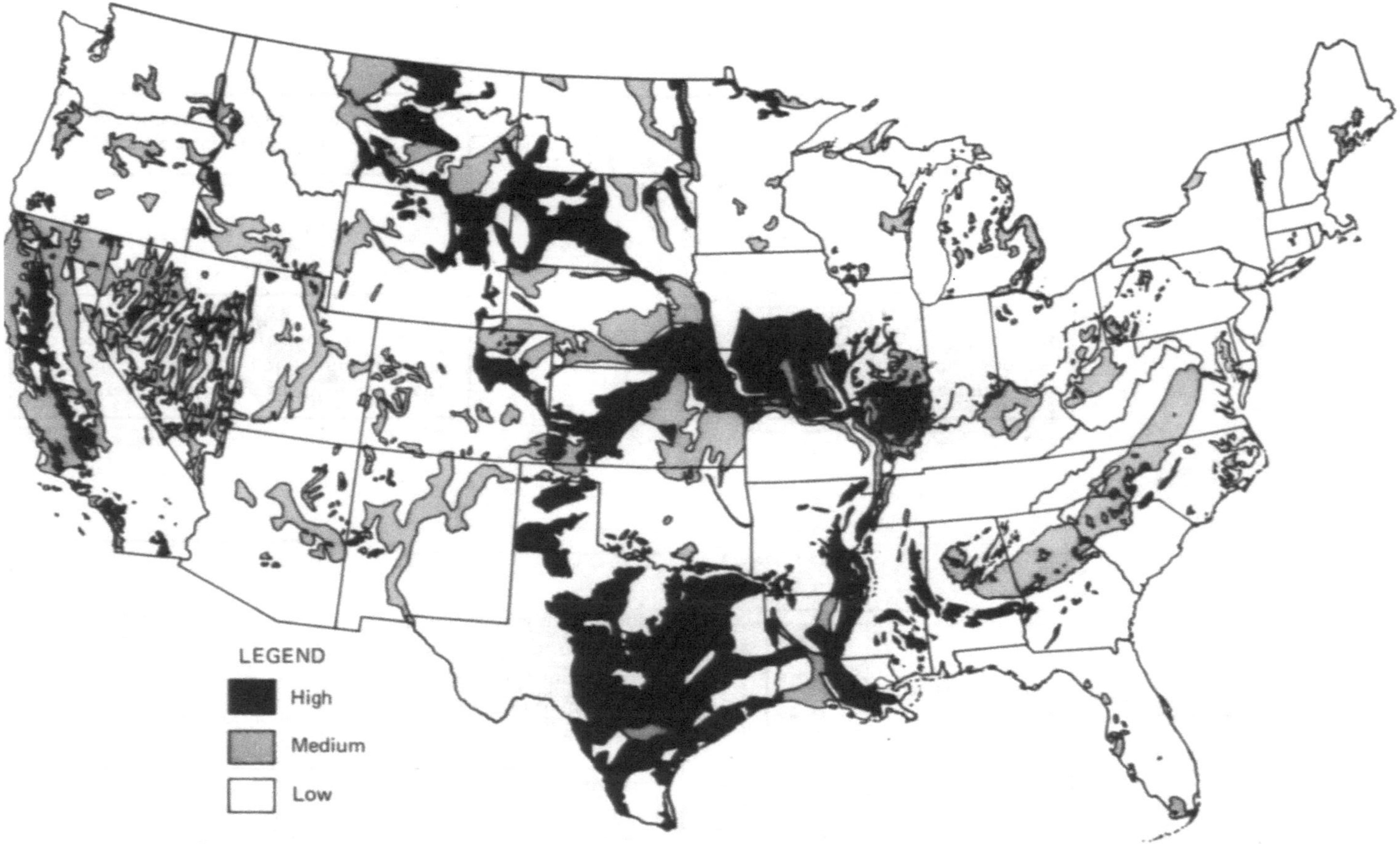

Figure 4-17. Expansive soils map. (From Wiggins et al., 1978 and Krohn and Slosson, 1980.)

Therefore, for purposes of this study, only those large mappable units have been considered. On most state maps, soils are depicted as soil "associations." Generally, these associations consist of one to three soil "series." A hypothetical soil association may contain three soil series (A-B-C), with A always representing the dominant soil series followed by B, then C. Therefore, for purposes of this study, if soils in position A, A-B, A-C, and B-C were montmorillonitic, then the entire soil association was given a moderate shrink–swell rating. In the case of the soil association consisting of one (A) or two (A-B) soil series, a high rating was assigned to the entire association providing the lone soil series (A) or one of the two soil series (A or B) contain montmorillonitic materials. For purposes of this analysis, the low expansive soil zone includes both low and nonexpansive soil areas (Krohn and Slosson, 1980).

After the map was completed and examined, all states were assigned to one of three groups: (1) no moderate or high expansive soils zones, (2) only moderate and low zones, and (3) all three zones. For the purpose of developing a county-level data base necessary to the generation of loss estimates, whole counties in states having only moderate or low expansive soil zones were assigned to such zones based on the latitude–longitude location of each county seat. If that latitude–longitude location fell within a map zone, then the whole of the county was assigned to that zone. A somewhat different procedure was used to develop loss estimates for states in which the map revealed all three soil expansivity zones. In those states, visual inspection of census maps was employed to assign appropriate fractions of counties to "high" and/or "moderate" zones. For the more populous counties, these estimates were based not on the proportion of the area of a county falling within the mapped zone, but on an estimate of the proportion of the county's total population residing within the mapped zone.

Since the damage algorithm used to estimate expansive soils losses was derived from the study of damage to single unit residential structures, the size of county populations exposed to expansive soils in an indicated category was reduced by the census-reported percentage of the county population residing in single family residential units.

The damage algorithm used to estimate expansive soils dollar losses was based entirely on two studies involving residential foundation data (Alfors, 1971; Smith and Allen, 1974). Smith and Allen report that homeowners in Dallas County, Texas, experience approximately 8470 residential foundation failures annually. The failures all involve expansive soil-related problems (e.g., cracked walls, foundations). It appears that 84% of the foundation failures occur within high expansive soil areas, 14% in moderate expansive soil areas, and 2% in low expansive soil areas. In 1970, based on census data, there were approximately 312,100 residential foundations in Dallas County (Smith and Allen, 1974). Of the total number of foundations, 42% exist on high expansive soil, 27% on moderate expansive material, and 31% on low expansive soil areas. Assuming that the average foundation repair cost is $1650 (Smith and Allen, 1974), the following can be inferred:

Expansive soil rating	Total no. foundations	Annual foundation failures	Cost per foundation ($)	Total cost of annual foundation failures ($)
High	130,400	7115	1,650	11,739,750
Moderate	83,600	1186	1,650	1,956,900
Low	98,100	169	1,650	278,850
Total	312,100	8470	1,650	13,975,500

$$\text{Damage loss per foundation per year} = \frac{\text{Total cost of annual foundation failures}}{\text{Total number of foundations per severity rating}} .$$

The average annual damage cost for residential foundations in Dallas County, Texas, for high expansive soil areas appears to be $90.00 per foundation per year. Foundation costs in moderate and low expansive soil regions are estimated to be $23.41 and $2.84 per foundation per year, respectively.

The average population rate per dwelling in Dallas County is estimated, by 1970 census data, to be three persons per housing unit. Therefore, estimated foundation failure losses per capita are:

Expansive soil rating	Per capita damage loss per foundation per year
High	30.00
Moderate	7.80
Low	0.95

Expansive soil related losses have also been calculated for California and discussed in the Open File Report 72-2, California Division of Mines and Geology. The method used to establish the map severity code used in the California study is somewhat similar to the method incorporated in the Smith and Allen (1974) report. Results of the California report (Alfors et al., 1971) are as follows:

Expansive soil rating	Per capita damage loss per year
High	$22.30
Moderate	$ 6.92
Low	$ 1.14

Averaging the California statistic with those of Dallas County, the following figures result:

Expansive soil rating	Per capita damage loss per year
High	$26.15
Moderate	$ 7.36
Low	$ 1.05

Accordingly, the damage algorithm used in this analysis assigned a loss of $26.00 per capita per year to single family residents in high zones; a damage loss of $8.00 per capita per year to single family residents in moderate zones; and a $1.50 damage loss per capita per year to single family residents in low zones.

Estimating Landslide Losses

Several assumptions were made during the development of the landslide intensity map, and were used to guide the map categorization:

1. It was assumed that adverse formations associated with previous landslide activity will continue to have a high potential for landslide activity in the future, providing all factors (topographic relief and precipitation) remain constant.
2. Due to size and scale, it was not possible to differentiate vertical and lateral formation variations within a given adverse formation. Therefore, if a formation was considered to be adverse in one area, all other locations containing the same formation were also considered adverse.
3. Although a given landslide potential rating was assigned to an area, local portions of that area may have ratings higher and/or lower than the ratings shown.

Recognizing the above limitations and constraints, a landslide intensity map was developed by Krohn and Slosson (1976) (Figure 4-18) to define landslide-prone areas within the continental United States. This map is based on three principal factors: topography, bedrock, and precipitation. Topography and rock type (Figure 4-19) were used to establish the high, moderate, and low landslide potential ratings defined on the landslide map. Mean annual rainfall data was consulted in order to complete the map. These factors, when considered for a given geographic area, provide the basis for development of a landslide intensity scale. These relative intensities are presented in Table 4-6.

The development of an exposure-vulnerability relationship for areas subjected to varying landslide intensities is difficult. Due to the limited data, it was necessary to develop building loss estimates from landslides using estimates of damage on a per capita basis as presented in the literature.

Exposure to landslide was determined, by county, from structure values, the population size, and a computation of average structure value per person. No

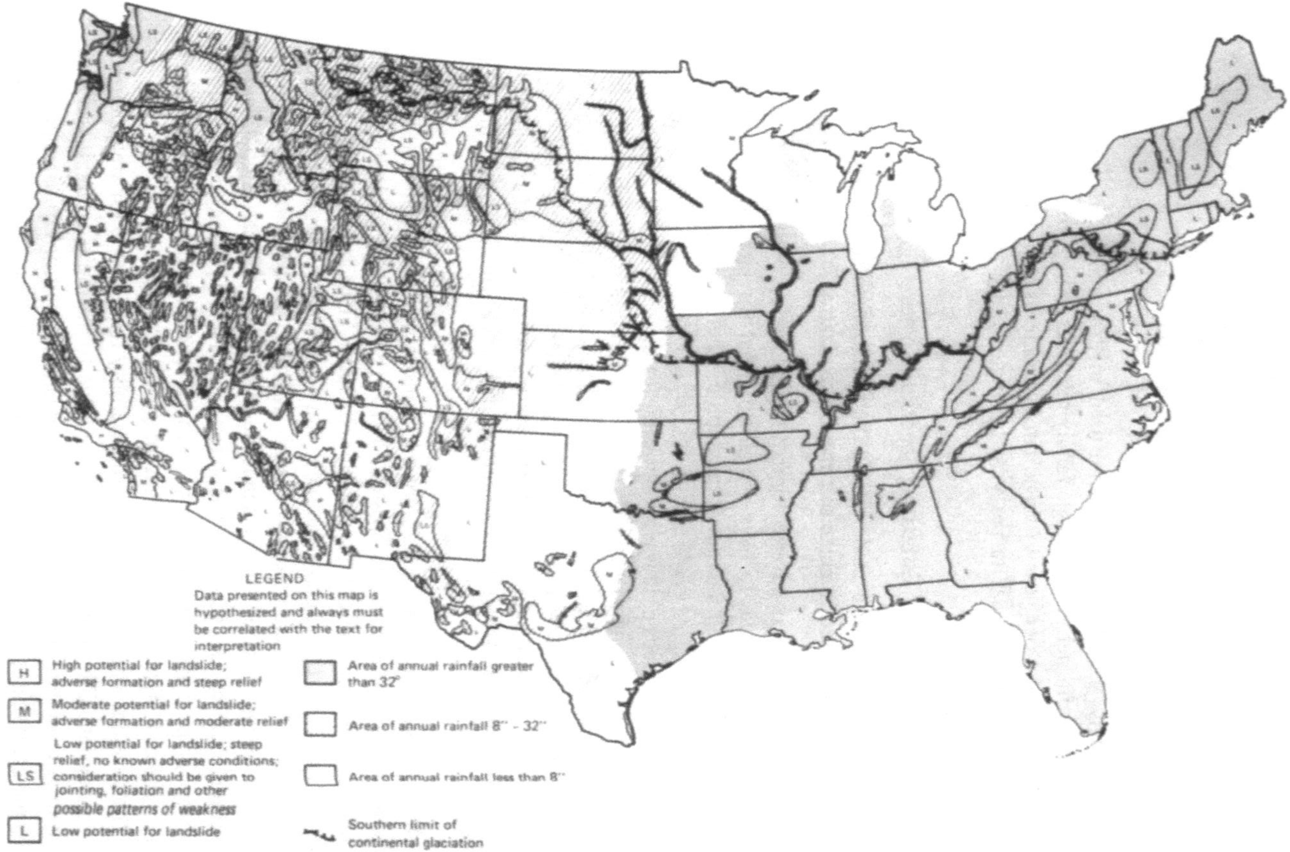

Figure 4-18. Interpretive landslide potential map. (From Krohn and Slosson, 1976.)

High (H): An area of steep topographic relief with a known landslide-prone bedrock formation (rock type).				
High	=	Adverse formation	+	Steep relief
Moderate (M): An area of moderate topographic relief with a known landslide-prone bedrock formation (rock type).				
Moderate	=	Adverse formation	+	Moderate relief
Low Steep (LS): An area of steep topographic relief without a known landslide-prone bedrock formation (rock type) of the two low designations, the low (steep) designation has the highest landslide potential.				
Low (steep)	=	No adverse formation	+	Steep relief
Low (L): An area of low topographic relief and may or may not contain a known landslide-prone bedrock formation (rock type); or moderate relief with no known landslide-prone bedrock formation.				
	=	No adverse formation	+	Moderate relief
Low	=	No adverse formation	+	Low relief
	=	Adverse formation	+	Low relief

Figure 4-19. Landslide potential rating criteria. (From Krohn and Slosson, 1976.)

attempt was made to distinguish between various buildings types such as residential, industrial, or commercial. It was assumed that dwellings usually suffer from landslides more frequently than do industrial-commercial properties; thus, loss estimates may trend toward the high side.

Using estimated costs of damages from landsliding in Allegheny County, Pennsylvania (Briggs et al., 1975) and estimated losses from landslides occurring in California (Brabb and Taylor, 1972; Slosson, 1969), Table 4-7 was developed to show landslide loss per capita as a function of intensity.

The landslide potential map and estimate of the precipitation made it possible to establish an intensity range for each U.S. county, and thus define a county risk zone. By knowing the population and structure value in a specific county and the risk zone for that county, it is possible to estimate the dollar loss resulting from landslides on an annual average basis. The risk determination procedure discussed above is summarized in Figure 4-20.

Table 4-6. Landslide Intensity Ranking Based on Rock Type, Topography and Precipitation

Landslide potential map rating	Rainfall (inches)	Relative intensity (highest ranking is the most severe)
High	32	XII
	8–32	X
	8	IV
Moderate	32	XI
	8–32	IX
	8	III
Low (steep)	32	VIII
	8–32	VI
	8	II
Low	32	VII
	8–32	V
	8	I

Table 4-7. Landslide Damage per Person Normalized to 1970 Dollars

Reference	Region	Intensity	Loss per person (1970 $)
1	California	XII	140.00
		IX	23.00
		V	1.10
2	California	XII	53.00
		IX	35.00
		V	1.00
3	Allegheny Co.	XII	1.20
4	Marin Co.	X 1/2	14.00
	San Mateo Co.	X 1/2	6.10
	Contra Costa Co.	IX	2.60
	Napa Co.	IX	1.80
	Sonoma Co.	XII	.88
	Alameda Co.	XII	.35
	Santa Clara Co.	V	.18
5	Marin Co.	X 1/2	5.30
	San Mateo Co.	X 1/2	7.40
	Contra Costa Co.	IX	11.00
	Napa Co.	IX	21.00
	Sonoma Co.	XII	39.00
	Alameda Co.	XII	5.30
	Santa Clara Co.	V	2.10
6	Los Angeles Co.	V	2.10

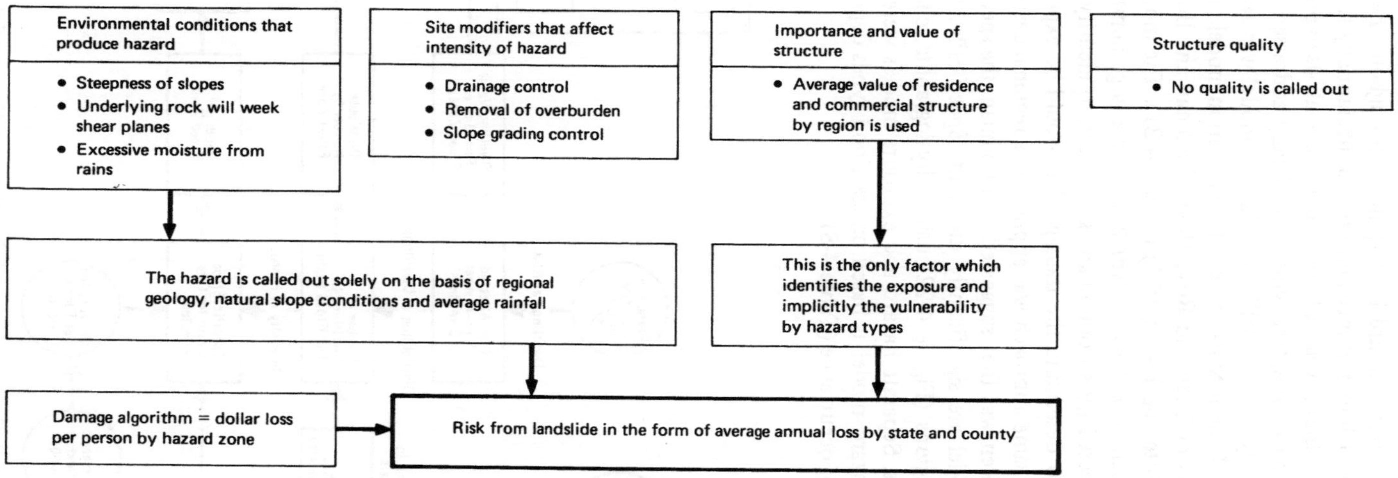

Figure 4-20. Landslide risk determination procedure.

Estimating Riverine Flooding Losses

Although the riverine flood hazard is ubiquitous throughout the United States, local variables are of such importance and vary so substantially from one riverine floodplain to another, that is was not possible in this analysis to present a national hazard model for riverine flooding which specifically addressed each individual floodplain. Accordingly, a "regional" flood damage model was developed by Lee et al. (1978) for the United States using the simulation model previously developed by Friedman (Friedman and Roy, 1966; Friedman and Boccacino, 1972) and reported by White and Haas (1975) (Figure 4-21). The model employed a Monte Carlo approach to determine which, among a large number of modeled cities, would be affected by a flood in any given year, to identify the magnitude of the flood, and to determine the damage that would be experienced by the affected cities. This same approach was applied on a regional basis. Although the general approach taken was that suggested by Friedman, the specific parameters of the required model are significantly different. Specifically, the nation was divided into ten regions (Figure 4-22) using physiographic, climatologic, and demographic factors. Separate hazard and exposure models were developed for each region. The hazard model is based on six flood types depending on their annual probability of occurrence (Table 4-8).

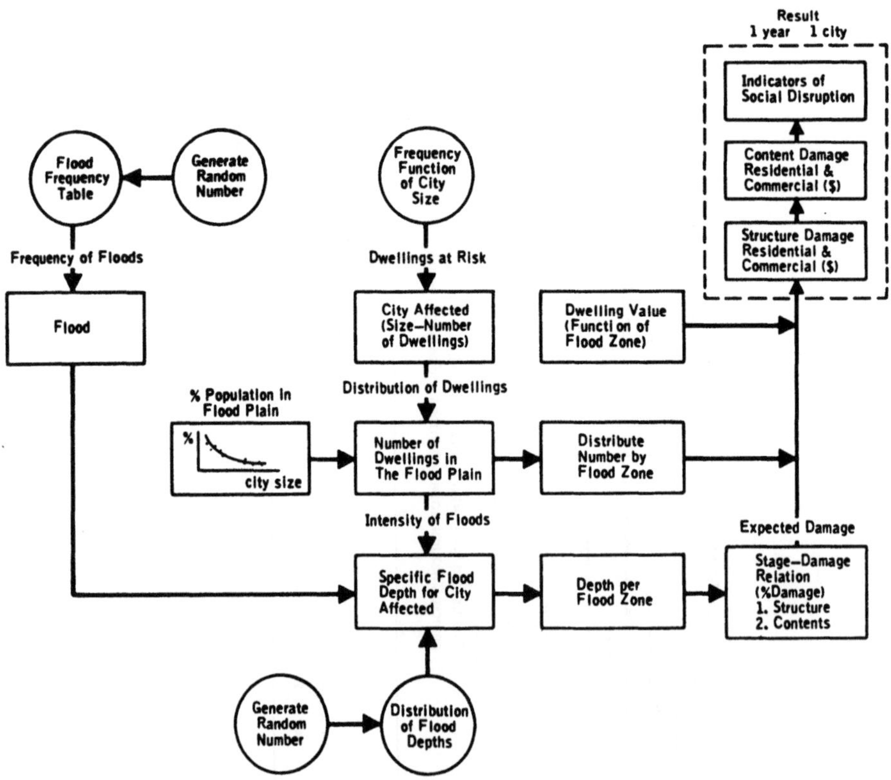

Figure 4-21. Simplified flow diagram of flood simulation model. (From White, 1975.)

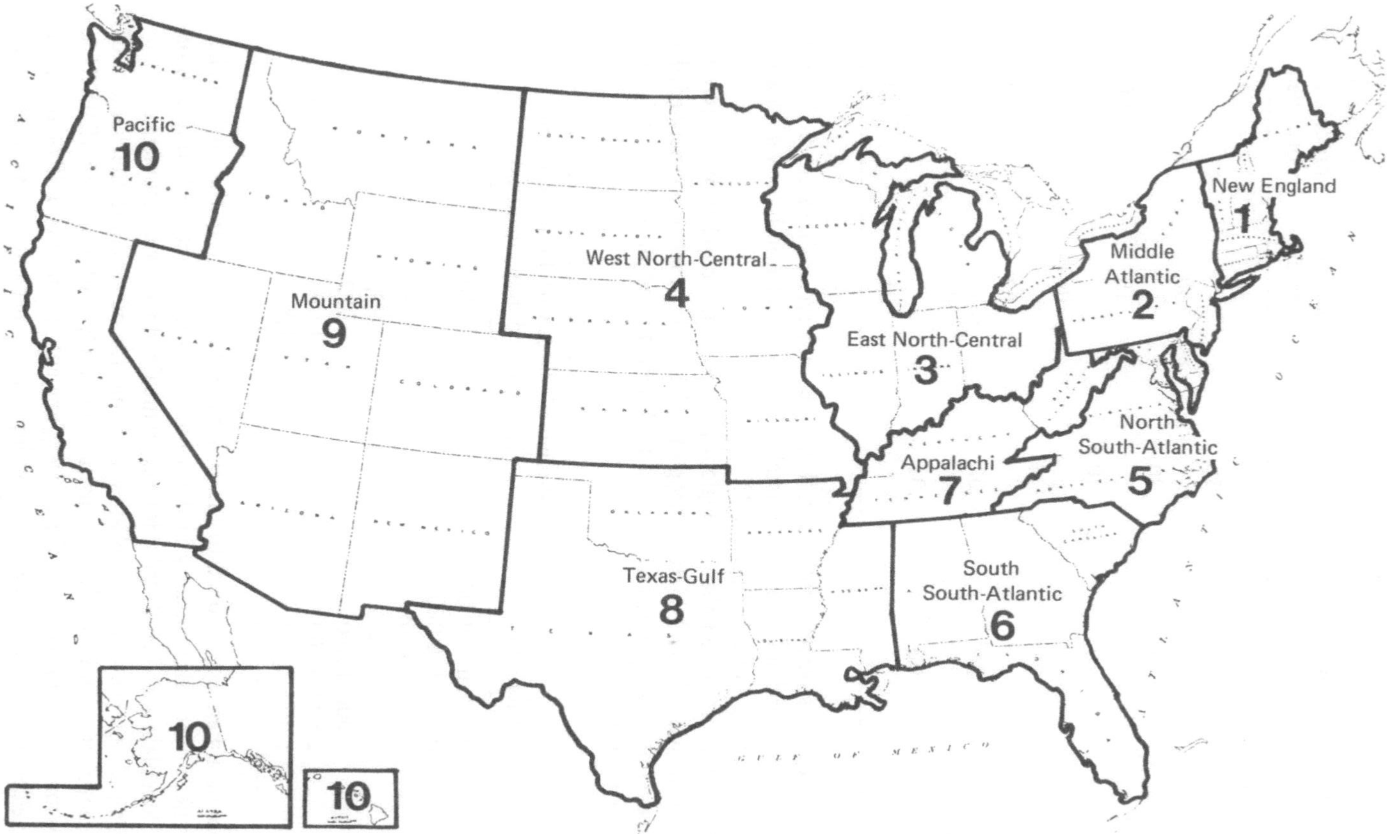

Figure 4-22. Regional model for riverine flooding. (From Lee et al., 1978.)

Table 4-8. Probability of Flood Occurrence

Flood type	Return period (years)	Probability of equaling or exceeding	Probability of occurrence	Cumulative probability of occurrence
No flood			.50	.50
	2	.50		
A			.30	.80
	5	.20		
B			.10	.90
	10	.10		
C			.06	.96
	25	.04		
D			.02	.98
	50	.02		
E			.01	.99
	100	.01		
F			.01	1.00

The depth associated with each flood type was regionalized on the basis of physiography and mean annual rainfall. Only four relationships were needed. A typical one is shown in Figure 4-23.

The exposure model also was regionalized. First the number of places (cities and towns with more than 1000 people) with a flood problem was defined. The following breakdown by city provided by Friedman and Bocaccino (1972) was used as the basis.

City size	Number of places with flood problems (incorporated municipalities)
Greater than 100,000	140
50,000–100,000	204
25,000–50,000	390
10,000–25,000	903
5,000–10,000	992
2,500–5,000	1,098
1,000–2,500	1,812

These data were apportioned to the 10 regions on the basis of the total number of cities of each size category in the region. Then the number of structures on the floodplain and their distribution across the floodplain were estimated.

Only 2 sources were immediately available for determining the number of residential structures on floodplains: a 1966 study performed for HUD (U.S. Department of Housing and Urban Development, 1966) and a 1974 study by Schneider and Goddard. Two relationships of the percent of dwellings on the 100 year floodplain as a function of city size were developed: one for the central

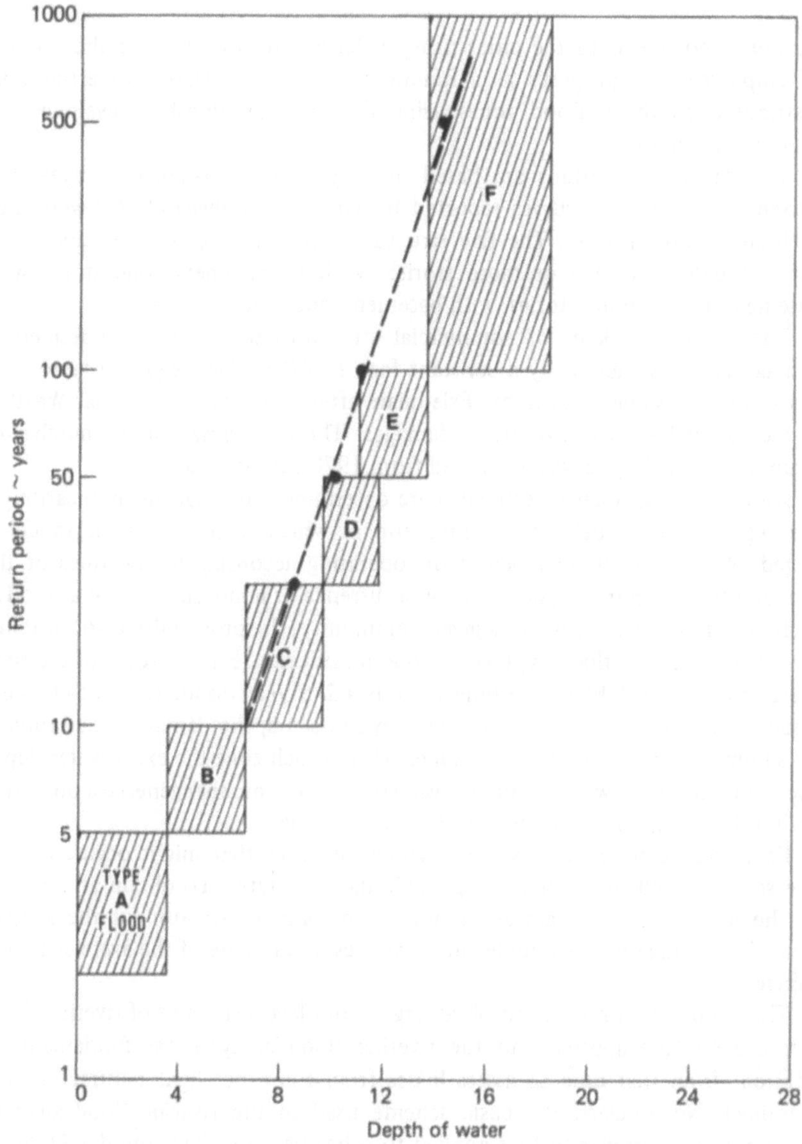

Figure 4-23. Hazard model–plains–high rainfall. (From Lee et al., 1978.)

and coastal plains and one for the western and eastern highlands. A further breakdown into the six specific flood zones was performed using the 1966 HUD data. The value of each residential unit was estimated and different values were obtained for different regions and city sizes.

An estimate of the total number of commercial-industrial-governmental properties in the United States was made in a somewhat coarser fashion. This national total was then apportioned to the 10 regions and within each region further apportioned to the cities using a derived relationship for the number of employees per property as a function of city size. These properties were distributed to the 6 flood zones using the same relationship developed for residential structures.

The damage algorithms are based on empirical depth-damage curves. The current depth-damage curves reported by Grigg and Helweg (1975) were used with slight modification. The five structure types considered were: one story, without basement; two or more stories, without basement; one story, with basement; two or more stories, with basement; and mobile homes.

The approach taken for commercial structures was to use the residential damage curves weighted by a constant factor of 0.3. The weighting factor was based on the value chosen by Friedman after a review of National Weather Service annual estimates of flood damages. The breakdown of the number of structures of each type was developed from 1970 census data.

Annual average damage estimates are determined through the application of the regionalized model by a sample run in which each city in the model is tested. A random number generator, operating according to the rules of the probability of occurrence, denotes the occurrence or nonoccurrence of a flood at each city. If a flood occurs for a given community, the probability-of-occurrence model specifies the flood type (i.e., flood magnitude). Fifty percent of the time there is no flood; 50% of the time there is a 2-5 year flood, 10%, a 5-10 year flood, etc. When a flood of a particular type occurs, the damage is determined by summing over all the affected zones. Within each zone the exact water depth above the first floor was determined using the random number generator operating within the depth ranges defined by the hazard model.

Each sample produces a set of regional damages that might occur, that is, one set out of all the millions of possible damage states. To obtain an estimate of the average annual damage, a number of samples are run for the subject year. The average of the sample runs provides an estimate of the average annual damage.

This model does not permit obtaining county-level estimates of riverine flood losses. Thus, the approach to the riverine flood analysis was fundamentally different from that used to assess losses from the other light natural hazards examined. Nonetheless, the basic scheme used in the riverine flood analysis (Figure 4-24) is similar to that applied to other hazards. Only the details of its application and the level of detail generated by the model differ.

Estimating Storm Surge Losses

The purpose of the hazard intensity model for storm surge is to associate probabilities of annual occurrence with given tide heights. This is done by developing plots of tide height versus frequency of occurrence for specific locations. Figure 4-25 shows a typical tide/frequency plot. This approach gives either the return period for exceeding a certain tide height, or the annual probability of exceeding that height. At the present time, 2 methods are used by analysts for generating

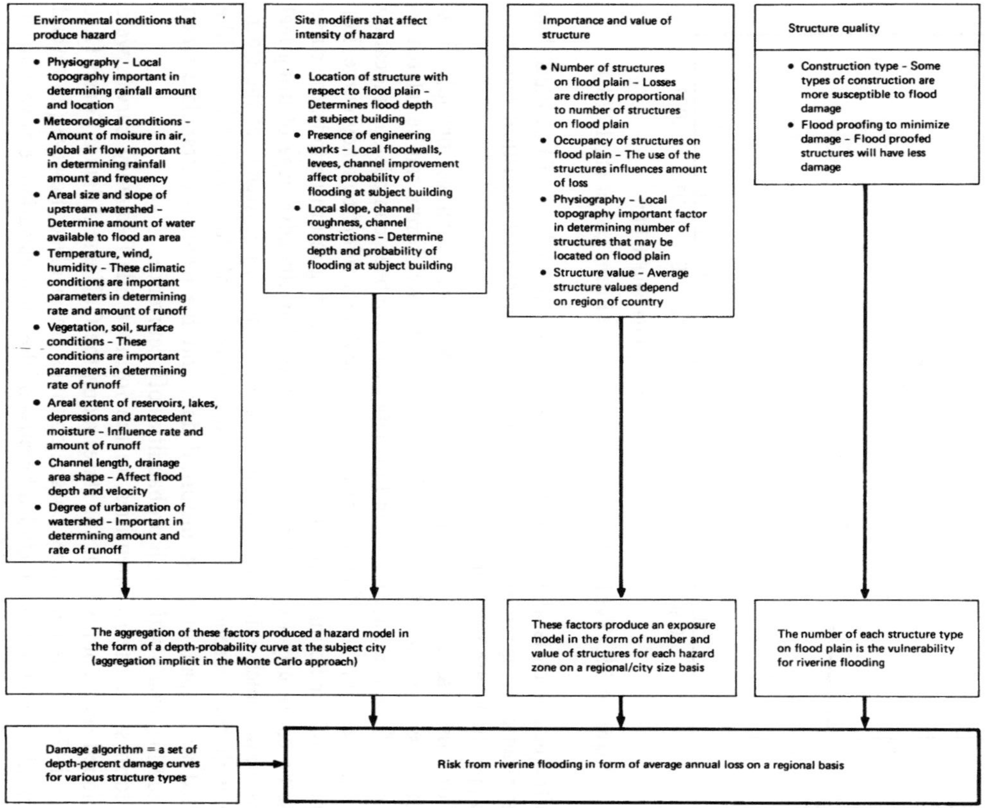

Figure 4-24. Riverine flooding risk determination process.

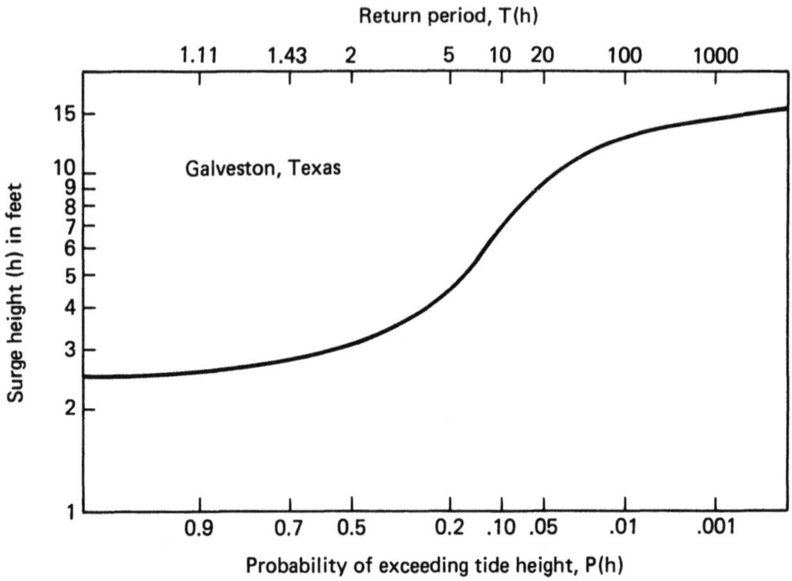

Figure 4-25. Typical tide frequency curve. (From Lee et al., 1978.)

tide/frequency plots at specific locations: statistical analysis of historical data or synthetically generated tide data based on generalized statistical hurricane data and the hydrodynamic surge model.

For this analysis, the historical method was used wherever possible. This method assumes that the tides of present and future storms can be related to the past. Therefore, actual storm tide data that have been compiled at a location for a limited number of years are used to calculate a tide/frequency curve. For purposes of predicting storm tides of events with long return periods, the curve must be cautiously extrapolated. This method has the advantage of using real storm data which inherently account for the random occurrence of the storm tide with the astronomical tide. The disadvantage of this method lies in the limited time span of most historical data and the need to extrapolate when considering events with small probabilities of occurrence.

Storm surge hazard intensities, in the form of tide/frequency curves, were developed on a county basis for states along the Gulf and Atlantic Coasts (Lee et al., 1978). In general, only coastal counties are involved; however, in several states inland counties are included because of their low elevation relative to mean sea level. The delineation of the hazard intensities on a county basis enables damage estimates to be produced.

Whether people are exposed to the storm surge hazard depends on their location relative to mean sea level. The largest surge height used in this study is 23 feet above mean sea level (msl). For this study, therefore, people living in coastal areas below 23 feet ms may be exposed to some degree of storm surge risk.

An estimate of the distribution of people and buildings by elevation intervals (i.e., one foot increments) was developed for each county along the Atlantic and Gulf Coasts. Only one other study to date (Friedman, 1971) has attempted to define such an extensive exposure breakdown. Data from that study were incorporated by Lee et al. (1978), particularly for regions where more current or accurate information was not available.

The exposure model provides only an approximation because of the following:

Distribution increments: Damage varies with water height. Thus, the population exposed must be distributed to relatively small elevation increments (one foot increment). This is a much finer increment than that for which population information is available.

Type of data available: Topographic and storm-evacuation maps at best only resolve elevation contours down to five-foot elevations. Thus, approximate relationships had to be developed to distribute the populace into one-foot elevation increments.

Quality of data: Analysis of all available topographic and storm-evacuation maps to estimate exposed population was not feasible within the constraints of this project.

The exposed population and their distribution must be considered in relation to man-made or natural barriers that might protect all or part of the county. Where large seawalls exist and protect all or nearly all of the county population to a specified height, the natural hazard model was modified by setting the surge probability for depths below the height to zero. These adjusted elevations are referred to in the analysis as adjusted mean sea level. In those cases where seawalls or levees protect only a limited area of a county, the exposure model for that county was reduced according to the percentage of the population protected. Table 4-9 summarizes the 1970 county exposure together with known information on the protective measures. It was assumed that the first floor of all structures would be two feet above grade due to the prevalence of foundations and basements.

The damage relations used for storm surge consist of a set of depth–damage curves, which provide an estimate of the percent of damage as a function of water height above the structure's first floor. There are 2 types of damage associated with storm surge. The first type is due to "still water" effects and is similar to that caused by riverine flooding except slightly more severe because of the corrosiveness of seawater. The second type is caused by wave action that imposes large forces on the structure.

Because the degree of damage is also a function of structure type, 10 separate structural categories were considered: one story residential without basement, one story residential with basement, two story residential without basement, two story residential with basement, one story commercial without basement, one story commercial with basement, two story commercial without basement, two story commercial with basement, industrial buildings, and mobile homes.

These damage curves were prepared for each structure type to reflect different levels of wave action: (1) still water, (2) light to moderate wave action,

Table 4-9. Summary of Storm Surge Exposures by State and County

State	County	1970 Exposed population (MSL–20)	Hazard (H) or exposure (E) reduction
Texas	Cameron	4,962	
	Willacy	99	
	Kenedy	0	
	Kleberg	15	
	Nueces	26,262	E 50% reduc. due to 13′ seawall at Corpus Cristi
	San Patricio	11,580	H 16″ cutoff–100-yr. wall at Freeport
	Aransas	5,379	
	Refugio	645	
	Calhoun	14,478	
	Jackson	813	
	Victoria	0	
	Matagorda	4,881	
	Brazoria	35,226	
	Galveston	140,319	E 95% reduc.–15′ seawall at Galveston, levee at Texas City
	Harris	29,451	
	Chambers	1,365	
	Jefferson	94,329	E 40% reduc.–100-yr. seawall at Pt. Arthur
	Orange	39,183	
Louisiana	Cameron	8,193	
	Calcasieu	143,520	
	Jefferson Davis	15,594	
	Acadia	13,275	
	Vermilion	42,999	
	Lafayette	0	
	St. Landry	2,184	
	St. Martin	24,300	
	Iberia	57,399	
	Pt. Coupee	0	
	St. Mary	46,740	E 30% reduc.–13′-14′ levee at Pt. Arthur
	Iberville	30,000	
	W. Baton Rouge	0	
	E. Baton Rouge	0	
	Terrebonne	46,899	
	Assumption	19,599	
	Ascension	36,999	

Table 4-9. (continued)

State	County	1970 Exposed population (MSL–20)	Hazard (H) or exposure (E) reduction
	Lafourche	32,199	
	St. Charles	21,840	H 9′ cutoff-storm evac. maps show levees
	St. John the Baptist	23,001	
	Livingston	1,734	
	Tangipamora	2,325	
	St. Tammany	57,999	H storm evac. maps and conversation show approx. 9′–10′ wall or levee protection use 9′ cutoff
	Jefferson	362,760	
	Orleans	578,000	
	Plaquemines	25,224	
	St. Bernard	51,186	
	St. James	19,731	
Mississippi	Hancock	6,843	H 6′ cutoff–levee from Waveland to Bay St. Louis
	Harrison	56,112	H 9′ cutoff–levee approx. 8′-12′
	Jackson	54,969	H 5′ cutoff–5′ beach erosion project
Alabama	Mobile	63,348	
	Baldwin	15,426	
Florida	Escambia	39,240	
	Santa Rosa	41,370	
	Okaloosa	38,046	
	Walton	591	
	Bay	16,338	
	Gulf	6,939	
	Franklin	5,934	
	Wakulla	1,455	
	Jefferson	24	
	Taylor	504	
	Dixie	522	
	Levy	7,752	
	Citrus	2,721	
	Hernando	285	
	Pasco	5,739	
	Pinellas	163,200	
	Hillsborough	98,700	
	Manatee	28,035	
	Sarasota	64,587	
	Charlotte	20,598	

Table 4-9. (continued)

State	County	1970 Exposed population (MSL–20)	Hazard (H) or exposure (E) reduction
	Lee	43,311	
	Collier	26,307	
	Monroe	43,443	
	Dade	750,000	E 30% reduc.–partial wall/ local protection
	Broward	463,200	
	Palm Beach	226,641	
	Martin	13,155	
	St. Lucie	34,278	
	Indian River	27,255	
	Brevard	137,064	
	Volusia	112,770	
	Flagler	2,052	
	St. Johns	14,325	
	Duval	23,487	
	Nassau	7,638	
Georgia	Camden	4,725	
	Glynn	26,739	
	McIntosh	3,075	
	Liberty	444	
	Bryan	900	
South Carolina	Chatham	64,716	E 20% reduc.–7′–9′ seawall at Savannah Beach
	Jasper	900	
	Beaufort	48,744	
	Colleton	900	
	Dorchester	0	
	Charleston	194,715	E 20% reduc.–7′ seawall fronting Charleston
	Berkeley	900	
	Georgetown	12,537	
	Horry	13,176	
North Carolina	Brunswick	3,261	
	New Hanover	28,809	
	Pender	681	
	Onslow	61,653	
	Jones	0	
	Carteret	31,602	
	Craven	17,646	
	Pamilico	9,468	
	Beaufort	35,649	

Table 4-9. (continued)

State	County	1970 Exposed population (MSL–20)	Hazard (H) or exposure (E) reduction
	Hyde	5,571	
	Dare	6,993	
	Tyrell	3,804	
	Washington	14,037	
	Martin	582	
	Bertie	2,613	
	Hertford	0	
	Gates	0	
	Chowan	9,174	
	Perquimans	6,174	
	Pasquotank	26,823	
	Camden	5,451	
	Currituck	6,975	
Virginia	Virginia Beach	185,493 }	E 20% reduc.–seawall on Atlantic side
	Chesapeake	17,250	
	Portsmouth	110,001	
	Norfolk	304,800	
	Nansemond	1,107	
	Isle of Wight	3,483	
	Newport News	102,240 }	E 10% reduc.–7′ seawall on Bay side
	Hampton	126,000 }	E 40% reduc.–3 seawalls along Chesapeake Bay
	York	7,770	
	Gloucester	2,526	
	Mathews	2,841	
	Middlesex	1,119	
	Lancaster	2,391	
	North Umberland	1,269	
	Westmoreland	2,757	
	Northampton	2,130	
	Accomack	3,933	
Maryland	St. Marys	2,400	
	Calvert	2,568	
	Anne Arundel	22,701	
	Baltimore City	135,000	
	Baltimore	71,751	
	Harford	4,731	
	Cecil	3,669	
	Kent	3,933	

Table 4-9. (continued)

State	County	1970 Exposed population (MSL–20)	Hazard (H) or exposure (E) reduction
	Queen Annes	3,000	
	Talbot	5,982	
	Dorchester	29,400	
	Wicomico	9,264	
	Somerset	17,775	
	Worcester	2,541	
Delaware	Sussex	6,438	
	Kent	708	
	New Castle	18,624	
New Jersey	Salem	26,577	
	Cumberland	6,624	
	Cape May	47,310	
	Atlantic	100,485	H 8′ cutoff–bulkheads at Atlantic City & Ocean City
	Burlington	501	
	Ocean	72,918	
	Monmouth	100,800	E 20% reduc.–seawalls at Keansburg and Monmouth Beach
	Middlesex	66,000	
	Union	50,700	
	Essex	102,999	
	Hudson	114,600	
New York	Richmond	25,380	
	New York	166,500	
	Kings	896,766	H 7′ cutoff–due to primary secondary beach dunes on Atlantic side (Kings, Queens, Nassau, Suffolk)
	Queens	297,750	
	Nassau	209,520	
	Suffolk	150,330	
	Bronx	170,895	
	Westchester	9,732	
Connecticut	Fairfield	111,000	E 30% reduc.–100-yr. wall at Stamford
	New Haven	72,498	
	Middlesex	10,320	
	New London	63,546	E 10% reduc.–100-yr. wall Powcatuck
Rhode Island	Washington	15,060	
	Kent	12,999	

Table 4-9. (continued)

State	County	1970 Exposed population (MSL–20)	Hazard (H) or exposure (E) reduction
	Providence	39,600 }	H 15′ cutoff–100-yr. wall fronting Providence
	Bristol	13,998	
	Newport	53,631	
Massachusetts	Bristol	60,000 }	H 14′ cutoff–100-yr. wall protecting New Bedford
	Plymouth	44,325	
	Barnstable	22,830	
	Dukes	2,943	
	Nantucket	10,920	
	Norfolk	17,499	
	Suffolk	50,001	
	Essex	88,182	
New Hampshire	Rockingham	10,743	
Maine	York	4,386	
	Cumberland	14,886	
	Sagadahoc	5,667	
	Lincoln	519	
	Knox	744	
	Waldo	168	
	Hancock	0	
	Washington	0	

and (3) moderate to heavy wave action. A typical damage-depth curve is presented in Figure 4-26. These curves represent average damage and only indicate how a large sample of structures would be damaged.

The risk model (Figure 4-27) uses the hazard, exposure, and vulnerability models to predict expected average annual losses from 1970 to 2000. For each county exposed to the storm surge hazard, the annual probabilities of different surge heights were applied to the exposed population and expected damages calculated from the damage curves and the vulnerability model.

Estimating Tsunami Losses

Although tsunamis have been recorded in the Atlantic Ocean and Carribean Sea, only one or two since the European Settlement began have ever caused any damage to the eastern edge of North America. One struck Puerto Rico in 1918, and another, the eastern coast of Canada in 1929 (Ayre, 1975). We therefore

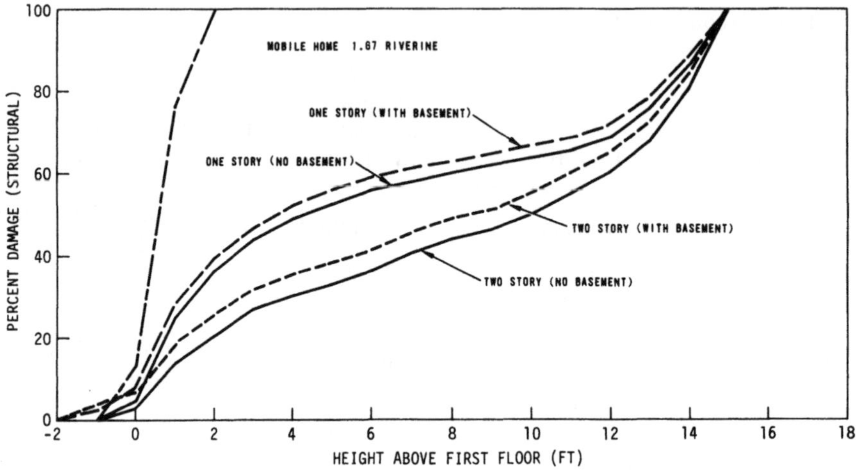

Figure 4-26. Typical storm surge depth versus damage curves (heavy wave action). (From Lee et al., 1978.)

know that the average annual risk is very small; however, an exact quantification will require a much larger data base than currently available. Consequently, the damage estimates developed by Lee et al. (1978) concentrated strictly on the Pacific Basin. A summary of the procedure used for assessing damage is presented in Figure 4-28.

Both the hazard and exposure estimates are organized on a community basis. A list was compiled of all recognizable communities, incorporated and unincorporated, along the U.S. Pacific Coast. Hazard parameters and an exposure estimate were then generated for each listed community. The results show that the critical factor in the determination of average annual damage was the exposure model (i.e., the number of people exposed to tsunami) and not the hazard model.

The hazard model for tsunami is similar to the storm surge model. A plot of runup height versus return period or annual probability of occurrence is prepared for each community. The primary data source for the hazard model for the coterminous United States was the work performed by Houston and Garcia (1974, 1975). This work is not based on historical runup data, since few exist, but on an inductive approach, using postulated source characteristics and propagation models. Only tsunamis of distant origin were considered, since the California coastal area does not have the characteristics of known tsunami-generating regions. Detailed height/frequency relationships were obtained for Hilo, Hawaii (U.S. Army Corps of Engineers, 1966), and Crescent City, California (California Office of Emergency Services, 1972). The Hilo curve formed the basis for the hazard intensity assessment in Hawaii and the Crescent City curve formed the basis for much of Alaska. (Figure 4-29) The uncertainty in the tsunami depth-probability relationships is estimated to range from ±40 to 75%.

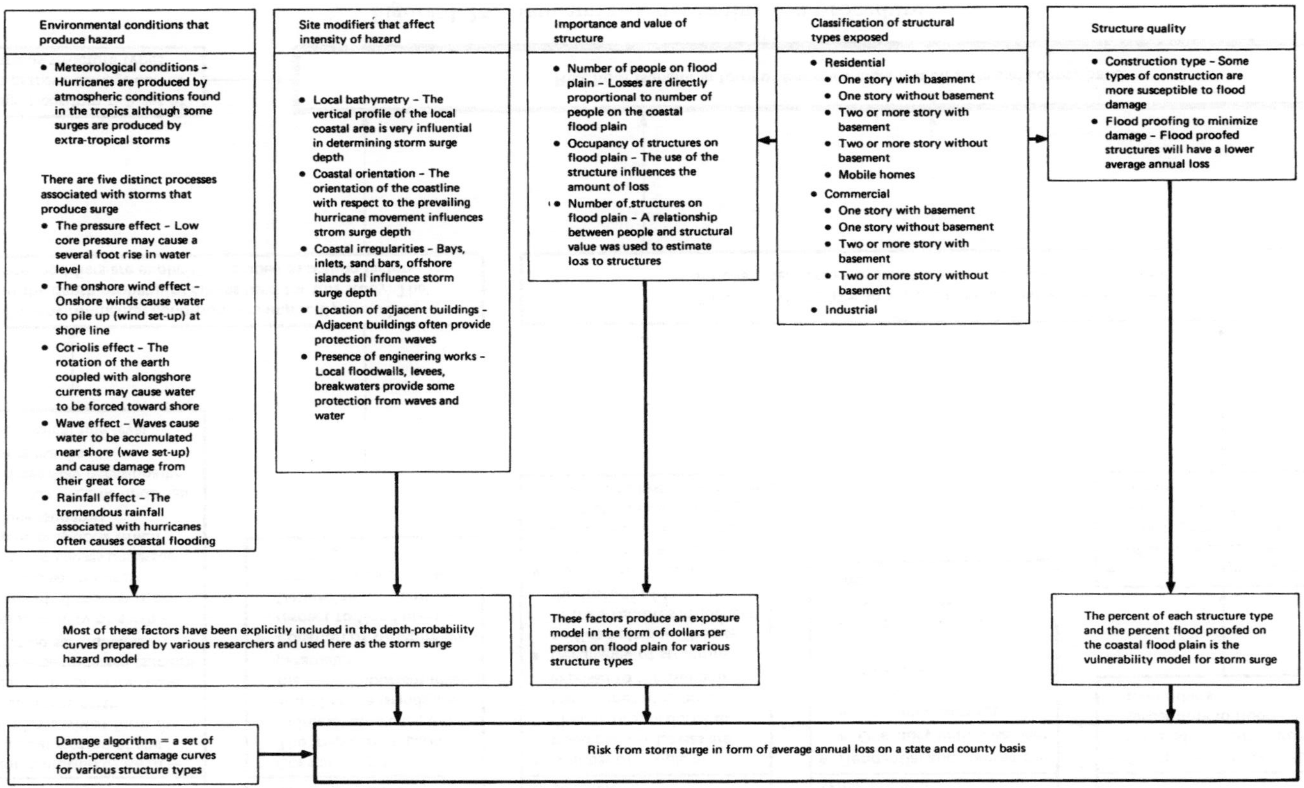

Figure 4-27. Storm surge risk determination procedure.

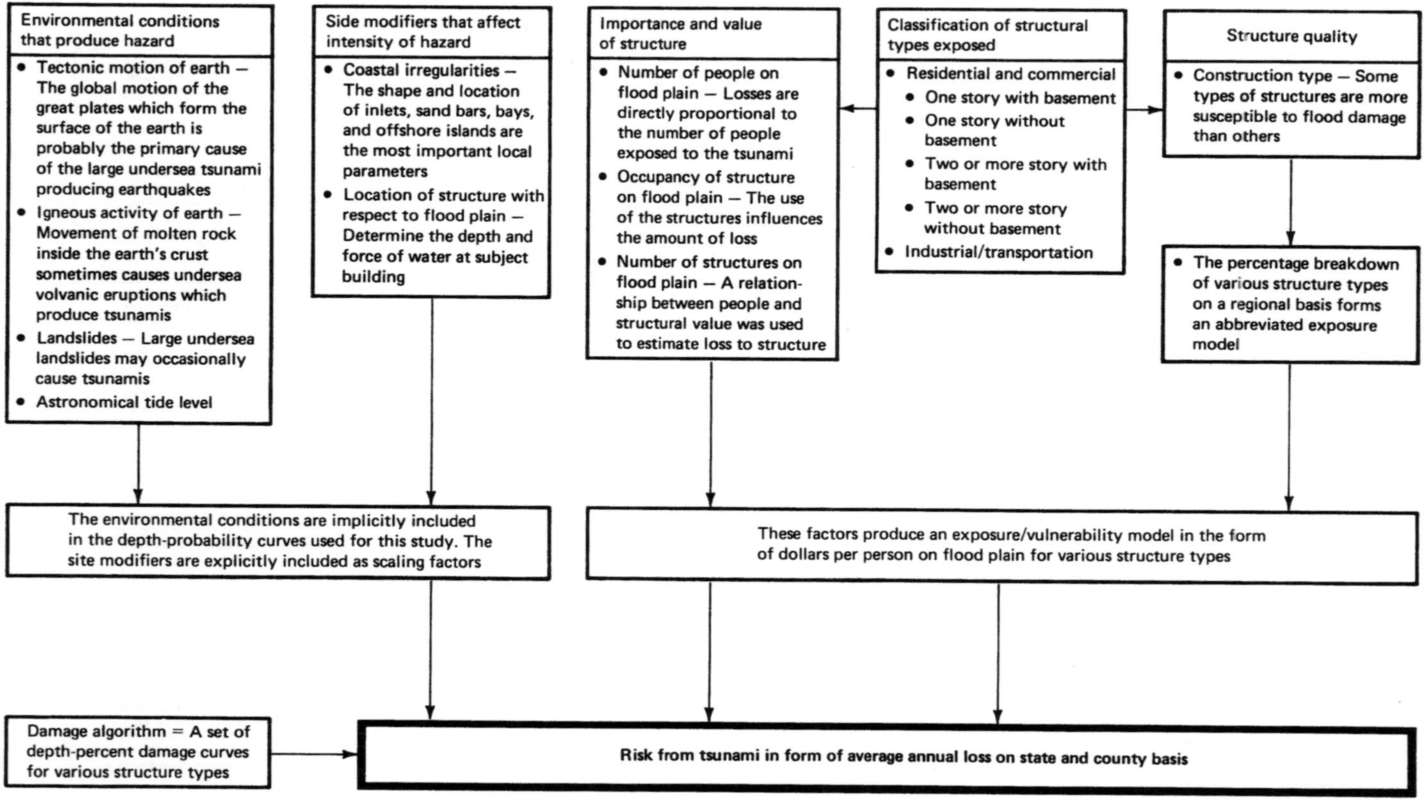

Figure 4-28. Tsunami risk determination procedure.

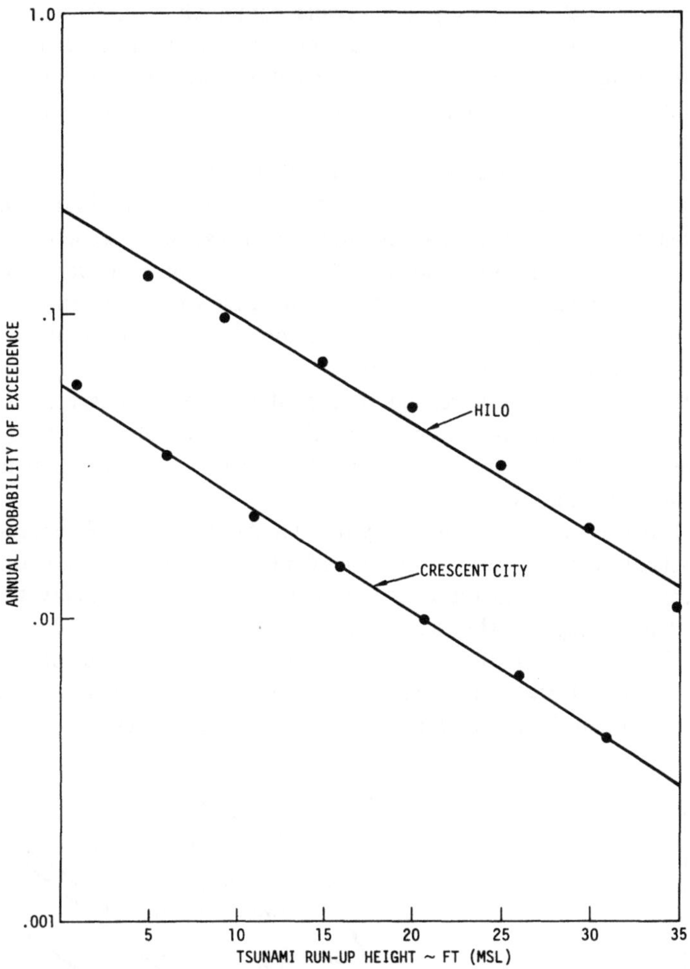

Figure 4-29. Typical tsunami runup/frequency relationships. (From Lee et al., 1978.)

The exposure model for tsunami was developed following four basic steps:

1. Determine (wherever possible and from whatever source available) an inundation area for the 100 year and/or 500 year events.
2. Assume a linear relationship between inundation area and tsunami height so that an inundation area can be associated with any return period.
3. Determine the population density of the inundated area using a relationship between city size and density and data (where available) on the specific type of land use present.
4. Limit the exposure at low elevations but providing that there be no structures exposed below the nearest integer height above the spring tide height (all heights measured from mean sea level).

There were 191 separately identifiable communities in California, 80 in Oregon, 182 in Washington, 197 in Alaska, and 147 in Hawaii that were catalogued as being potentially subject to tsunami. Inundation areas for 78% of the communities potentially subject to tsunami were developed. In many cases, this area was determined to be zero.

Fourteen different sources (Lee et al., 1978) were used to estimate inundation areas. Most data sources provided maps which outlined inundation areas associated with specific recurrence intervals. These areas were measured to quantify the size of the inundation area. When inundation areas were not already defined, topographic maps were utilized. In these cases, the inundation area was defined as that area lying between the shoreline and the elevation contour equal to the tsunami recurrence height.

As for the other water related hazard, the tsunami damage relationships consist of a set of depth–damage curves. The structural categories are similar to those used for storm surge, except for the expansion of industrial buildings to include port facilities.

Depth–damage curves developed by the U.S. Army Corps of Engineers (1966) were available for one and two story residential structures without basements. These data, together with the Federal Insurance Administration curves for riverine flooding (Grigg and Helwig, 1975), formed the basis for the depth-damage curves (Figure 4-30) adopted for this analysis. Since only limited data are available on tsunami depth-damage relationships, engineering judgment was a fundamental component in the development of these damage curves.

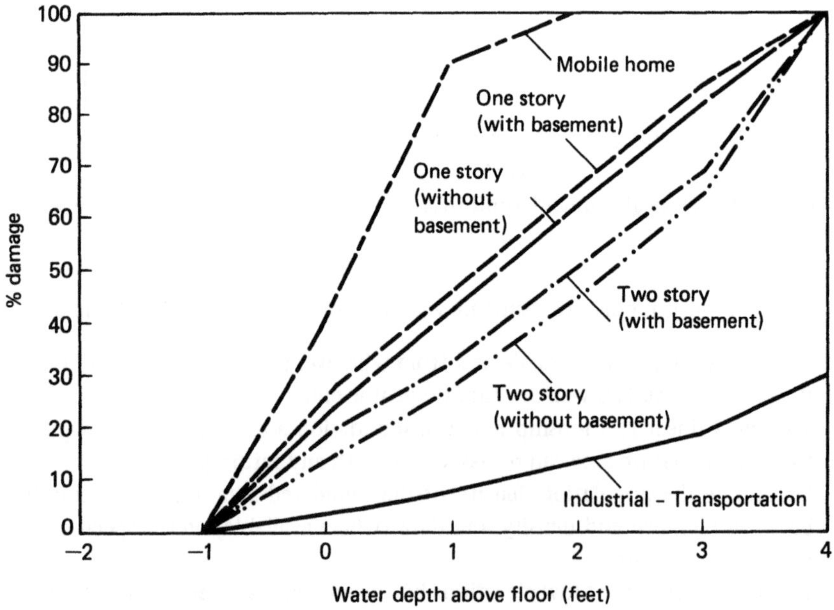

Figure 4-30. Tsunami depth–damage relationships. (From Lee et al., 1978.)

The risk model uses the hazard, exposure, and vulnerability models to estimate expected average annual losses for the year 1970 to the year 2000. For each county, as well as for each community within that county, the annual probability of different tsunami heights was applied to the exposed population and expected damages were calculated using the damage curves and the vulnerability model.

Estimating Tornado Losses

Although a tornado is a complex meteorologic system with as yet no universally accepted description, the distribution, occurrence, and intensity for these phenomena have been fairly well established. The work of two researchers in particular (Thom, 1968 and Pautz, 1969) was applicable to this effort due to the convenient data format they used. They performed similar studies of tornado occurrence and intensity during the years 1955–1967. Both used a 1° by 1° regionalization for tornado occurrence. The total number of tornadoes recorded by them in each 1° X 1° cell for the period from 1955 to 1967 are shown in Figure 4-31.

The environmental conditions that produce tornadoes and the site modifiers that affect their intensity are interrelated. The dominant characteristic of a tornado, however, is the maximum wind speed. Fujita (1970a, b) developed a scheme whereby tornadoes are classified by their maximum wind speeds, which were used for this study. It was also assumed that the probability of a tornado striking an area is uniform over each 1° by 1° geographic grid. Thus, each county in the United States was assigned the tornado probability for the 1° grid in which its county seat is located. Risk of damage was estimated using the Thom and Pautz data bases for tornado occurrence and the tornado intensity distribution shown in Table 4-10. The risk determination procedure used for the tornado analysis is presented in Figure 4-32.

Estimating Hurricane Losses

Hurricanes are large cyclonic storms with wind velocities in excess of 73 miles per hour. Although hurricanes differ from tornadoes, the risk associated with them is determined in a similar fashion. The procedure presented in Figure 4-32 applies to hurricane with the only differences occurring in the area of environmental conditions that produce the hazard. The most significant differences between hurricane and tornado are:

1. Hurricane maximum wind speed considered to be 150 miles per hour is usually less than that of a tornado.
2. The rotating hurricane air mass covers a much greater land area than a tornado.
3. The lifetime of a hurricane is measured in days, whereas a tornado is measured in minutes or hours, thus the structure exposure to the wind forces is longer.

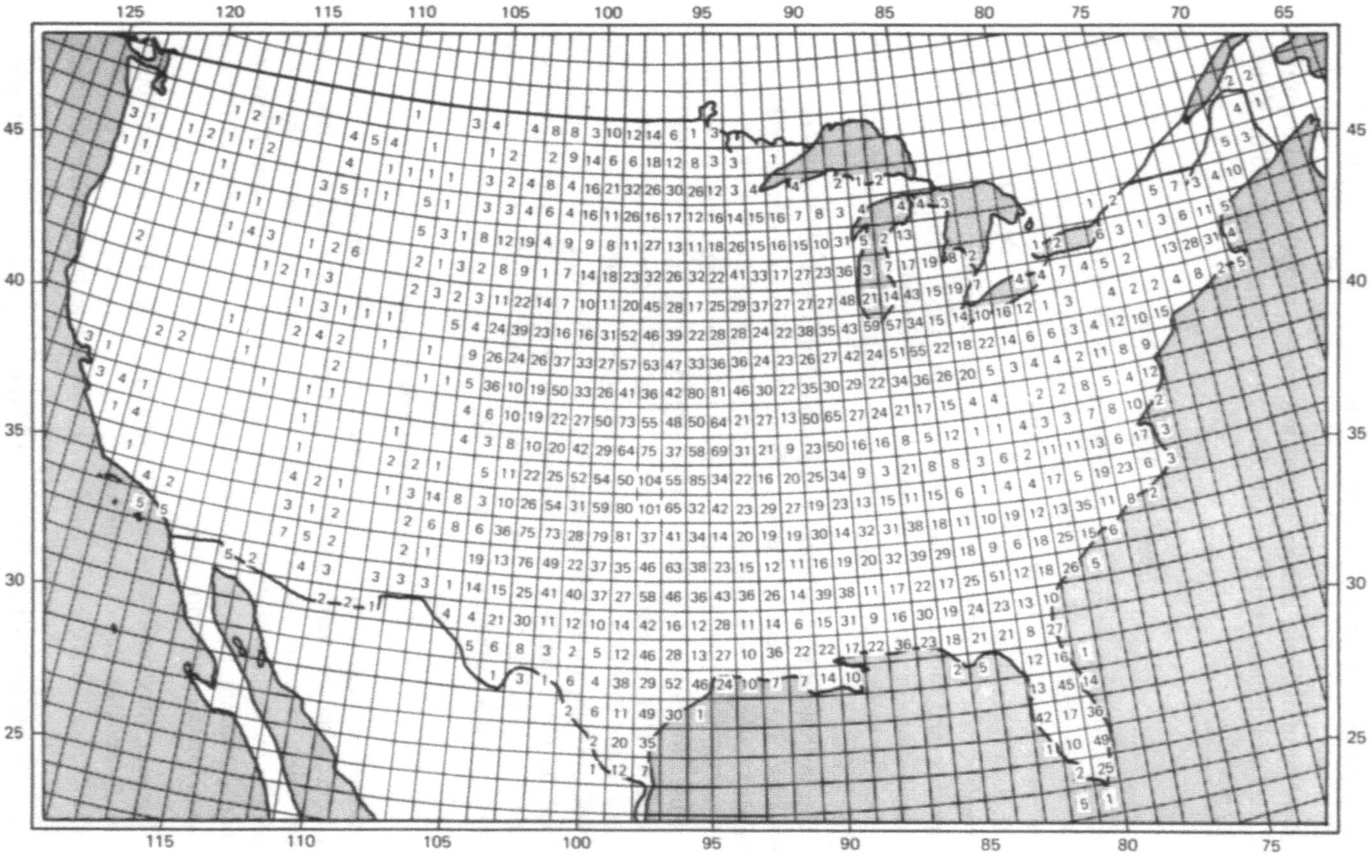

Figure 4-31. Tornadoes in 1° by 1° grid cells throughout the U.S. (From U.S. Department of Commerce, 1969.)

Table 4-10. Fujita Classification and Tornado Occurrence

Wind state	Fujita classification	Maximum wind velocity (mph)	Percent of observed tornadoes in classification
1	–	–	not applicable to tornado
2	F0	40 - 72 (56)[a]	19.9[b]
3	F1	73 - 112 (93)	44.0
4	F2	113 - 157 (135)	26.6
5	F3	158 - 206 (182)	7.2
6	F4	207 - 260 (234)	2.1
7	F5	261 - 318 (290)	0.2

[a]simple average wind speed; wind speeds are at 1/4 mile.
[b]Fujita and Pearson, 1973.

4. The pressure differential associated with a hurricane is more gradual than with a tornado.
5. Hurricanes are less prone to carry heavy missiles than tornadoes, due to the lower wind speeds.

Using the extreme wind speed model proposed by Thom (1968), a 1° by 1° grid was overlaid on the coastal area of the United States where hurricane winds occur. With Thom's wind speed maps, it was possible to establish the probability of hurricane winds occurring at varying wind speeds for each grid cell and thus each county. The maximum wind speed ranges and the representative wind speed for each wind state are presented in Table 4-11. As with tornado, the probability of occurrence is assumed to be uniform over the entire area included in each 1° grid cell.

Estimating Severe Wind Losses

Severe winds ranging in speed up to 150 mph are common to all states. The risk determination procedure for severe winds is similar to the tornado and hurricane wind hazards. Figure 4-32 summarizes the general risk assessment procedure applied to the severe wind analysis. The environmental conditions that produce the severe wind hazard are: (1) a maximum wind speed of 150 miles per hour; and (2) structure exposure to wind speeds lasting several hours. Thom (1968) developed an extreme wind characterization of the United States. However, his wind speed maps are not appropriate for damage estimation because they are based only on the single most severe wind speed each year. Therefore, the occurrence statistics for severe winds based on maximum hourly wind speed observations and maps developed by Valley (1965) were used in this analysis. These data permit one to calculate the probability of occurrence of multiple severe winds in any single month.

Consistent with the procedures used in estimating losses from tornado and hurricane winds, we assumed severe wind occurrences to be uniform over each

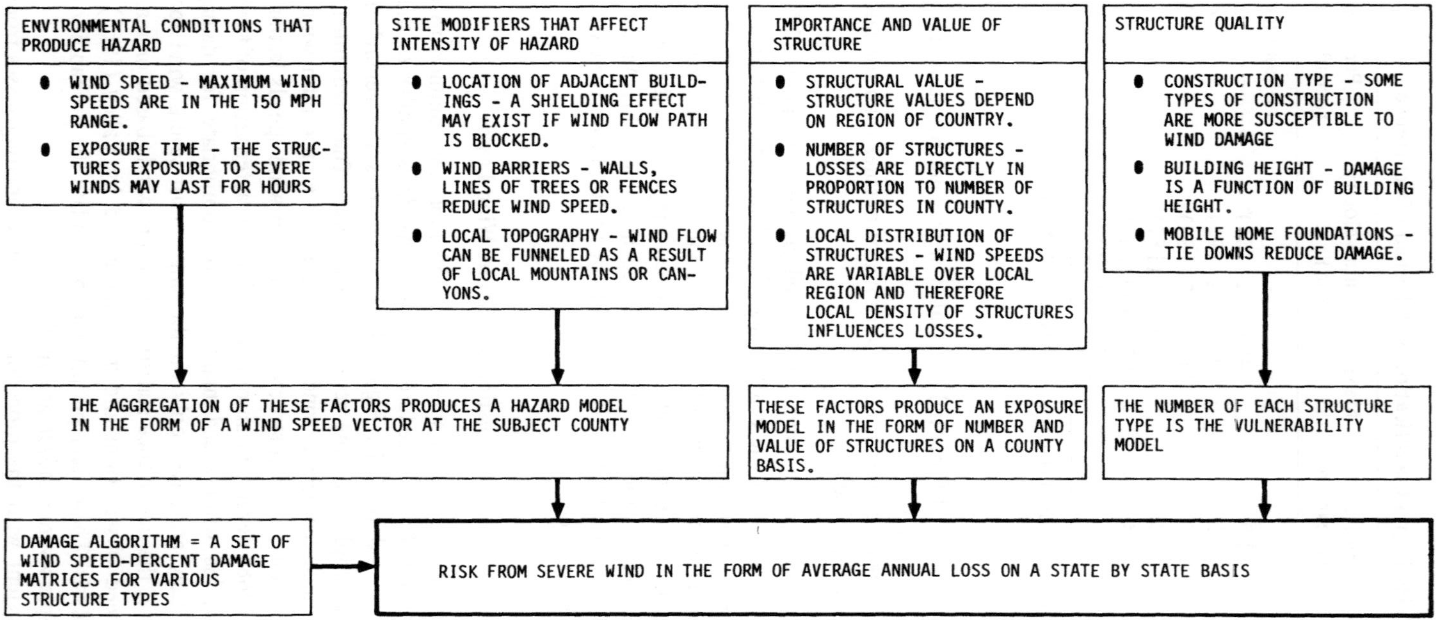

Figure 4-32. Tornado risk determination procedure.

Table 4-11. Hurricane Representative Windspeeds and Corresponding Windspeed Range

Wind state	Representative windspeed (mph)	Range of windspeed (mph)
1	75	73.0 - 87.5
2	100	87.5 - 112.5
3	125	112.5 - 137.5
4	150	137.5 - 175.0
5	200	175.0 - 225.0
6	250	225.0 - 275.0
7	300	275.0 - 325.0

1° by 1° grid. Although this uniformity is not always true, a severe wind micronization was not deemed warranted, or practical, for the broad purposes of this analysis.

Wind Damage Relationships

The degree of damage that can be expected for a building exposed to a particular wind state depends on the type of construction (Hart, 1976). Consequently, eight separate structural categories were considered: one to three story wood frame residential; one to three story wood frame commercial, financial, and service; one to three story concrete and masonry wall, residential; one to three story metal commercial, financial, and service; one to three story concrete and metal commercial and industrial; four or more story steel ductile frame; mobile homes with engineered tie-downs; and mobile homes without engineered tie-downs. Also, window damage was estimated using algorithms related to two categories: windows in structures of less than four stories and windows in structures of four or more stories.

Structure type, structure value, numbers of structures, and the distribution of structures, all determined on a county basis, provide the exposure model used to assess the risk of buildings to wind hazards. Damages were calculated for each of the building and window categories using a wind state-percent damage matrix (Hart, 1976). Using the probabilistic wind speed intensity grids developed for each wind type, the damage matrices, and estimates of the United States building population, it was possible to estimate the risk (i.e., average annual loss) for each state and county.

The quantitative description of damage resulting from wind is of significant engineering importance and can lead to new design requirements for more wind resistant structures. There is, however, a shortage of quantitative knowledge concerning the relationship between wind damage and wind velocities (Hart, 1976). Since empirically based predictions of building damage under specific wind states was not within the present state of the art, it was necessary to develop a means to relate wind state (i.e., speed) to building damage rates. The damage matrix approach developed by Whitman et al. (1974) for quantifying earthquake damage was found to be applicable. Damage matrices were developed for each structure type and for windows. A panel of wind engineering experts was then asked to estimate the probable degree of damage as a function of wind speed for each structure type. These data then were combined to produce a set of probabilistic damage matrices. A typical damage rate versus wind speed plot is presented in Figure 4-33. These curves represent average damage and are indicative of how a large sample of structures would be damaged when subjected to winds of varying intensities (Hart, 1976).

Cost Analysis Methods

For each hazard, a set of possible mitigations was identified and estimates of potential damage reductions for selected mitigation methods were prepared. For most of these mitigations, the investment costs associated with application of the mitigation were determined. These estimates were expressed either as unit costs (viz., cost to provide 50 year flood protection to one community, cost to floodproof one single family dwelling unit) or as a percent of the value of the structure(s) to which the mitigation was intended to be applied. Typically, these data then were utilized to calculate 50 year annual amortized mitigation costs over a range of *true* interest rates (viz. actual interest minus the rate of inflation). The rates of annual mitigation costs to annual dollar loss reductions predicted for the mitigation then were determined and annual net savings (annual amortized mitigation costs minus annual loss reduction) were calculated. Also, for all hazards, the building life projections of "net savings" have been discounted over a range of appropriate rates.

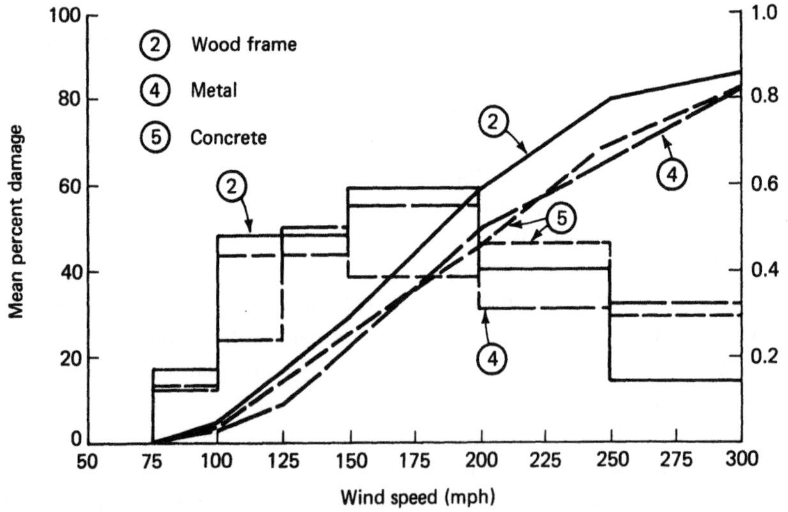

Figure 4-33. Typical mean damage versus wind speed (one to three story commercial-industrial). (From Hart, 1976.)

Mitigation Cost Estimates

Estimates of mitigation costs were developed through use of data presented by the Naval Facilities Engineering Command (1972), Carson (1975), U.S. Army Corps of Engineers (1975), Leslie and Biggs (1972), U.S. Department of Housing and Urban Development (1976), and Structural Engineers Association of California (SEAOC) (1970). All estimates include the cost of both labor and materials. Since some of the source data were available only to specific regions, all cost data have been adjusted to account for the regional differences in construction costs depicted in Figure 4-34.

Earthquake Mitigation Cost Data

Cost estimates were developed for the following earthquake mitigations:

1. Modifying or designing structural and nonstructural systems to UBC Seismic Zone 1.
2. Modifying or designing structural and nonstructural systems to UBC Seismic Zone 2.
3. Modifying or designing structural and nonstructural systems to UBC Seismic Zone 3.
4. Modifying or designing structural and nonstructural systems to UBC Seismic Zone S (2X UBC Zone 3).
5. Modifying or designing nonstructural systems to UBC Seismic Zone 1.
6. Modifying or designing nonstructural systems to UBC Seismic Zone 2.
7. Modifying or designing nonstructural systems to UBC Seismic Zone 3.
8. Modifying or designing nonstructural systems to UBC Seismic Zone S (i.e., 2X UBC Zone 3).

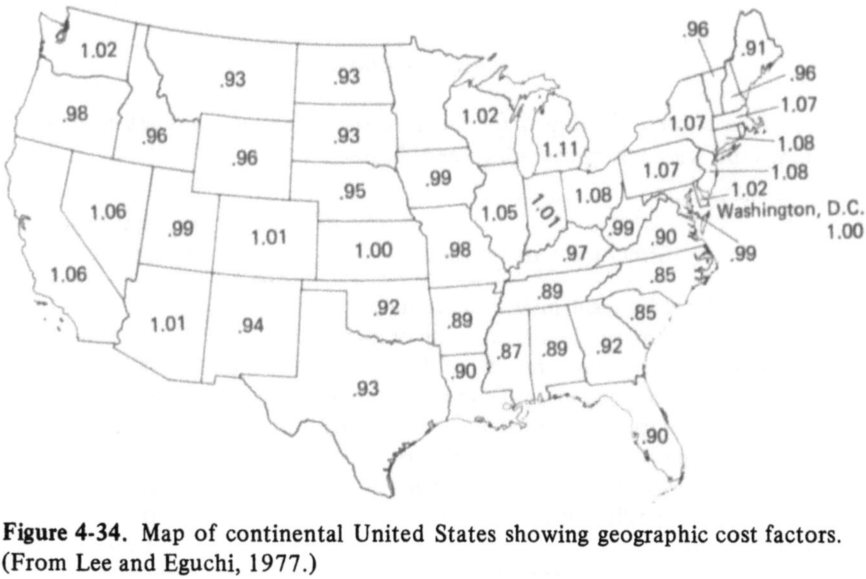

Figure 4-34. Map of continental United States showing geographic cost factors. (From Lee and Eguchi, 1977.)

Nonstructural systems are defined as those elements of a building that are not integral to its construction, such as the architectural treatment, mechanical and electrical facilities, partitions, glass panels, etc. Figure 4-35 provides estimates for existing and proposed high-rise structures based on a study performed by Leslie and Biggs (1972), and Figure 4-36 gives estimates for proposed low-rise structures based on the report by SEAOC (1970). These estimates represent the percent increase over the total original costs. For high-rise buildings, estimates are provided for reinforced concrete structures (frame and nonstructural) and steel structures (frame and nonstructural). The higher costs shown for reinforced concrete structures are a result of several factors: (1) higher design and detailing cost for a structurally more complicated resisting frame; (2) stringent design criteria (ductile, seismic resistant structures); and (3) for an equal volume structure, a concrete structure is stiffer than a steel structure and so attracts higher seismic design forces. From Figure 4-35 the highest cost increases are reported for concrete block buildings. In fact, these structures are highly unsuited for high lateral forces, and as a result, need extensive reinforcing.

The cost estimates assume that the original structure is of quality design and meets minimum code design force (wind forces), and that the structural design for each seismic zone is the most economic for the given force constraints. The curves are not based on increasing the lateral force-resisting system that was adequate for the wind loads. Rather, the cost curves include the change of frame type to accomodate changing force constraints (e.g., change from nonductile reinforced concrete vertical load frame in Zone 1 to a ductile reinforced concrete moment-resistant frame in Zone 3). All cost curves in Figures 4-35 and 4-36 are for new buildings. Since the cost of modifying an existing structure to the current levels is considerably higher than that for a new structure, mitigation costs (as percent of building value) for modification of existing structures are assumed to be eight times those identified above.

Wind Mitigation Cost Data

Cost estimates were developed for building strengthening measures capable of yielding UBC wind resistance ratings, as indicated:

Structural classification	Structural design level (UBC)
One to three story residential wood frame	0.75, 1.0, 1.5, 3.0
One to three story residential concrete or masonry	0.75, 1.0, 1.5, 3.0
One to three story commercial wood frame	0.75
One to three story commercial concrete or masonry	0.75
One to three story commercial metal (nonsteel)	0.75
Four or more story, concrete or masonry	0.75
Four or more story, concrete shear wall	0.75
Steel frame, regardless of height	0.75
Mobile home	1.00

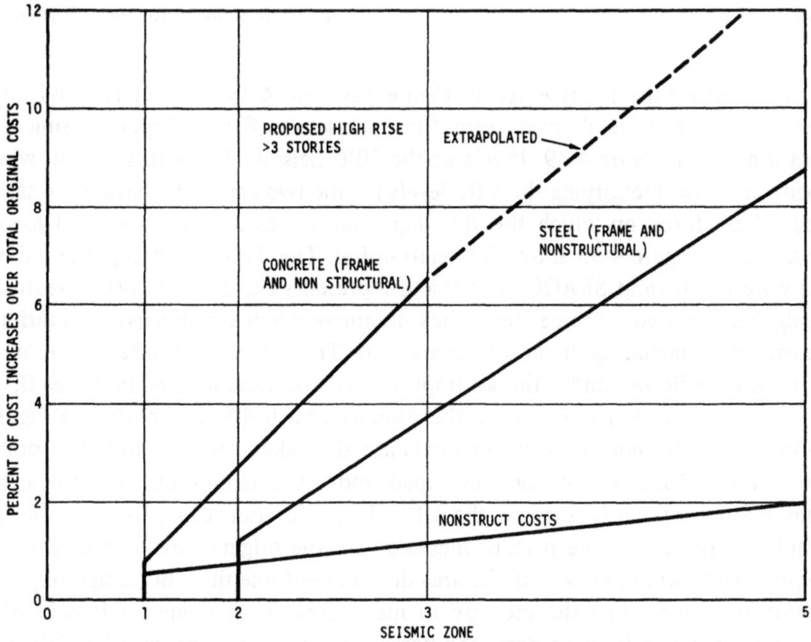

Figure 4-35. Earthquake cost data—high rise. (From Lee and Eguchi, 1977.)

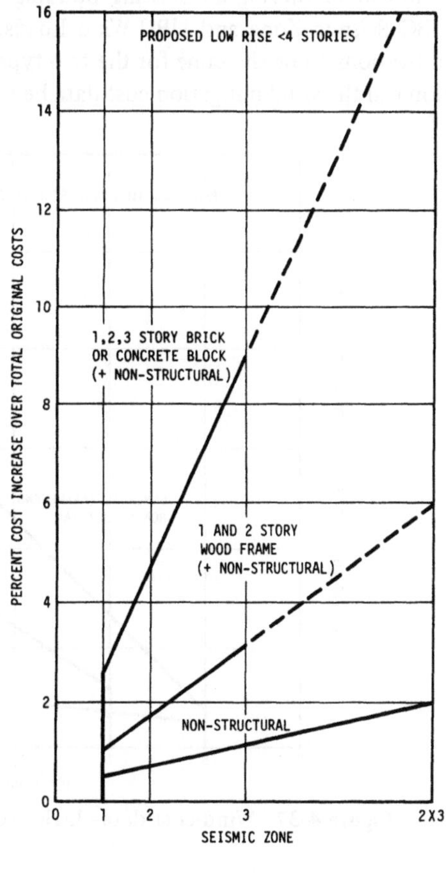

Figure 4-36. Earthquake cost data —low rise. (From Lee and Eguchi, 1977.)

The cost estimates presented in Figure 4-37 and 4-38, for high-rise and low-rise structures, were derived from those presented for earthquake using the relationship of Figure 4-39. It relates the UBC Seismic Zones to the UBC wind loads based on identifying the UBC levels for the two categories having the same base shear force on which the Building Code is based. The upper and lower abscissas of Figure 4-39 show this relationship. The data points (represented by ▲) were taken from SEAOC (1970) and represents the percent cost increase in designing one, two, or three story brick or concrete block buildings to the various design levels excluding the nonstructural part. The data point for Seismic Zone 2 cost was developed under the assumption that for areas located in Zones 0, 1, and 2, the percent increase over the original cost is 4%, as reported in Figure 4-38, where the nonstructural elements are also taken into account. The point for Seismic Zone 3 cost was developed under the assumption that for areas that do not presently enforce the UBC design for hurricane, tornado, severe wind, or earthquake, the percent increase over the original cost to meet Zone 3 requirements would be 9%. If the area does not enforce the UBC design requirements for wind, then the increase to meet Zone 3 requirements is only 4%.

As with the earthquake data, the percent increase for a proposed building is multiplied by a factor of eight to derive the cost (as a function of replacement value) of modifying an existing building. In working the comparison between UBC Seismic Zone and UBC Wind Loads, we consider the effectiveness of the mitigations to be the same for the two types of hazards. Only under this assumption can the wind mitigation cost data be considered reliable.

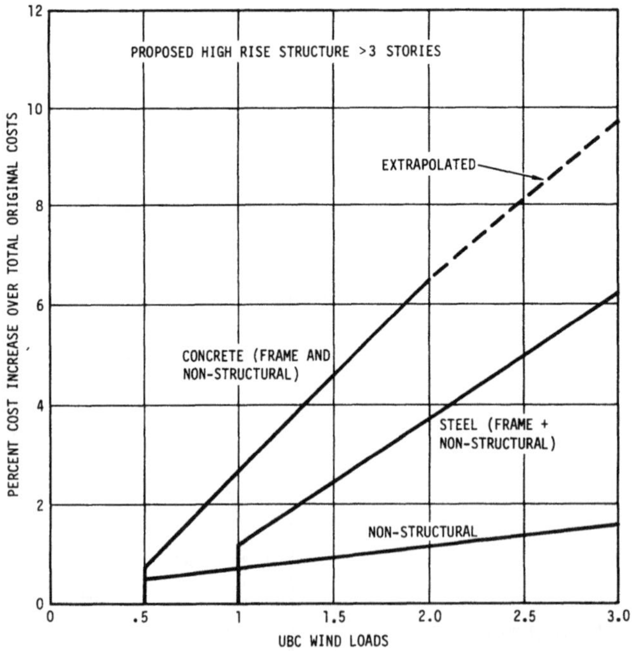

Figure 4-37. Wind-cost data–high rise. (From Lee and Eguchi, 1977.)

Figure 4-38. Wind-cost data–low rise. (From Lee and Eguchi, 1977.)

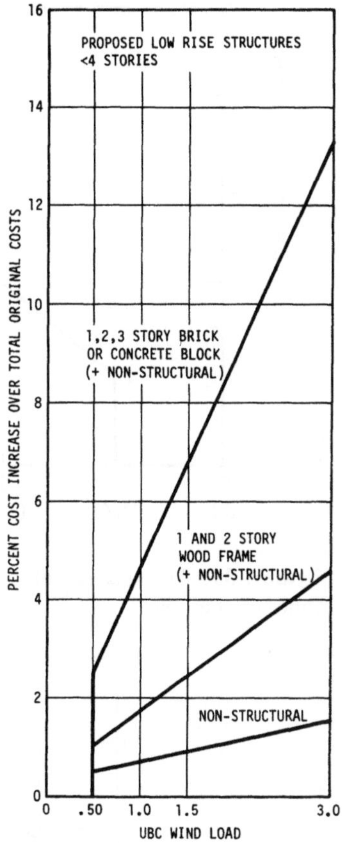

The cost of anchoring a mobile structure was found to be approximately \$500 per unit from a telephone survey of several mobile home dealers. The cost of strengthening windows for wind loading was found to be \$0.62 per square foot from Building Construction Cost Data (1972). Overall, the percent cost increase over the original cost is about 75% higher for concrete than for steel structures in high-rise buildings, and about 200% higher for brick or concrete than it would be for wood frame structures for low-rise buildings.

Flood Mitigation Cost Data

Cost estimates were developed for two classes of mitigations: (1) floodproofing to exclude water (for residential and commercial structures); (2) elevating building (on "fill" and "other than fill") for masonry, wood frame, and wood frame on slab foundation. The cost data are summarized in Figures 4-40 through 4-42. These figures are a rough approximation of data provided in Carson (1975), U.S. Department of Housing and Urban Development (1976), and U.S. Army Corps of Engineers (1975), using an assumed baseline construction cost of \$25/square foot. Figure 4-40 shows the percent increase in cost at different

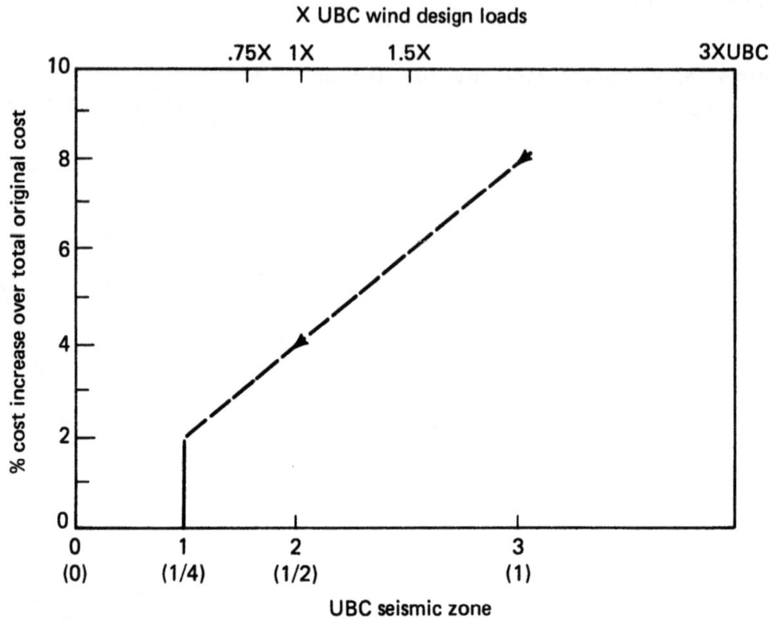

Figure 4-39. The relationships between cost factors for wind loads and Seismic Zones of the UBC.

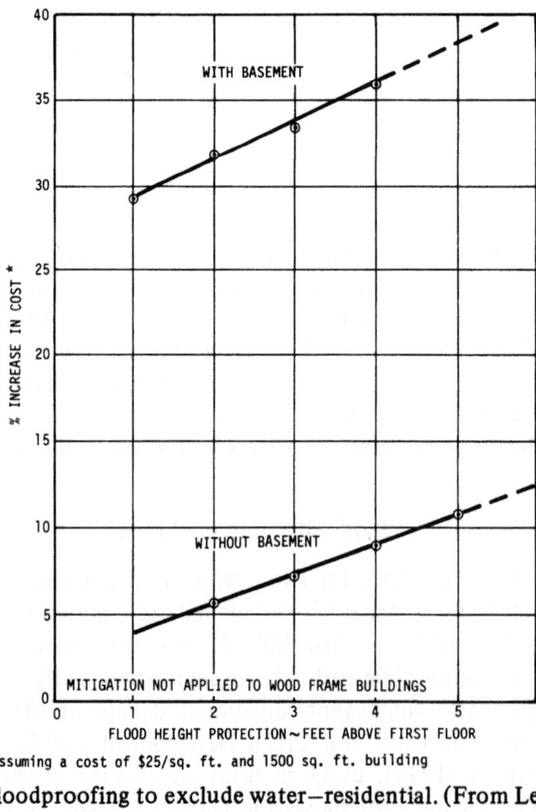

Figure 4-40. Floodproofing to exclude water–residential. (From Lee and Eguchi, 1977.)

levels of mitigation, represented as flood height protection for residential structures excluding wood frame buildings. It shows that floodproofing structures with basements would cost 3.5 times the same mitigation applied to structures without the basement. For commercial buildings the difference drops to 0.60 times (Figure 4-41).

Figure 4-42 summarizes the costs of elevating structures by means other than fill, usually through pilings or walls. The increasing cost for this mitigation is 60% more for masonry buildings than for wood frame buildings, mainly due to the higher weight of masonry structures. An average decrease of 70% is obtained when passing from "raising by means other than fill" to "raising on fill" up to four feet. At higher levels there is a more pronounced increase of costs of elevation on fill because the initial construction costs for foundation and walls in the A and B type of mitigation (Figure 4-42) are amortized for higher elevations.

Tabular expressions of the earthquake, wind, and flood mitigation cost curves are presented in Tables 4-12, 4-13, and 4-14.

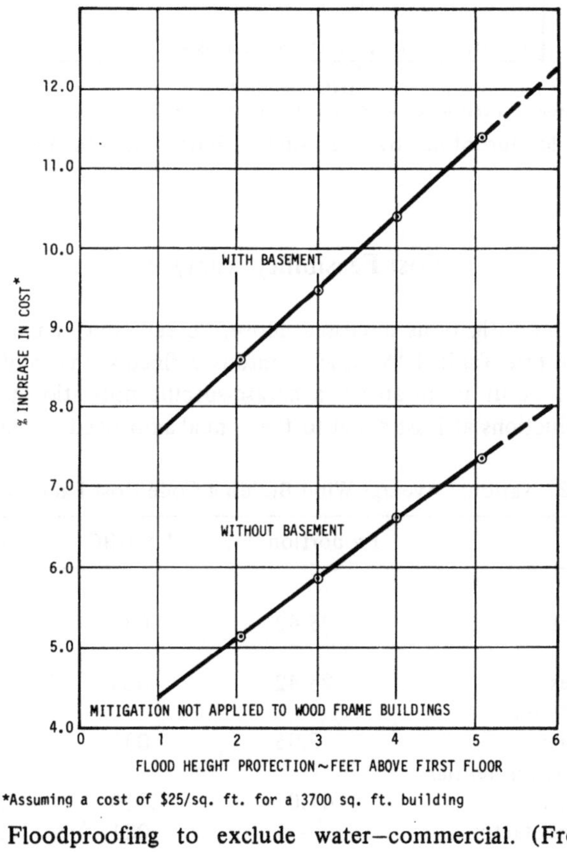

Figure 4-41. Floodproofing to exclude water–commercial. (From Lee and Eguchi, 1977.)

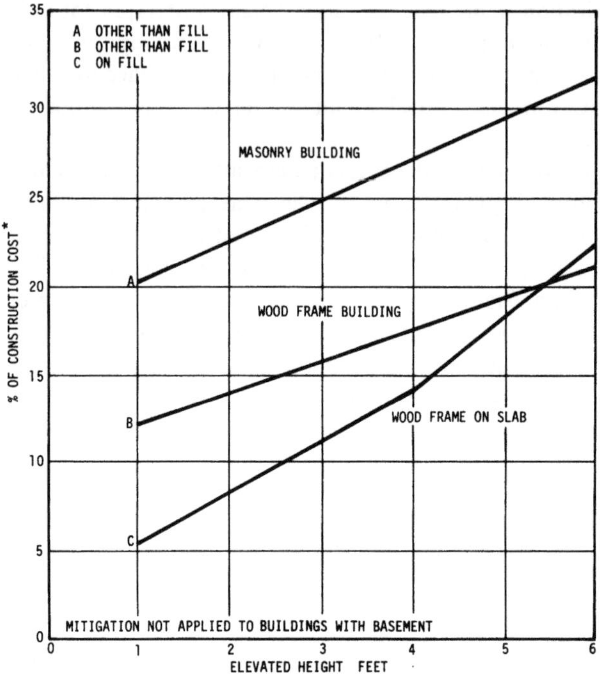

*Assuming a cost of $25/sq. ft. and 1500 sq. ft. building

Figure 4-42. Floodproofing by elevating building. (From Lee and Eguchi, 1977.)

Cost Feasibility Analysis

For each of the mitigations discussed above, "break-even" damage rates also were calculated (see Table 4-15). Such a rate is defined as that building damage rate that must occur in an area if a hazard-specific mitigation is to produce annual loss reductions at least equal to the annual amortized cost of the mitiga-

Table 4-12. National Average Wind Building Code Cost Factors

	Proportion (%)	1.5 UBC	3.0 UBC
Concrete			
+3 stories	25.42	.031	.069
Steel			
+3 stories	25.42	.025	.062
Brick residential			
1-3 stories	8.05	.033	.095
Wood frame residential			
1-3 stories	41.05	.010	.030
Average cost factor	–	.021	.051

Table 4-13. National Average Earthquake Building Code Cost Factors

	Proportion (%)	Zone 3	2 X Zone 3
Concrete			
+3 stories	25.42	.038	.130
Steel			
+3 stories	25.42	.027	.088
Brick residential			
1-3 stories	8.05	.065	.180
Wood frame residential			
1-3 stories	41.05	.020	.060
Average cost factor	–	.030	.095

tion. Break-even damage rates were used to identify counties in which specific mitigations might be considered cost feasible.

Five values are necessary for determination of "break-even" damage rates: (1) the initial cost of the mitigation; (2) the annual expected loss reduction associated with the mitigation; (3) the period of time over which costs are to be amortized and loss reductions are to be experienced; (4) the total estimated loss reduction that will be produced by the mitigation over the lifetimes of buildings on areas to which the mitigation is applied; (5) *either* the discount rate that is applied to building-life loss reductions, *or* the building life accumulated annual amortized costs of the mitigation at a specified interest rate.

Table 4-14. National Average Storm Surge Building Code Cost Factors

Floodproofing	Proportions (%)	2 Foot protection	4 Foot protection
Residential			
with basement	27.52	.333	.373
Residential			
without basement	27.08	.053	.093
Commercial and industrial			
with basement	25.82	.074	.103
Commercial and industrial			
without basement	22.58	.038	.066
Average cost factor		.132	.166
Masonry	16.40	.227	.272
Wood frame	17.73	.139	.176
Wood frame or slab	65.87	.083	.136
Average cost factor		.117	.165

Table 4-15. Average Damage Rates Associated with a Range of Cost/Loss-Reduction Ratios for National Average Building Type Mix Assuming a Building Life of 50 Years and a Discount Rate of 6.1% ($\alpha = 0.64$)

	Wind hazards	
Cost/loss-reduction ratio	1.5 UBC (cost factor = .021)	3.0 (cost factor = .051)
4:3	.0135	.0218
1:1	.0180	.0290
4:5	.0225	.0362
2:3	.0270	.0434
4:7	.0315	.0506
1:2	.0350	.0580

	Earthquake	
Cost/loss-reduction ratio	Zone 3 UBC (cost factor = .030)	2 X Zone 3 (cost factor = .095)
4:3	.0662	.0587
1:1	.0882	.0782
4:5	.110	.0978
2:3	.132	.117
4:7	.154	.137
1:2	.176	.156

	Storm surge		
Cost/loss-reduction ratio	4 ft. elevation floodproofing (cost factor = .166)	2 ft. elevation (cost factor = .117)	2 ft. flood-proofing (cost factor = .132)
4:3	.0568	.0392	.0470
1:1	.0757	.0549	.0627
4:5	.0946	.0706	.0784
2:3	.114	.0863	.0941
4:7	.132	.102	.110
1:2	.152	.118	.126

The "break-even" damage rate is that specific rate for a given hazard with the following relationship:

$$\overline{DR}_{hrkl} = \frac{C_k}{(1 + C_k)} \cdot \frac{1}{S_h - MS_h} \cdot \alpha_{ik}$$

where

$\overline{DR}_{hrkl}$ = the annual proportion of value lost due to the hazard h, in the region r, to structure type k, to wealth type l (building wealth)

C_k = the proportional increase in construction costs due to the mitigation on building type k (Tables 4-12, 4-13 or 4-14)

$(1 + C_k)$ = the multiplier to determine new value of the building type k, after mitigation

S_h = the portion of the annual structural losses saved due to the mitigation to hazard h (Table 4-15)

MS_h = the ratio of total economic losses to building losses for hazard h, nationally. (Table 4-16)

α_{ik} = value based on discount rate, building life, and building type (Table 4-17)

Analysis of Social Impacts

The impacts of natural hazards discussed so far have been solely economic impacts that lend themselves to quantification for an overall picture of the effects of hazards on the population. However, there are other "social" impacts accompanying the occurrence of natural disasters. The social impacts treated here do not comprise an exhaustive list, but rather they represent an attempt at a preliminary description of the pertinent social indicators that may be of interest to the decision makers in their efforts to assess alternative mitigation policies and methods. The indicators are estimates of mortality, housing loss, homelessness, and unemployment.

Table 4-16. The Ratios of Total Economic Loss to Building Loss

		1970 \$ 10^6	MS ratio 1970	2000 \$ 10^6	MS ratio 2000	MS ratio average
Earthquake	buildings	655.2	1.192	1177.0	1.32	1.256
	Total	781.1		1553.7		
Tornado	buildings	879.8	1.882	2058.0	2.536	2.209
	Total	1656.0		5219.1		
Hurricane	buildings	685.5	1.541	1742.0	2.024	1.782
	Total	1056.0		3526.3		
Severe wind	buildings	11.4	1.579	24.8	2.153	1.866
	Total	18.0		53.4		
Storm surge	buildings	441.6	1.452	1176.0	1.992	1.722
	Total	641.2		2342.9		
Tsunami	buildings	8.7	1.724	19.8	2.040	1.882
	Total	15.0		40.4		
Wind average	buildings	1576.7	1.731	3824.8	2.300	2.016
	Total	2730.0		8798.8		

Table 4-17. Values of α Based on Different Discount Rates, and Building Lives

Discount rates (i)						Building lives
0.0	2.9	6.1	10.0	15.0	20.0	(L)
0.050	0.066	0.086	0.116	0.159	0.204	20.0
0.040	0.056	0.078	0.109	0.154	0.200	25.0
0.033	0.050	0.072	0.105	0.152	0.200	30.0
0.025	0.042	0.067	0.102	0.149	0.200	40.0
0.020	0.037	0.064	0.101	0.149	0.200	50.0
0.017	0.035	0.063	0.100	0.149	0.200	60.0
0.013	0.033	0.062	0.100	0.149	0.200	75.0
0.010	0.031	0.061	0.100	0.149	0.200	100.0

Estimate of Mortality

The number of lives lost in an extreme natural event is represented as a function of economic loss to buildings from that event. A relationship between damages (in constant 1970 dollars) and the number of lives lost was estimated from data available for past natural disasters. This relationship was based on assumptions that the dollar amounts of damage recorded for an event reflected only building damage, the lives lost were directly caused by the event, the vulnerability of the populations was constant, and the value of all structures in all events was the same.

Dollar amounts of damage of past occurrences of natural hazards were supplied by insurance company records and include many types of damage (e.g., contents losses). However, building losses that are not paid often go unnoted. For instance, most homeowner insurance policies for earthquake have a 5% deductible provision; thus, ceiling cracks and broken windows will generally not be reported by insurance companies. In addition, many businesses and homes are not insured for earth movement or earthquake damage.

The modeled mortality rate represents a simplification of the actual occurrence. Deaths may be partly related to the level of success of rescue operations as well as the level of physical destruction or structure characteristics. Population density, amount of warning, foreknowledge of the effects associated with the event, psychological impressions of the population, availability of shelter, and time of day of the occurrence certainly influence the number of casualties. No estimate of mortality can take all these factors into account.

A further complication in the construction of a life loss model is the relationship between value at risk and the value of potential losses. A disaster in an area with a high value of building wealth per person will incur greater economic loss per capita in a given event than an area exhibiting lower wealth per capita. Accordingly, life loss estimates based on per capita dollar value of building damage losses will be effected by the widely varying levels of building wealth per capita in different counties and regions of the nation.

Several authors (Dacy and Kunreuther, 1969; U.S. Department of Commerce, 1972) have observed a temporal variation in the relationship between hazard induced deaths and the economic value of hazard induced damages to property. The choice of time as the single independent variable in the functional form of such a ratio may be weak, but having the relationship in this form is desirable and, as demonstrated in the data plots in Figures 4-43 and 4-44, a trend does exist. The data are shown with a vertical scale of the logarithm (base 10) of the ratio life loss to structural damage (LL/SD) plotted as a function of time from 1929 to 1970, and forecasted to the year 2000. In forecasting to the year 2000, both the choice of the model and the fit of the data influence the result. In this case it appears that it is possible that the projections are very uncertain and could vary by an order of magnitude.

The forecasts of deaths for storm surge and hurricane wind impacts are not distinguishable for estimation of separate life loss relationships. Thus, they are both assumed to be equally represented in the data. Flood losses were assumed to exhibit the same characteristics as those for storm surge. The estimates of life loss due to severe wind were also estimated from the tornado data, since no source of data dealt solely with severe wind. Since, by definition, no water-

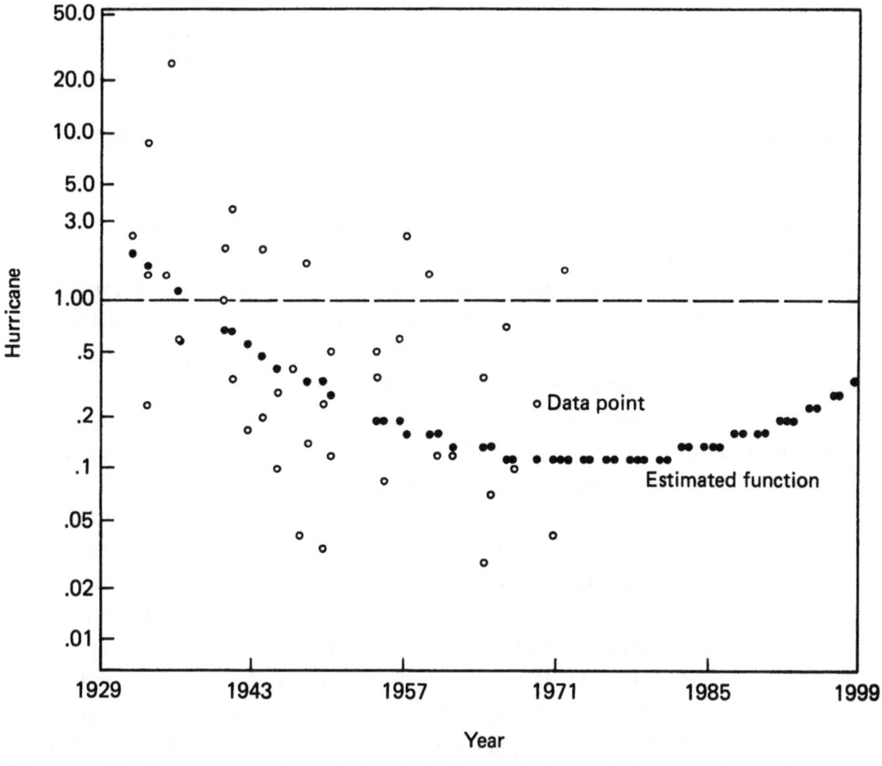

Figure 4-43. Regression plot of deaths/million dollars of damage due to hurricane. (From Hirschberg et al., 1978.)

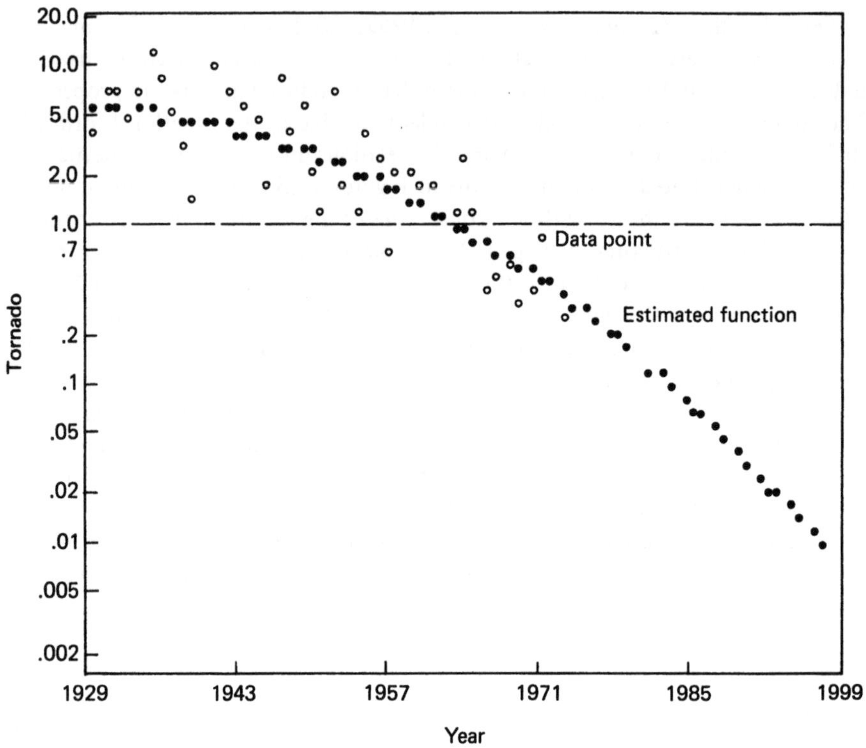

Figure 4-44. Regression plot of deaths/million dollars of damage due to tornado. (From Hirschberg et al., 1978.)

related deaths could be due to severe wind, the hurricane relationship is not applicable for this hazard.

Consequently, for hurricane and storm surge .0956 deaths per million dollars damage, and for tornado and severe wind .488 deaths per million dollars damage, were used to calculate life loss estimates for the projected years. The death/damage ratio for tsunami was 2.432 deaths per million dollars of building damage and was calculated from the average of three tsunamis. Because a warning system was not present for those tsunamis, but is now available, the predicted deaths due to tsunami should be taken as an upper bound.

Dacy and Kunreuther (1969), have suggested that deaths per million dollars damage declined over the period 1930–1964 because of warnings, stronger structures, and a higher wealth per capita. In order to examine this contention it was necessary, by estimating the wealth per capita exposed, to normalize the effect of the change in structure value to determine the impact of time-dependent technologic change. Due to data limitations, this was only performed with U.S. hurricanes and earthquakes beginning in 1933. For these hazards and time period, well-defined areas of impact and reliable economic estimates were available for analysis. Because no earthquake in the sample had a prior warning, while hurricanes have been predicted with increasing accuracy, a comparison of these two hazards can provide an indication of the effects of warnings versus building code requirements.

To account for the differences in structure value per capita in different locations at different periods in time, a relative value of structures per capita was computed to scale the dollar losses reported for each event. To account for the regional effect, the relative income per capita was used. The annual structure values per person were estimated for private structures, both residential and commercial. Using the estimates from Musgrave (1974) and from Young (1971) (in constant 1967 dollars), a yearly estimate of total private structure wealth was made. The estimated total population (U.S. Bureau of the Census, 1971) for corresponding years was then used to compute a private structure value per capita (Figure 4-45). These annual national estimates were then weighted by the interpolated relative per capita income in the area of primary effect for the year of the event, available by SMSA and BEA areas for years back to 1929 (U.S. Water Resources Council, 1974).

For example, in the case of the 1933 earthquake in Long Beach, California, the relative per capita income for the Long Beach-Los Angeles SMSA was

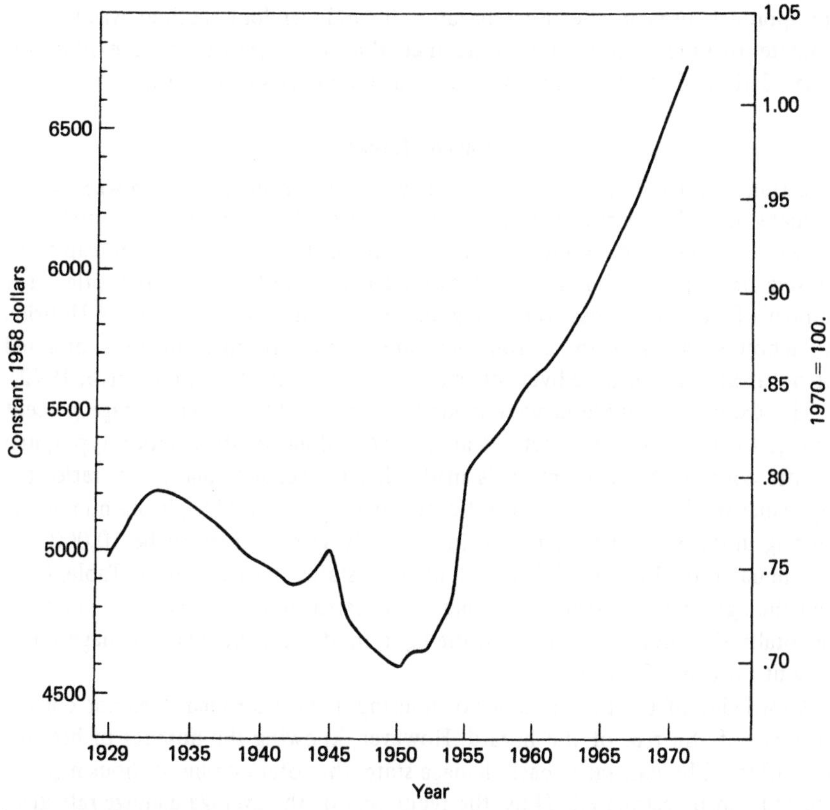

Figure 4-45. Private structure value per capita in the United States 1939–1971. (From Hirschberg et al., 1978.)

estimated as 1.41 times the national average in 1933. Therefore, the value of private structures was estimated using the 1.41 factor. Figure 4-45 shows that the relative structure value per capita in 1933 compared to 1970 is approximately 0.79.

Thus, the relative structure value per capita in Long Beach, California, in 1933 compared to the 1970 national average is $1.41 \times .79 = 1.12$. To normalize the dollar losses reported for the 1933 Long Beach earthquake, the dollar losses of $81.8 million is multiplied by 1.12 to yield a normalized dollar value of $91.6 million (in 1967 constant dollars). Table 4-18 lists the life loss, the normalized dollar losses, and the value per capita relative to the 1970 national average.

By weighting the damages by the ratio of the historical per capita value to the 1970 per capita structure value, the effect of the regional and time-dependent factors of the variations in wealth at risk can be factored out. Then the time effect is a better indicator of society's changing ability to cope with the occurrence of the disaster.

Table 4-19 lists the life loss totals by hazard for 1970 and 2000 as calculated by the model described above. Expansive soil is limited to property damage. Because the model utilized in estimating riverine flooding losses is different than the approach used in the other hazards, no independent method was used to estimate riverine deaths. It was assumed that such deaths would exhibit the same relationship to dollar losses as that calculated for storm surge.

Housing Losses

To estimate dollar losses to housing, the value of housing per capita was used to estimate the value at risk of residential structures. In estimating the number of houses destroyed, the value at risk was changed from dollars per capita to housing units per capita. Thus, the estimated values of housing loss reflect the number of such units lost. A housing unit is defined as in the Census of Housing as: "a house, an apartment, a group of rooms, or single room occupied or intended for occupancy as separate living quarters . . . " (U.S. Dept. of Commerce, 1972).

The damage calculations were made by applying the annual average percent damage to the values of structures at risk, according to construction types, in a specific county. An assumption is made that the repair/replacement ratios are applicable to the determination of housing damage rates. Thus, if the number of housing units of a particular type in a county was assumed to be 10,000 and the annual probabilities of different damage states are as given in Table 4-20, then the aggregate number of housing units destroyed is 90.2. Of those, only 10 are totally destroyed, and those in the light, moderate, and heavy damage states account for over 40% of the total.

Knowledge of the total number of housing units destroyed does not convey the scale of the risk of the hazard. However, knowing the number of housing units affected by damage in each damage state, the total number of housing units affected can be estimated. Thus, the reciprocal of the average damage rate gives the number of houses affected for each house totally destroyed, at each damage state. In the example community shown in Table 4-20, with 10,000 housing units at risk, there would be 16.56% chance that any one housing unit would be affected in any year. But of those affected, about 60% would be in the category of light damage, and less than 10% would have damage ratios above .075. Table 4-21 extends the example in Table 4-20 to show the number of housing units affected.

Table 4-18. Hurricane and Earthquake Data Used in Time Effect Analysis

	LOCATION		YEAR	DEATHS	NORMALIZED(1) IN 1967$	RELATIVE STRUCTURE(2) VALUE/CAPITA
DATA FOR HURRICANE	KEY LARGO	FLORIDA	29.0	3.00	3554.47	.438
	COLUMBIA	TEXAS	32.0	40.0	60550.4	.300
	BROWNSVIL	TEXAS	33.0	40.0	86705.9	.356
	JUPITER	FLORIDA	33.0	2.00	20829.7	.507
	CPHATTERA	NOCAROLINA	33.0	21.0	9563.15	.269
	FLA KEYS	FLORIDA	35.0	408.	27559.5	.525
	MIAMI	FLORIDA	35.0	19.0	14978.9	.886
	CPHATTERA	NOCAROLINA	36.0	2.00	13294.2	.287
	PROVIDENC	RHODE IS	38.0	600.	723662.	.994
	CHARLESTO	SOCAROLINA	40.0	34.0	56527.6	.293
	MATAGORDA	TEXAS	41.0	4.00	16955.5	.802
	MIAMI	FLORIDA	41.0	5.00	1824.66	.869
	PT LAVACA	TEXAS	42.0	8.00	133950.	.402
	TORTUGAS	FLORIDA	44.0	18.0	190529.	.593
	PT LAVACA	TEXAS	45.0	3.00	92298.1	.402
	HOMESTEAD	FLORIDA	45.0	4.00	165404.	.606
	HILLSBORO	FLORIDA	47.0	51.0	277901.	.587
	MIAMI	FLORIDA	47.0	1.00	38363.5	.773
	KEY WEST	FLORIDA	48.0	3.00	41475.3	.581
	JUPITER	FLORIDA	49.0	2.00	124301.	.581
	FREEPORT	TEXAS	49.0	2.00	23539.3	.413
	CEDAR KEY	FLORIDA	50.0	2.00	7744.94	.586
	MIAMI	FLORIDA	50.0	4.00	52571.7	.733
	LONG IS	NEW YORK	54.0	21.0	67729.7	.740
	CAPE FEAR	NOCAROLINA	54.0	98.0	829235.	.373
		NOCAROLINA	55.0	7.00	202315.	.538
	GRAND ISL	LOUISIANA	56.0	15.0	67879.6	.447
	OAK GROVE	LOUISIANA	57.0	390.	398468.	.443
	BEAUFORT	NOCAROLINA	59.0	22.0	36819.8	.432
	FLA KEYS	FLORIDA	60.0	50.0	637354.	.753
	PT LAVACA	TEXAS	61.0	46.0	822160.	.550
	MIAMI	FLORIDA	64.0	3.00	161203.	.851
	ST AUGUST	FLORIDA	64.0	5.00	458374.	.582
	FRAKLIN	LOUISIANA	64.0	38.0	232768.	.573
	FLA AND L	FLORIDA	65.0	75.0	.185013E+07	.812
	PORTRICHE	FLORIDA	66.0	6.00	18243.1	.566
	COLUMBIA	MISSISSIPP	69.0	256.	.219346E+07	.589
	RAYMONDVI	TEXAS	67.0	15.0	450784.	.443
	CORPUSCHR	TEXAS	70.0	11.0	721315.	.540
DATA FOR EARTHQUAKE	LONG BEACH	CALIFORNIA	33.0	115.	91590.7	1.12
	HELENA	MONTANA	35.0	4.00	13380.4	.722
	IMPERIALVA	CALIFORNIA	40.0	9.00	18525.9	.767
	SANTABARBA	CALIFORNIA	41.0	*.	547.776	1.03
	PUGET SOUN	WASHINGTON	49.0	8.00	50613.0	.686
	KERN CO	CALIFORNIA	52.0	14.0	105041.	.713
	EUREKA	CALIFORNIA	54.0	1.00	2973.35	.871
	FALLON	NEVADA	54.0	*.	1025.59	.842
	OAKLAND	CALIFORNIA	55.0	1.00	1184.89	1.04
	SANFRANCIS	CALIFORNIA	57.0	*.	1112.00	1.06
	HEBGEN LAK	MONTANA	59.0	28.0	16958.9	.737
		HAWAII	60.0	61.0	38590.0	.744
	WEST COAST	WEST COAST	64.0	125.	546684.	.977
	PUGET SOUN	WASHINGTON	65.0	7.00	15749.7	.839
	SANTA ROSA	CALIFORNIA	69.0	*.	7901.30	.836
	SANFERNAND	CALIFORNIA	71.0	58.0	345011.	1.14

(1) NORMALIZED TO THE 1970 STRUCTURE VALUE/CAPITA.

$$\text{NORMALIZED DAMAGES} = \frac{(\text{DAMAGES})}{(\text{RELATIVE STRUCTURE VALUE/CAPITA})}$$

(2) RELATIVE TO 1970 NATIONAL STRUCTURE VALUE/CAPITA.

*NO DATA AVAILABLE

Table 4-19. Life Loss by Hazard

	1970	2000
Earthquake	273	400
Tornado	392	920
Hurricane	62	153
Storm surge	37	103
Tsunami[a]	20	44
Severe wind	5	11
Expansive soil[b]	NA	NA
Riverine flooding	190	159
Total	979	1790

[a]These estimates assume no warning prior to event.

[b]Expansive soil does not have associated life loss estimates.

Table 4-20. Example of Estimates of the Distribution of Housing Unit Losses

Damage state	Range of damage ratios	Geometric means of damage ratios	Hypothetical annual probability of occurrence	Effective annual damage ratios	Housing units at risk	Housing units damaged
Light	.0005– .0125	.0025	.100	.00025	10,000	2.5
Moderate	.0125– .0750	.0306	.050	.00153	10,000	15.3
Heavy	.0750– .6500	.2208	.010	.00221	10,000	22.1
Severe	.6500–1.0000	.8062	.005	.00403	10,000	40.3
Collapse	1.0000	1.0000	.001	.00100	10,000	10.0
Total				.00902		90.2

Table 4-21. Example Computation of Number of Housing Units Affected

Damage state	Mean damage ratio	Number of housing units affected per housing units damaged	Housing units damaged	Housing units affected	Percentages in each damage state (%)
Light	.0025	400.0	2.5	1000.0	60.4
Moderate	.0306	32.6	15.3	499.0	30.2
Heavy	.2208	4.5	22.1	9.9	6.0
Severe	.8062	1.2	40.3	48.0	2.8
Collapse	1.0000	1.0	10.0	10.0	.6
Total			90.2	1656.0	100.0

Homelessness

In order to quantify homelessness, a time element and an occupancy factor were used to estimate the number of persons displaced by given levels of housing losses. From Whitman et al. (1974), estimates were made by damage state of building use time lost due to repair and restoration, i.e., "time of inoperation."

These estimates can then be applied to the number of housing units damaged at each level. Table 4-22 gives the times of inoperation, the number of housing units damaged, and the computed homelessness, assuming a national average of 3 persons per housing unit in the example county. In this example, the number of people displaced in one year would amount to 92.3 or about 3% of the county's population of 30,000 living in the structure type. The national average percent of "year round vacant for rent" housing units as recorded in the 1970 Census of Housing (U.S. Bureau of the Census, 1972) is 2.46%; thus, the number of vacant housing units in this example community is estimated as 246, or housing for 738 persons.

The annual rates assume that every year a little of each possible disaster occurs, but in reality this is not the case. When a disaster occurs, the effects can be distributed over an area so wide that local resources may be overwhelmed, with the possible exception of tornado, storm surge, and tsunami, which may be of relatively local scale. Also, the type of housing units destroyed and those already vacant may be sufficiently dissimilar in type and rental costs that direct tenant transfer may be impossible.

Table 4-23 lists the expected number of homes destroyed, homes affected, and homelessness of hurricane, tornado, severe wind, tsunami, storm surge, earthquake, landslide, and expansive soil. (Data for riverine flood was not comparable.) On a national scale the gradual effect of the annual probabilities for different levels is more realistic, due to the aggregation of the probabilities of many occurrences over a year.

The values of losses and of housing units at risk are taken from the structure value calculations discussed by Hirschberg et al. (1978). The average number of persons per housing unit varies by hazard, and is based on the census of housing data. It should be noted that the structure value is per person, based on the 1970 stock of buildings, in 1970 dollars.

Table 4-22. Example Estimation of Homelessness

Damage state	Housing units affected	Time of inoperation in years	Persons per housing unit	Homelessness (in people-years)
Light	2.5	.00000	3.0	0.0
Moderate	15.3	.00055	3.0	0.0
Heavy	22.1	.02600	3.0	1.8
Severe	40.3	.50000	3.0	60.0
Collapse	10.0	1.00000	3.0	30.0
Total	90.2			92.3

Table 4-23. Housing Unit Losses by Hazard

Hazard	No. of HUs totally destroyed (100% lost)	No. of HUs affected by some damage	HUs out of service for one year	People displaced for one year (homelessness)	HUs at risk × 10^6	Avg. value HUs at risk (structure only) in 1970$	Avg. value of HUs affected by some damage in 1970$	Avg. value of loss of HUs affected in 1970$	Avg. no. of people/HU in areas at risk	% chance of being affected if in exposed area	% chance of being hit on national scale
Hurricane	28,190.	984,684.	10,388.0	34,508.0	20.76	17,400.0	16,000.0	980.0	3.34	2.210	0.680
Severe wind	485.	16,314.	255.6	767.4	67.56	16,372.0	15,670.0	463.0	3.00	.024	0.024
Tornado	30,970.	45,486.	25,004.6	88,144.0	60.01	16,081.1	16,215.0	11,548.5	3.52	.072	0.064
Tsunami	233.	1,369.	100.0	345.0	6.34	17,296.8	19,407.7	3,303.1	3.45	.022	.002
Earthquake	20,510.	2,805,925.	254.7	735.9	47.79	16,632.3	17,269.6	126.0	2.89	5.870	4.150
Storm surge	24,672.	515,595.	2,532.1	7,208.5	12.89	17,879.7	13,486.0	645.0	2.85	4.000	0.760
Landslide	—[a]		—[a]	—[a]	67.39	16,316.8	16,316.8	1.7	3.00	—[a]	—[a]
Expansive soil	—[a]		—[a]	—[a]	67.39	16,316.8	16,316.8	8.9	3.00	—[a]	—[a]

[a] Expansive soil and landslide did not have a damage state breakdown, thus only the averages are available for reporting

Unemployment

Using the income losses calculated by Hirschberg et al. (1978), the number of person-years of unemployment can be calculated from the ratio of payroll to employees for each economic sector. From the data in the Multi-Regional Input-Output Model for the United States (Polenske, 1970), estimates are available of the number of employees and payrolls by sector (Rodgers, 1972).

The income per employee ratios are available by state and were used to weight the income losses to derive the total unemployment by sector. The unemployment computed here is not the statistic often cited as a percentage of the work force, but is an unemployment statistic with a time dimension. For example, a loss in income to a basic sector (farming and mining) is given as $3.0 million for a year due to a particular hazard's annual risk. From Table 4-24 we calculate an average income of $3063 per employee annually. Thus, the loss of $3 million of income could result in approximately 980 person-years of unemployment, in the form of either 980 persons unemployed for a year or 11,760 persons unemployed for a month.

Table 4-25 lists the unemployment estimated for each sector due to hurricane wind, severe wind, tornado, storm surge, tsunami and earthquake. Again, due to limitations in the expansive soil and landslide models, unemployment estimates were not calculated. The unemployment distribution follows that of income loss. Thus, all employees in the farming and mining sector are exposed to severe wind hazards, yielding 19.1 person-years lost in 1970. About 5% of the total employed suffered some joblessness in this sector from severe winds. Again, the usual cautions pertaining to this study's county level data should be observed. The percentage of the total loss listed by sector for each hazard demonstrates the regional variation of the hazards and the types of exposure.

Table 4-24. Average Income Per Employee in 1970 Dollars in 1963

Economic sectors	Average income
Household	1,681.5
Mining, farming	3,063.2
Construction	6,716.5
Manufacturing	6,712.6
Transportation and utilities	7,520.9
Wholesale and retail trade	4,646.0
Finance, insurance, etc.	5,867.1
Services	4,464.8
Federal government	6,463.6
State and local government	5,860.5

Table 4-25. Distribution of Annual Unemployment by Economic Sector and Hazard in Person Years Unemployed for 1970

	Household	Farming and mining	Construction	Manufacturing
Estimated employee years	1,456.0	7,283.0	4,548.0	20,630.0
percent distribution of total	1.6	8.1	5.1	23.1
Severe wind				
percent of exposure	100.0	100.0	100.0	100.0
person-years	37.9	19.1	5.6	28.1
percent distribution of total	10.2	5.1	1.5	7.5
Hurricane				
percent of exposure	32.9	21.5	34.4	30.5
person-years	1,841.0	1,411.0	583.0	2,408.0
percent distribution of total	8.8	6.7	2.8	11.5
Tornado				
percent of exposure	92.7	94.0	90.6	91.6
person-years	2,058.0	4,289.0	1,507.0	8,626.0
percent distribution of total	3.6	7.5	2.6	15.0
Tsunami				
percent of exposure	8.0	3.3	7.6	7.2
person-years	7.2	.3	.8	1.5
percent distribution of total	7.4	.3	.8	1.5
Earthquake				
percent of exposure	80.7	71.9	70.1	67.8
person-years	21.4	61.3	12.5	55.8
percent distribution of total	5.2	14.9	3.0	13.5
Storm surge				
percent of exposure	20.8	14.2	21.2	17.7
person-years	47.7	48.4	8.4	28.0
percent distribution of total	12.9	13.1	2.3	7.6

[a]The percent of the total number employed who would lose their jobs.
[b]This estimate is 11.9% greater than the total employment estimates for 1970 in the OBERs report (U.S. Water Resources Council, 1974) and 12.2% greater than the employment reported in the Statistical Abstract for 1970 (U.S. Bureau of the Census, 1972b.)

Transportation and utilities	Retail and wholesale trade	Finance and insurance	Services	Federal government	State and local government	Total
4,711.0	17,685.0	4,332.0	16,845.0	3,265.0	8,888.0	89,643.0[b]
5.3	19.7	4.8	18.8	3.6	9.9	100.0
100.0	100.0	100.0	100.0	100.0	100.0	100.0
8.4	95.0	77.0	77.7	9.5	14.6	372.9
2.3	25.5	20.6	20.8	2.5	3.9	100.0
34.0	32.6	39.2	35.7	36.3	32.7	32.9
810.0	4,254.0	3,997.0	3,818.0	653.0	1,228.0	21,003.0
3.9	20.2	19.0	18.2	3.1	5.8	100.0
85.2	88.4	82.6	87.4	87.8	87.5	88.5
2,147.0	12,980.0	10,190.0	10,290.0	1,977.0	3,476.0	57,540.0
3.7	22.6	17.7	17.9	3.4	6.0	100.0
10.7	9.4	10.8	9.4	11.3	8.4	8.6
23.7	21.4	15.4	22.9	2.8	1.6	97.6
24.3	21.9	15.8	23.5	2.9	1.6	100.0
72.3	70.8	74.2	73.0	76.0	71.2	71.4
24.5	61.4	59.0	47.5	35.1	35.4	413.9
5.9	14.8	14.3	11.5	8.5	8.6	100.0
24.3	21.0	27.7	23.2	18.1	19.9	20.9
15.2	77.5	57.0	59.4	9.6	17.5	368.7
4.1	71.1	19.5	16.1	2.6	4.7	100.0

Conclusion

The foregoing discussion of the analysis methods is intended to provide a brief summary of a complex set of models employed to develop the results presented in Chapters 5 and 6. A more complete and thorough discussion of the socioeconomic models may be found in Hirschberg et al. (1978); the earthquake, landslide, and expansive soil models are developed in Wiggins et al. (1978); the riverine flooding, storm surge, and tsunami models are presented in Lee et al. (1978); and the hurricane, tornado, and severe wind models are presented by Hart (1976).

It should be emphasized and understood that the results derived from the risk analysis models are not to be considered "fact." Much uncertainty is associated with the findings generated by the models. However, the findings are useful in framing a complex problem, providing a basis for assessing the consequences of natural hazard losses and mitigations, and providing a basis for performing subsequent problem and policy analyses.

References

Alfors, J. T., J. L. Burnett, and T. E. Gay, Jr. 1971. Urban Geology–Master Plan for California–Phase 1. A Method for Setting Priorities. Open File Report 72-2. California Division of Mines and Geology, Sacramento.

Algermissen, S. T., and D. M. Perkins. 1976. A Probabilistic Estimate of Maximum Acceleration in Rock in the Contiguous United States. U.S. Geological Survey Open File Report 76-416. U.S. Department of the Interior, Washington, D.C.

Ayre, R. S. 1975. Earthquake and Tsunami Hazards in the United States: A Research Assessment. Monograph No. NSF-RA-E-75-005. University of Colorado, Boulder.

Barnes, S. B., and C. Pinkham. 1972. Damage Probability Matrix. Prepared for Massachusetts Institute of Technology, Cambridge, Massachusetts.

Brabb, E. E., and F. A. Taylor. 1972. U.S. Geological Survey Field Studies Map MF-327 (BDC 37). U.S. Geological Survey, Washington, D.C.

Briggs, R. P., J. S. Pomeroy, and W. E. Davies. 1975. Landsliding in Allegheny County, Pennsylvania. U.S. Geological Survey Circular 728. U.S. Geological Survey, Washington, D.C.

Building Construction Cost Data. 1972. Building Cost File. Construction Publishing Company, Inc., New York.

California Office of Emergency Services. 1972. California Emergency Plan. COES.

Carson, W. D. 1975. Estimating Costs and Benefits for Non-structural Flood Control Measures. The Hydrologic Engineering Center, U.S. Army Corps of Engineers, Washington, D.C.

Culver, C., H. S. Lew, G. C. Hart, and C. W. Pinkham. 1975. Natural Hazard Evaluation of Existing Buildings. Building Science Series 61. U.S. Department of Commerce, Washington, D.C.

Dacy, D. C., and H. Kunreuther. 1969. The Economics of Natural Disasters, Implications for Federal Policy. Free Press, New York.

Environmental Science Service Administration. 1969. Studies in Seismicity and Earthquake Damage Statistics. Prepared for the Department of Housing and Urban Development, Washington, D.C.

Friedman, D. G. 1971. The Storm Surge Along the Gulf and South Atlantic Coastlines. Travelers Insurance Company, Hartford, Connecticut.

Friedman, D. G., and M. Bocaccino. 1972. Computer Simulation of the Effects of Adjustments to the Inland Flood Hazard. Travelers Insurance Company, Hartford, Connecticut.

Friedman, D. G., and T. S. Roy. 1966. Simulation of Total Flood Loss Experience of Dwellings on Inland and Coastal Flood Plains. Travelers Insurance Company, Hartford, Connecticut.

Fujita, T. T. 1970a. Estimate of Areal Probability of Tornadoes from Inflationary Reporting of Their Frequencies. SMRP Research Paper No. 89. Department of Geophysical Sciences, University of Chicago, Chicago, Illinois.

Fujita, T. T. 1970b. A Proposed Characterization of Tornadoes and Hurricanes by Area and Intensity. SMRP Research Paper No. 91. Department of Geophysical Sciences, University of Chicago, Chicago, Illinois.

Fujita, T. T., and A. D. Pearson. 1973. Results of FPP Classification of 1971 and 1972 Tornadoes. Presented at the Eighth Annual Conference on Severe Local Winds, Denver, Colorado, October 1973.

Goldsmith, R. W. 1971. Institutional Investors Study Report of the Securities and Exchange Commission. National Bureau of Economics Research. Columbia University Press, New York.

Grigg, N. S., and O. J. Helweg. 1975. State-of-the-Art of Estimating Flood Damage in Urban Areas. AWRA Water Rsour. Bull. pp. 370–390.

Hart, G. C. 1976. Natural Hazards: Tornado, Hurricane, Severe Wind Loss Models. Prepared for the National Science Foundation (NTIS No. PB 294594/AS). J. H. Wiggins Company, Redondo Beach, California.

Hirschberg, J., P. Gordon, and W. Petak. 1978. Natural Hazards: Socio-economic Impact Assessment Model. Prepared for the National Science Foundation (NTIS No. PB 294681/AS). J. H. Wiggins Company, Redondo Beach, California.

Houston, J. R., and A. W. Garcia. 1974. Type 16 Flood Insurance Study: Tsunami Predictions for Pacific Coastal Communities. U.S. Army Engineer Waterways Experiment Station. Report H-74-3, May 1974.

Houston, J. R., and A. W. Garcia. 1975. Type 16 Flood Insurance Study: Tsunami Predictions for Monterey Bay, San Francisco Bay and Puget Sound. U.S. Army Engineer Waterways Experimental Station, Report H-75-17, November 1975.

Krohn, J. P., and J. E. Slosson. 1976. Landslide potential in the United States. California Geology, 29(10):224–231. California Division of Mines and Geology, Sacramento, California.

Krohn, J. P., and J. E. Slosson. 1980. Assessment of expansive soils in the United States. Proceedings of the 4th International Conference on Expansive Soils, 1:596–608. Am. Soc. Civ. Eng., New York.

Lee, L. T., and R. Eguchi. 1977. Capital Investment Program for Mitigation of Risks from Natural Hazards. Prepared for the U.S. Naval Facilities Command. J. H. Wiggins Company, Redondo Beach, California:

Lee, L. T., J. Chrostowski, and R. Eguchi. 1978. Natural Hazards: Riverine Flooding, Storm Surge, Tsunami Loss Models. Prepared for the National

Science Foundation (NTIS No. PB 294682/AS). J. H. Wiggins Company, Redondo Beach, California.

Leslie, S. K., and J. M. Biggs. 1972. Earthquake Code Evaluation and Effects of Seismic Design on the Cost of Buildings. Seismic Design Decision Analysis. Report No. 2, MIT-CE-R72-20. Massachusetts Institute of Technology, Cambridge, Massachusetts.

Loftus, S. F. 1972. Stocks of business inventories in the United States, Surv. Curr. Business 52(3):29–32. U.S. Govt. Printing Office, Washington, D.C.

Manvel, A. D. 1968. Trends in the value of real estate and land, 1956 to 1966. *In*: Three Land Studies. Prepared for the National Commission on Urban Problems, Research Report No. 12. U.S. Govt. Printing Office, Washington, D.C.

McClure, F. E. 1973. Performance of Single Family Dwellings in the San Fernando Earthquake of February 9, 1971. Prepared for Housing and Urban Development and National Oceanographic and Atmospheric Administration, Washington, D.C.

Musgrave, J. C. 1974. New estimates of fixed non-residential business capital in the United States, 1925–73. Surv. Curr. Business 54(3):23–27. U.S. Govt. Printing Office, Washington, D.C.

Naval Facility Engineering Command. 1972. Conceptual Military Construction Cost Engineering Data (NAVFAC P-448. NFEC, Washington, D.C.

Pautz, M. E. 1969. Severe Local Storm Occurrences, 1955–1967. ESSA Memo WBTM FCST 12. Environmental Science Services Administration, U.S. Department of Commerce, Washington, D.C.

Polenske, K. R. 1970. A Multi-Regional Input-Output Model for the United States. Harvard Economic Research Project, EDA 71-070-6-7-35212, NTIS No. OOM-71-00943. Harvard University, Cambridge, Massachusetts.

Rodgers, J. M. 1972. State Estimates of Outputs, Employment, and Payrolls, 1974, 1958, 1963. Lexington Books, Lexington, Massachusetts.

Schneider, W. J., and J. E. Goddard. 1974. Extent and Development of Urban Flood Plains. Geol. Surv. Circular 601-J. U.S. Geological Survey, Washington, D.C.

Shavell, H. 1971. The stock of durable goods in the hands of consumers, 1946–1969. Proc. Business Econ. Sect. Am. Stat. Assoc., 1970. pp. 241–248.

Slosson, J. E. 1969. The role of engineering geology in urban planning: The Governor's conference on environmental geology. Colo. Geol. Surv. Spec. Publ. No. 1, pp. 8–15.

Smith, D. M., and P. M. Allen. 1974. A Cost-Benefit Analysis of Housing Foundation Failures. Department of Economics, Southern Methodist University, Dallas, Texas.

Steinbrugge, K. V. 1971. San Fernando Earthquake, February 9, 1971. Pacific Fire Rating Bureau. San Francisco, California.

Structural Engineers Association of California (SEAOC). 1970. Report of the Ad Hoc Committee on Cost of Design for Earthquake. O.S.T. Task Force on Earthquake Hazard Reduction, San Francisco, California.

Sugg, A. L., L. G. Pardue, and R. L. Carrodus. 1971. Memorable Hurricanes of the United States 1873. NOAA Technical Memorandum WS SR-56. Southern Region, Forth Worth, Texas.

Thom, H. C. S. 1968. New distribution of extreme winds in the United States. J. Struct. Div., Am. Soc. Civ. Eng. 87(No. ST7):1787–1801.

U.S. Army Corps of Engineers. 1966. Flood Insurance Study, Hilo, Hawaii. U.S. Army Engineer District, Honolulu.

U.S. Army Corps of Engineers. 1967. List of Urban Places with Information About Flood Problems. Department of the Army, Washington, D.C.

U.S. Army Corps of Engineers. 1975. Cost of Placing Fill in a Flood Plain. Hydrologic Engineering Center, Washington, D.C.

U.S. Bureau of the Census. 1971. Current Population Reports. Series P-25, No. 470. Projections of the United States, by age and sex: 1970 to 2020. U.S. Govt. Printing Office, Washington, D.C.

U.S. Bureau of the Census. 1972a. Census of Housing: 1970 Detailed Housing Characteristics. Final Report HC (1)-B1. U.S. Summary. U.S. Govt. Printing Office, Washington, D.C.

U.S. Bureau of the Census. 1972b. Statistical Abstract of the United States: 1972. 93rd edition. U.S. Govt. Printing Office, Washington, D.C.

U.S. Bureau of the Census. 1973. County and City Data Book 1972. U.S. Govt. Printing Office, Washington, D.C.

U.S. Bureau of Economic Analysis. 1974a. Revised deflators for new construction, 1947–1973. Surv. Curr. Business 54(8):18–27. U.S. Govt. Printing Office, Washington, D.C.

U.S. Bureau of Economic Analysis. 1974b. The input-output structure of the United States economy: 1967. Surv. Curr. Business 54(1):24–29. U.S. Govt. Printing Office, Washington, D.C.

U.S. Department of Agriculture. 1971. Guide for Interpreting Engineering Uses of Soils. Soil Conservation Service, USDA, Washington, D.C.

U.S. Department of Commerce. 1969. Severe Local Storm Occurrences, 1955–1967. M. E. Pautz, ESSA Memo WBTM FCST 12. Environmental Sciences Services Administration, U.S. Department of Commerce, Washington, D.C.

U.S. Department of Commerce. 1972. An Inquiry into the Long-Term Economic Impact of Natural Disasters in the United States. Prepared by Harbridge House, Inc. for Office of Technical Assistance, Economic Development Administration. U.S. Department of Commerce, Washington, D.C.

U.S. Department of Commerce, National Bureau of Standards. 1974. Design, Siting, and Construction of Low-Cost Housing and Community Buildings to Better Withstand Earthquakes and Windstorms. Building Science Series 48. U.S. Govt. Printing, Office, Washington, D.C.

U.S. Department of Housing and Urban Development, Federal Housing Administration. 1966. Minimum Property Standards for One and Two Living Units. Report No. 300. Housing and Urban Development, Federal Housing Administration, Washington, D.C.

U.S. Department of Housing and Urban Development, Federal Insurance Administration. 1976. Elevated Residential Structures–Reducing Flood Damage Through Building Design: A Guide Manual. Housing and Urban Development, Federal Insurance Administration, Washington, D.C.

U.S. Department of the Interior. 1970. The National Atlas of the United States of America. Department of the Geological Survey, Washington, D.C.

U.S. Office of Emergency Preparedness, Executive Office of the President. 1972. Disaster Preparedness. U.S. Govt. Printing Office, Washington, D.C.

U.S. Water Resources Council. 1974. OBERS Projections, Regional Economic Activity in the United States, Series E, Populations. Vols. 1–7. U.S. Govt. Printing Office, Washington, D.C.

Valley, S. L. 1965. Handbook of Geophysics and Space Environments. McGraw-Hill, New York.
Wiggins, J. H. 1975. Seismic Hazard Maps of the United States. Prepared for Massachusetts Institute of Technology. J. H. Wiggins Company, Redondo Beach, California.
Wiggins, J. H., J. Hirschberg, and A. Bronowicki. 1974. Budget Justification for Earthquake Engineering Research. Prepared for the National Science Foundation. J. H. Wiggins Company, Redondo Beach, California.
Wiggins, J. H., J. E. Slosson, and J. Krohn. 1978. Natural Hazards: Earthquake, Landslide, Expansive Soil Loss Models. Prepared for the National Science Foundation (NTIS No. PB 294686/AS). J. H. Wiggins Company, Redondo Beach, California.
White, G. F., and E. J. Haas. 1975. Assessment of Research on Natural Hazards. MIT Press, Cambridge, Massachusetts.
Whitman, R. V., J. M. Biggs, J. Brennan, R. de Neufville, and E. H. Vanmarcke. 1974. Methodology and Pilot Application. Seismic Design Decision Analysis Report No. 10. Department of Civil Engineering, Massachusetts Institute of Technology, Cambridge, Massachusetts.
Whitman, R. V., S. T. Hange, and J. W. Reed. 1973. Damage Statistics for High Rise Buildings in the Vicinity of the San Fernando Earthquake. School of Engineering, Massachusetts Institute of Technology, Cambridge, Massachusetts.
Young, A. H., J. C. Musgrave, and C. Harkins. 1971. Residential capital in the United States, 1925–70. Surv. Curr. Business 5(11):16–27. U.S. Govt. Printing Office, Washington, D.C.

5

Natural Hazard Exposure and Loss Analysis

Introduction

Numerous analysts have attempted to measure the annual magnitude, geographic distribution, and impacts of natural hazard exposure in the United States. Although most have focused on single hazards or related clusters of hazards, a few comprehensively oriented assessments have been published, and for at least sixteen natural hazards, various estimates of annual losses are available.

In 1969, one team of analysts estimated that the 40 year costs (1925-1965) of hurricanes, floods, tornadoes, and earthquakes had reached 21,467 million dollars (in constant 1970 dollars) and 15,768 deaths (Dacy and Kunreuther, 1969).

In 1975, another team of researchers estimated that direct property losses from fifteen natural hazards equaled \$23 per capita, or 0.05% of per capita income (White and Haas, 1975). The hazards inventoried were avalanche, coastal erosion, drought, earthquake, flood, frost, hail, hurricane, landslide, lightning, tornado, tsunami, urban snow, volcano, and windstorm. Another set of authors has placed annual losses for expansive soils at 2.3 billion dollars per year, or about \$11 per capita (Jones and Holtz, 1973).

Other authors such as Bates (1963) and Cochrane (1975) have examined the distribution and impacts of hazard losses within the populations of communities that have experienced major natural disasters. From these studies a consistent finding has emerged: the persons most adversely affected are the poor, the elderly, and members of minority groups. Thus, in Bates's study, the most severe effects of a hurricane on the population of a Louisiana parish were found to be concentrated among the community's black population. Nearly half the black population was killed in the disaster, but only 4% of the whites were killed. Similarly, Cochrane's examination of eight communities that had experienced disasters revealed that "the lower income groups consistently bear a disproportionate share of the losses; they receive, in most instances, the smallest proportion of disaster relief; they are the least likely to be insured . . . ; and they live in dwellings which are of the poorest construction and most subject to damage." Also, the elderly consistently are overrepresented among the dead and injured of disaster-struck communities.

Whether the occurrence of a disaster has long-term effects on the economic vitality of a community has been examined by a number of investigators, including Dacy and Kunreuther (1969), Haas, et al. (1977), and Wright et al. (1979). Surprisingly, these investigators answer this question in the negative. The first group (Dacy and Kunreuther) concluded that a community actually "may benefit economically from a disaster through a rapid inflow of capital for rebuilding purposes and the adoption of technological innovations to meet crisis situations." Haas and his team concluded that predisaster trends, rather than the impacts of the disaster itself, determine the future fortunes of a community. "Simply stated, rapidly-growing cities recover rapidly; stable, stagnant or declining cities recover slowly and may have their decline accelerated." In the study by Wright et al., a sophisticated analysis of data for 1612 counties that had experienced flood, hurricane, or tornado damage between 1960 and 1970 failed to find any "consistent and significant" effects of such disasters.

This analysis differed in several significant ways from the other investigations. First of all, it did not seek to tally the losses inventoried over any given time frame by field investigators in past disaster reports. Instead, it used a probabilistic approach that resulted in the generation of annual estimates of losses that may be expected to occur over a very long time frame as a result of both frequent and infrequent hazardous occurrences. This required the use of several sets of data, respectively, concerned with such subjects as (a) the vulnerability of exposed building mixes to a range of hazard and hazard intensities; (b) the geographic distribution of specific hazards; and (c) the probability of hazard occurrence, by intensity level, in specific hazard exposure areas or types of areas. Although comprehensive in its treatment of natural hazard exposure, the anlysis did not fully express all the major probable natural hazard losses in the United States. Instead, the analysis focused on nine hazards and did not consider such major sources of loss as nonriverine flash flooding; headwater flooding; shore erosion; drought; forest, field, and grass fires; hail, snow, and ice storms; and subsidence. Also, there was no attempt to inventory such important potential losses as damage to automobiles, boats, ornamental vegetation, agricultural crops, and community infrastructure (streets, sidewalks, utility lines, etc.). Although focusing, in part, on potential expansive soil losses, the analysis did not consider possible expansive soil damage to structures other than detached one and two unit single family residences. This omission was due to a paucity of empirical data necessary to the development of supportable damage algorithms. Finally, the riverine flood parts of the study are limited to an urban-oriented model in which the number of flood-prone cities in the United States was assumed to range from 5539 to 6455 between 1970 and 2000. However, by mid-1976 more than 16,000 United States communities had been identified by the Federal Insurance Administration as flood-prone. Thus, the analysis understates the real magnitude of natural hazard losses in the United States.

The loss analysis focused on several objectives: (1) identification of the primary effects of the exposure of people and buildings to nine natural hazards and of selected secondary and higher order effects; (2) determination of the characteristics of the U.S. human and building populations exposed to various intensity levels of each of these hazards based on 1970 estimates and expected exposure to the same phenomena in the year 2000; and (3) quantification of the magnitude and geographic occurrence of seven major effects associated with the exposure of buildings, building contents, and building occupants to these same hazardous natural events. In most cases, data on hazard effects were probabilistically derived, calculated as annual expected or "annualized" estimates. Except for riverine flooding, estimates were made at the scale of individual counties, states, and the whole nation.

The loss analysis also included the development of an annual expected hazard-specific and aggregated *building damage rate* for each state and county in the United States, and a hazard-specific rate for other selected hazard zones or areas. The *building damage rate* for any area is the annual expected dollar value of hazard-induced damage to buildings divided by the total value of all buildings within the area and is a crude indicator of the level of hazard-related risk associated with a particular place. The rate is a function of two sets of factors: the probability of hazard occurrence within the area by various intensity levels, and the level of vulnerability of area buildings and property to hazard-induced damage.

For any given area, the annual expected dollar losses derived from the loss analysis are expressed as the sum of the natural hazard dollar losses (in 1970 dollars) expected to be incurred within the area over a period of years divided by the number of years in the period. Accordingly, this calculation treats all expected losses over a large frame of years as though the losses were sustained in equal amounts during each of the several years of exposure. Dollar losses inventoried include those resulting from damage to buildings, damage to building contents, supplier losses due to increased transportation costs, and income reduction sustained as a result of the closure or slowdown of economic activities conducted in damaged buildings. Single event losses resulting from a major event are analyzed in Chapter 7 using earthquake and hurricane scenarios.

Natural Hazard Effects

Whenever persons or property are exposed to hazardous natural events, the exposure may result in chains of effects, some of which may be disvalued by subsets of the human population. Among such consequences, or impacts, are those shown in Table 5-1.

The *primary effects* of a natural hazard are those instantly and directly

Table 5-1. Taxonomy of Natural Hazard Effects

Primary effects	Secondary effects	Higher order effects
1. Injury or death to human beings 2. Injury or death to livestock and domestic animals 3. Damage to structures and their contents 4. Damage to community facilities and infrastructure 5. Damage to automobiles, boats, and other personal property 6. Damage to crops, forests, and ornamental vegetation 7. Alteration of valued geophysical configurations 8. Psychological trauma 9. Loss of community housing stock	1. Homelessness 2. Shutdowns or slowdowns in business and industry 3. Disruption of utility service 4. Financial expenditures by communities, families, and businesses for clean-up and recovery operations 5. Financial expenditures by communities, families, and businesses for repair or replacement of damaged structures, building contents and other property 6. Financial expenditures by government for repair and/or replacement of damaged community facilities and infrastructure 7. Insurance payouts to policy holders	1. Unemployment 2. Loss of personal family income 3. Loss of business/ industrial income 4. Diversion of investment capital to rehabilitation and recovery projects 5. Alteration of land and property values 6. Alteration of area population growth trends 7. Alteration of population migration patterns 8. Alteration of community overhead burdens 9. Increased tax burdens to finance recovery operations 10. Alteration in community socio-economic viability 11. Alterations in familial socioeconomic viability 12. Depletion of personal or business savings or capital 13. Financial cost to suppliers

resulting from the occurrence of the hazard and the exposure of people or property to that hazard. *Secondary effects* are the instant and direct outcome of primary effects. Similarly, secondary effects are proximate causes of still other consequences, referred to as *higher-order* effects. The hazard-induced effects or consequences quantified through performance of the loss analysis included (1) economic losses to buildings (building losses); (2) economic losses to the contents of buildings (contents losses); (3) economic losses in the impacted area due to increased transportation costs of goods (supplier losses);

(4) economic losses in the form of lost take-home pay (income loss); (5) years of unemployment induced by hazard-caused damage to buildings and their contents (unemployment); (6) years of homelessness induced by damage to private residences (homelessness); and (7) loss of life resulting from the occurrence of hazardous natural events (life loss).

Of course, several other major effects also are associated with the exposure of people, buildings, and other properties to hazardous natural events. Although not considered in this analysis, some of the most significant of these include (1) damage to community infrastructure, including utility lines, pipelines, railroads, streets, highways, and other similar facilities; (2) damage to personal property not countable as building contents, such as automobiles, boats, and recreational equipment; and (3) damage to growing crops, livestock, and ornamental vegetation. From an economic point of view, the annual expected dollar losses associated with these types of hazard effects may well be equal to or greater than those inventoried in this analysis.

Being guided by a primary focus on hazard-induced damage to buildings, the analysis revealed considerable differences in the type, magnitude, and frequency of damage to buildings as a function of the type of hazard. Taking into account the geographic areas in which each type of hazard occurs and the vulnerability of buildings to the hazard, the computer models disclosed that light to moderate damage would be sustained by 61.6% of the buildings experiencing damage from earthquakes, by 26.7% of those damaged by storm surge, by 18.3% of those damaged by tsunami, and by about 16% of those damaged by either hurricane winds or severe winds. Structural collapse or severe damage is expected to be experienced by 92.7% of those buildings damaged by tornadoes, by 50.7% of those damaged by hurricane winds, and by 36.1% of those damaged by tsunami (Table 5-2 and Figure 5-1).

Table 5-2. Damage State Description

Damage state (%)	Structure condition
None (0–0.5)	No damage.
Light (0.5–1.25)	Minor ceiling tile or partition cracking; possible damage due to missiles.
Moderate (1.25–1.75)	Many partitions cracked or ceiling tiles fallen down; a few structural members appear to be stressed beyond yield level.
Heavy (7.5–65)	Significant number of structural members with structural damage, or damage to a structural system, roof having major damage.
Very severe (65–100)	Major damage; structure standing but will probably be taken down.
Collapse (100)	Structure does not remain standing.

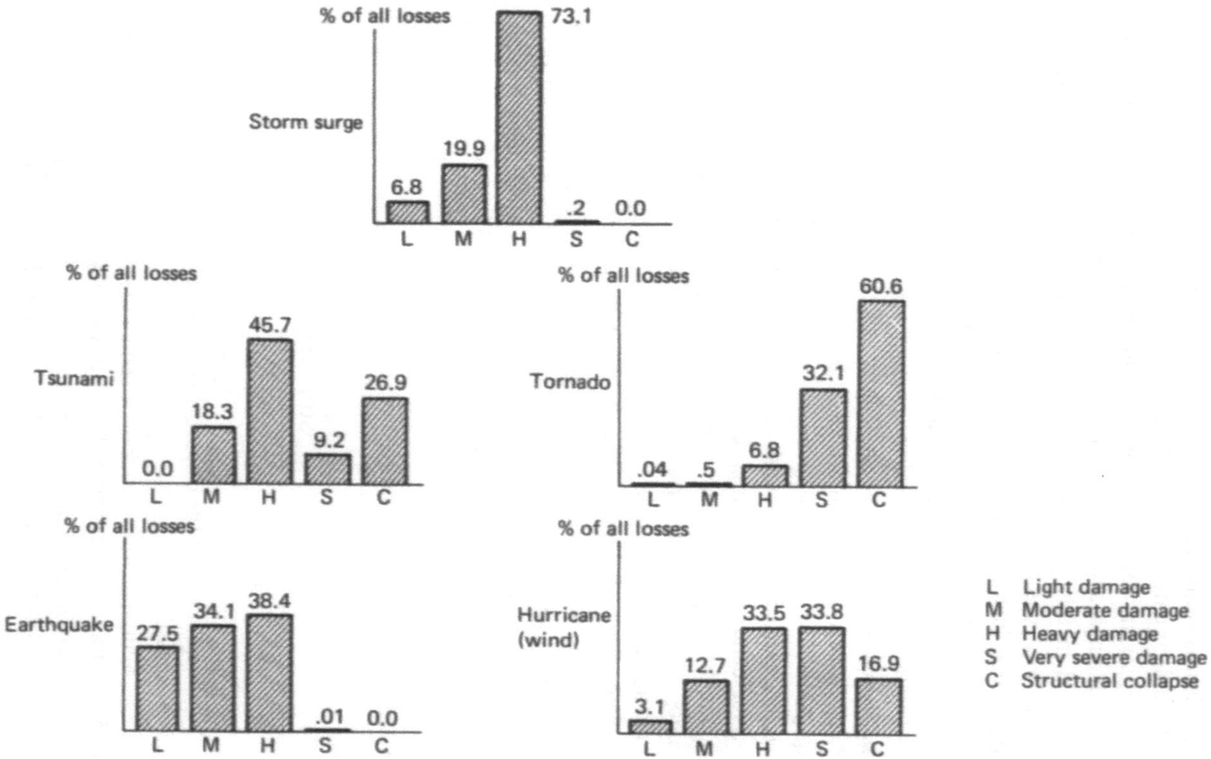

Figure 5-1. Average proportionate distribution of building damage by selected hazard type in the United States and by level of severity of hazard effects.

Natural Hazard Loss Analysis (1970–2000)

Based on the estimating procedures used in this analysis the effects produced by natural hazard exposures in the United States in 1970 yielded total annual expected losses of 8.1 billion dollars for 1970, as well as the loss of 113.9 thousand housing units and nearly one thousand lives (Table 5-3). In addition 1970 natural hazards exposures also resulted in such annual expected impacts as 130 thousand person-years of homelessness and 80 thousand person-years of unemployment. In descending order of importance, the death-inducing hazards were tornado (392), earthquake (273), riverine flood (190), hurricane (62), storm surge (37), tsuanmi (20), and severe wind (5). It is these same low probability but high consequence hazards that result in the collapse of buildings and therefore in significant destruction of housing units and the related production of homelessness and unemployment.

In descending order of total economic importance, the nine natural hazards examined were expected to be responsible for total dollar losses as follows: riverine flooding ($2758 million), tornado ($1656 million), hurricane ($1056 million), expansive soil ($798 million), earthquake ($781 million), storm surge ($641 million), landslide ($370 million), severe wind ($18 million), and tsunami ($15 million). Damage to buildings accounted for over 71% of all annual expected dollar losses under 1970 exposures, while income loss and damage to building contents accounted for 5% and 24% respectively, of all losses.

For those hazards for which appropriate data are available, damage to housing structures was found to account for 50% of all building losses in 1970; damage to all governmental structures produced 11% of such losses, and damage to transportation and utility structures yielded 10% of such losses (Table 5-4).

Annual expected dollar losses resulting from 1970 exposures to all nine natural hazards averaged $39.76 per capita and were equivalent to 1.00% of national per capita income (see Table 5-5). As indicated in Table 5-6, building losses from all hazards were equal to .279% of the value of all buildings exposed. However, both annual per capita hazard costs and building damage rates vary substantially between counties, states, and other identifiable populations at risk.

Expressed as a percentage of per capita income, 1970 building losses from all hazards except riverine flooding range from a low of 0.40% to a high of 5.82%. If building contents and other losses are considered, these hazard-related demands on per capita income would be 30% higher.

At the state level, per capita losses of all types (building, contents, income and supplier losses) range from $16.86 (Vermont) to $110.32 (Florida) and per capita losses as a percentage of per capita income range from 0.40% (D.C.) to 2.95% (Florida) (Table 5-6). At a finer scale, the 500 most damage-prone counties of the United States exhibited building damage rates that varied from 0.287 to 3.51% of building value (Table 5-7) and per capita losses that ranged from $21.18 to $219.89 (Table 5-5). When these building losses are expressed as a percent of per capita income, they varied from 0.63 to 5.82%.

In terms of the states, the aggregate of damage rates induced by all nine

Table 5.3 Expected Annual Losses from Natural Hazard Exposures in the United States, by Type of Hazard and Type of Loss, 1970 Exposure

	Expected annual losses								
	Dollar losses resulting from indicated factor (millions of 1970 $)					Other losses			
Hazard	(1) Building damage	(2) Contents damage	(3) Income loss	(4) Supplier loss	(5) Total (1–4)	(6) Number of deaths	(7) Housing units lost	(8) Person-years of home-lessness	(9) Person-years of unem-ployment
1. Earthquake	655.2	123.23	2.651	0.030	781.1	273	20,485	736	413.5
2. Expansive soil	798.1	–	–	–	798.1	–	–	–	–
3. Hurricane	685.5	267.57	101.803	1.092	1056.0	62	31,885	34,505	21,003.7
4. Landslides	370.3	–	–	–	370.3	–	–	–	–
5. Riverine flooding	1901.0	847.02	10.166	0.120	2758.3	190	–	–	–
6. Severe wind	11.4	4.47	2.090	0.022	18.0	5	547	852	373.1
7. Storm surge	441.6	197.24	2.367	0.028	641.2	37	24,521	7,290	369.7
8. Tornado	879.8	469.93	302.821	3.451	1656.0	392	36,212	86,122	57,541.6
9. Tsunami	8.7	5.54	0.727	0.012	15.0	20	234	345	97.5
Totals	5751.6	1915.0	422.625	4.755	8094.0	979	113,884	129,850	79,799.1

hazards ranged from 0.087 to 0.825%; annual expected *total* per capita losses varied from $16.86 to $110.32; and total per capita losses as a percent of per capita income ranged from 0.40 to 2.95%.

Estimates of natural hazard losses and other effects also were developed for the year 2000. These estimates were based on projected increases in building and human populations in the several hazard exposure areas (counties, states, and flood control regions) and generally assumed no change in land-use zoning, building codes, and new construction practices. For the 30 year period, the loss analysis revealed that (1) annual expected losses will jump to $17,779 million in the year 2000, an increase of $9685 million; (2) hazard-related deaths will climb to 1790 per year; and (3) that other annual expected adverse effects will record a similar sharp increase (Table 5-8). It should be noted that the 1970-2000 increase in total expected losses, ($9685 million) exceeds the total estimated loss for 1970 ($8094) million).

By the year 2000, the annual per capita loss from natural hazard exposures is expected to nearly double. The hazards of principal importance to doubling the per capita losses are tornado, hurricane, landslide, storm surge, and earthquake (Tables 5-9 and 5-10). The combination of storm surge and hurricane wind per capita losses will total $22.92 in 2000 and will make hurricanes the leading source of national per capita loss, followed closely by tornadoes ($20.38 per capita). Because of the expected benefits from area flood control projects and floodplain zoning, per capita losses from riverine flooding in 2000 are expected to drop below 1970 levels and reduce the comparative loss-producing importance of this hazard.

Two factors are of particular importance to the 1970-2000 escalation of total hazard losses: (1) the predicted increase in the ratio of building content value to building value, and (2) the predicted large rates of migration of population and investment capital into hazard-prone areas.

As shown in Table 5-11, the 13 states that, in 1970, ranked among the upper quarter of all states in terms of building damage rates will experience 31.54% of all the building wealth additions projected for that 30 year period, and will account for 50.21% of the estimated increase in total dollar losses for all natural hazards within that same period. Although a lower than expected proportion of high damage rate states exhibit high rates of building wealth addition between 1970 and 2000, the observed differences in proportion of high and low growth rate states among the several classes of damage rate states lack statistical significance. Based on available data, therefore, it appears that the hazard-proneness of states does not influence rates of migration to, and capital investment within, those states. Indeed, it is the large 30 year projected increase in building wealth additions in high damage rate states, and counties, principally along the Gulf Coast and South Atlantic, that accounts for much of the projected increase in annual expected natural hazard losses between 1970 and 2000. The proportion of the total national value of buildings and building contents in low damage rate states will decline between 1970 and 2000 while the proportion in high damage rate states will increase.

Table 5-4. Distribution of Building Values and Losses by Type (1970 Exposures)

	Building types				
	Housing	Farming and mining	Construction	Manufacturing	Transportation and Utilities
Total estimated value exposed in 10^6 1970$	1,106,073.0	21,026.0	27,058.1	109,897.4	235,109.4
Percent of total	53.7	1.0	1.3	5.3	11.4
Severe wind					
Exposure total = 100	100.	100.	100.	100.	100.
Est. loss 10^6 1970$	7.55	.07	.07	.26	.64
Percent of total	66.6	0.6	0.6	2.3	5.17
Hurricane wind					
Exposure total = 100	32.6	19.2	33.3	28.9	33.5
Est. loss 10^6 1970$	451.0	3.4	5.3	16.0	45.9
Percent of total	65.9	0.5	0.8	2.3	6.7
Storm surge					
Exposure total = 100	20.8	13.7	21.1	17.2	24.0
Est. loss 10^6 1970$	333.	3.1	2.9	7.2	28.9
Percent of total	75.4	0.7	0.7	1.6	6.5
Tornado					
Exposure total = 100	87.2	92.2	88.5	91.2	84.5
Est. loss 10^6 1970$	502.	7.2	8.8	36.8	75.0
Percent of total	57.1	0.8	1.0	4.2	8.5
Tsunami					
Exposure total = 100	10.3	3.9	9.7	8.5	11.5
Est. loss 10^6 1970$	4.52	.004	.02	.04	3.3
Percent of total	51.8	0.0	.2	.5	
Earthquake					
Exposure total = 100	71.9	72.8	71.3	68.5	72.7
Est. loss 10^6 1970$	354.2	5.7	7.2	24.8	77.9
Percent of total	54.0	0.9	1.1	3.8	11.9

Building types					
Retail and wholesale trade	Finance	Services	Federal government	State and local	Total value
85,075.9	104,911.4	93,039.0	128,561.01	153,753.8	2,064,507.5
4.1	5.1	4.5	6.2	7.4	100.0
100.	100.	100.	100.	100.	100.
.60	.69	.64	.38	.42	11.32
5.3	6.1	5.7	3.4	3.7	100.0
32.6	39.5	36.5	36.2	31.2	33.1
33.6	45.7	37.9	20.9	25.6	685.3
4.9	6.7	5.5	3.0	3.7	100.0
21.5	28.5	24.4	18.0	19.0	21.3
12.9	15.4	13.2	11.6	13.6	441.8
2.9	3.5	3.0	2.6	3.1	100.0
87.1	81.1	85.0	87.9	84.9	86.7
53.9	59.5	53.8	37.6	44.9	879.5
6.1	6.8	6.1	4.3	5.1	100.0
10.6	11.6	11.6	11.1	11.5	10.5
.17	.18	.24	.14	.11	8.72
1.9	2.1	2.8	1.6	1.3	100.0
71.8	75.1	74.6	75.9	72.7	72.4
26.0	34.1	29.1	46.5	49.7	655.2
4.0	5.2	4.4	7.1	7.6	100.0

Table 5-5. Annual Expected Total and Per Capita Losses for Selected Populations at Risk of Exposure to Nine Natural Hazards (1970 Exposure)

		Per capita loss	
Population at risk	Total losses (millions of 1970$)	Total	As % of per capita income in state
1. Highest damage rate county[a]	5.7	219.89	5.82
2. Fifth-ranked county[a]	7.2	123.14	3.98
3. High damage-rate state (FL)	749.07	110.32	2.95
4. Fiftieth-ranked county[a]	3.2	62.97	1.68
5. Group of thirteen highest damage-rate states	3745.95	58.83	1.48[b]
6. Total national population	8081.42	39.76	1.00[b]
7. Group of twenty-five medium damage-rate states	3109.33	35.00	0.88[b]
8. Median damage-rate state (Michigan)	295.11	33.23	0.80
9. Group of thirteen low damage-rate states	1226.14	24.17	0.61[b]
10. Low damage-rate state (Washington, D.C.)	15.25	20.07	0.40
11. County ranking 500th in damage rate[a]	0.62	21.18	0.63

[a]Includes only the losses resulting from damage to buildings and does *not* include losses from riverine flooding. In contrast, data for states includes all losses from all hazards.
[b]Expressed as percent of national per capita income.

The populations at risk of experiencing the adverse effects of natural hazards and exposures may be viewed as consisting of three broad classes of individuals and groups:

1. Locational Risk Takers. These populations are located in direct contact or immediate proximity to specific natural hazards, and include such persons as those who construct structures within a 50 year floodplain, on steep hillsides, or on earthquake faults. These risk takers consist of two groups: (a) those who take no special measures to protect themselves, and (b) those who take one or more special steps to protect themselves, such as structurally strengthening their buildings or purchasing insurance. Each of these groups may also be classified as voluntary or involuntary risk takers. The size of locational risk-taking groups usually can be determined only by fine-grained analysis of a specific local land area. Accordingly, the size of locational risk-taking groups was not determined for all hazards in this analysis. Per capita and other estimates of the losses incurred by locational risk takers are presented for a few hazards.

2. Area Risk Takers. This class includes persons and properties located in geopolitically identifiable local areas within which segments of the land area available for development are known to exhibit particularly high hazard potentials. Thus, a person constructing a dwelling unit in the cities of Dallas or Houston, Texas, could be designated an area risk taker because a high proportion of these land areas exhibits a high potential for expansive soil-induced damage to foundations and structures. Similarly, a person establishing residence in the San Fernando Valley of Los Angeles County or in the City of San Francisco could be declared to be an area risk taker because of the known earthquake hazard.

3. Communities at Interest. This class includes the individuals, groups, and aggregations of building wealth comprising the "environmental surrounds" of locational and area risk takers, the populations of communities that bear primary power for controlling hazard exposures or for mitigating the risks induced by such exposures, and further includes the groups within which the costs of natural hazard exposures incurred by locational and area risk takers may be distributed. Communities at interest may be viewed as consisting of several groups, the largest being the population of the United States.

In most cases it was possible to identify populations at risk only at the scale of *area risk takers* and at the scale of *communities at interest*. Generally, references to county-level populations should be viewed as area populations at risk while references to state or national populations should be viewed as communities at interest. In this analysis, the finest-grained hazard exposure areas defined for selected hazards are listed in Table 5-12. Data for all hazards except riverine flooding also were aggregated by county and state.

Expressed as a percentage of the total U.S. population, the combined populations of natural hazard exposure areas ranged from 100% for severe winds to only 0.05% for 100 year tsunami inundation areas (Table 5-13). Moreover, comparatively small fractions of the total U.S. population account for the bulk of the annual expected natural hazard losses.

Illustratively 89% of the U.S. population is exposed to tornadoes at some risk level, while the major burden of the losses is borne by minor fractions of the exposed population. For example, in the contiguous United States, 14 states and the District of Columbia experience fewer than one tornado per year per 10,000 square miles of land area. In the remaining 34 states, strike occurrences per 10,000 square miles of land area per year are as follows:

Strike occurrences per year	Number of states
1–1.99	10
2–3.99	16
4–4.99	4
5+	4

Table 5-6. Per Capita Natural Hazard Expected Losses (Building, Contents, Income and Supplier Cost) for Nine Hazards, by State (1970 Exposure)

State	Building damage rate	Ratio to national average	Rank	Expected per capita loss	Rank	Ratio to national average	Per capita income	Per capita loss as % of per capita income	Rank
FL	.825	2.96	1	110.32	1	2.77	3741	2.95	1
LA	.721	2.59	2	86.07	2	2.16	3097	2.78	2
MS	.477	1.71	3	44.94	13	1.13	2630	1.71	3
KS	.423	1.52	4	57.74	3	1.45	3857	1.50	4
WA	.404	1.45	5	52.43	5	1.32	4053	1.29	9
TX	.398	1.43	6	51.20	7	1.29	3600	1.42	7
MO	.396	1.42	7	55.08	4	1.39	3775	1.46	6
NE	.391	1.40	8	49.82	8	1.25	3786	1.32	8
CA	.375	1.35	9	52.08	6	1.31	4496	1.16	11
OK	.338	1.21	10	49.62	9	1.24	3381	1.47	5
CO	.319	1.15	11	41.87	14	1.05	3851	1.09	14
SD	.311	1.12	12	34.15	21	.86	3101	1.10	13
SC	.311	1.12	13	33.92	22	.85	2992	1.13	12
NC	.310	1.11	14	35.50	18	.89	3256	1.09	15
AL	.309	1.11	15	35.31	19	.89	2947	1.20	10
AR	.299	1.07	16	29.33	28	.74	2886	1.02	20
IA	.291	1.04	17	38.49	16	.97	3755	1.03	19
UT	.286	1.02	18	34.18	20	.86	3218	1.06	17
RI	.284	1.02	19	40.12	15	1.01	3960	1.01	21
CT	.284	1.02	20	48.98	10	1.23	4923	.99	23
ND	.277	.99	21	29.87	27	.75	3191	.94	24
IL	.274	.98	22	48.79	11	1.23	4504	1.08	16
MA	.272	.98	23	45.96	12	1.16	4347	1.06	18
IN	.252	.90	24	37.92	17	.95	3768	1.01	22

ME	.233	.84	25	27.70	35	.70	3309	.84	28
MI	.230	.83	26	33.23	23	.84	4175	.80	30
GA	.223	.80	27	28.83	32	.73	3357	.86	26
VA	.220	.79	28	32.63	25	.82	3720	.88	25
NH	.220	.79	29	33.05	24	.83	3850	.86	27
MN	.215	.77	30	31.55	26	.79	3839	.82	29
WI	.213	.76	31	28.97	31	.73	3809	.76	31
TN	.210	.75	32	22.86	38	.57	3124	.73	32
MT	.202	.73	33	22.20	41	.56	3504	.63	36
WY	.200	.72	34	22.55	39	.57	3816	.59	42
OH	.194	.70	35	28.43	33	.71	4011	.71	33
KY	.192	.69	36	21.15	44	.53	3118	.68	35
DE	.184	.66	37	27.93	34	.70	4527	.62	39
NM	.182	.65	38	21.43	43	.54	3092	.69	34
ID	.179	.64	39	20.13	45	.51	3294	.61	41
WV	.166	.60	40	17.60	49	.44	3070	.57	43
NY	.166	.59	41	29.17	30	.73	4712	.62	40
AK	.164	.59	42	29.19	29	.73	4632	.63	37
NJ	.159	.57	43	27.05	37	.68	4705	.57	44
MD	.155	.56	44	27.23	36	.68	4318	.63	38
NV	.153	.55	45	22.27	40	.56	4563	.49	47
OR	.151	.54	46	18.56	48	.47	3717	.50	46
VT	.137	.49	47	16.86	51	.42	3328	.51	45
PA	.136	.49	48	18.73	47	.47	3970	.47	48
AZ	.130	.47	49	17.30	50	.44	3665	.47	49
HI	.114	.41	50	21.44	42	.54	4623	.46	50
DC	.087	.31	51	20.07	46	.50	5036	.40	51
U.S.	.279			39.76			3966	1.00	

Table 5-7. Proportion of Building Value Expected To Be Lost, Annually, in Several "Most Damage Prone" U.S. Counties as a Result of Exposure to Selected Natural Hazards (1970 Exposure)

	Expected annual building loss for county of indicated rank[a]				
Hazard types	Rank 1 (%)	Rank 10 (%)	Rank 50 (%)	Rank 100 (%)	Rank 500 (%)
1. Earthquake, tsunami, storm surge, tornado, and hurricane wind	3.43	1.40	0.743	0.405	0.14
2. Expansive soil, landslide, and severe wind	0.462	0.262	0.260	0.234	0.113
3. All hazards, except riverine flooding	3.51	1.444	0.784	0.483	0.287

[a]Represents 500 most damage-prone counties of U.S., ranked in descending order by damage rate (e.g., proportion of building value expected to be lost, annually for indicated hazard category).

Thus, about 63% of all U.S. tornado strikes per year are experienced by less than 20% of the land area, which represents a slightly lower fraction of the population (Hart, 1976).

The extent to which various proportions of the U.S. population are exposed to differing levels of risk arising from exposure to natural hazards may be suggested through examination of the *building damage rates* for various areas and 1970 exposures. Accordingly, building damage rates are presented for each hazard and state in Table 5-14.

In terms of U.S. land area, building damage losses for 1970 exposures were estimated to be equal to .279% of the nation's building wealth. At the level of whole states, only 2 states (Vermont and Arizona) and the District of Columbia were found to experience expected annual building damage rates below the national rate for *all* nine hazards. Twenty states recorded annual expected building damage rates for all hazards equal to or greater than the aggregated national rate for all hazards. All Gulf Coast states, 4 of the Atlantic seaboard states, and 2 of the Pacific coast states were in this group, but only 9 of the interior, land-locked states (Colorado, Utah, South Dakota, Kansas, Missouri, Nebraska, Arkansas, Oklahoma, and Iowa). These 20 states claimed a total population of 81.98 million persons and a combined value of buildings in excess of $785 billion.

These same 20 states accounted for 54.6% of the nation's annual expected hazard-induced dollar losses to buildings for exposures in 1970. Also, 15 states experienced an all-hazard annual expected building damage rate of 0.3% or higher. Containing a combined population of over 72 million persons, these "high" damage-rate states included Florida (0.825%), Louisiana (0.721%),

Table 5-8. Expected Annual Losses from Natural Hazard Exposures in the United States, by Type of Hazard and Type of Loss in the Year 2000.

	Expected annual losses								
	Dollar losses resulting from indicated factor (millions of 1970$)					Other losses			
	(1)	(2)	(3)	(4)	(5)	(6)	(7)	(8)	(9)
Hazard	Building damage	Contents damage	Income loss	Supplier loss	Total (1–4)	Number of deaths	Housing units lost	Person years of homelessness	Person years of unemploy.
1. Earthquake	1177.0	372.76	3.906	0.048	1553.7	400	22,868	648	634.9
2. Expansive soil	997.1	–	–	–	997.1	–	–	–	–
3. Hurricane	1742.0	1504.98	276.191	3.095	3526.3	153	52,237	48,271	58,223.7
4. Landslide	871.2	–	–	–	871.2	–	–	–	–
5. Riverine flooding	1594.0	1572.54	8.68	.105	3175.33	159	–	–	–
6. Severe wind	24.8	23.8	4.696	0.051	53.4	11	748	1,014	850.9
7. Storm surge	1176.0	1160.43	6.407	0.007	2342.9	103	43,757	10,330	1,018.3
8. Tornado	2058.0	2401.32	750.780	9.042	5219.1	920	52,119	107,650	146,568.5
9. Tsunami	19.8	19.10	1.479	0.027	40.4	44	335	389	195.9
Totals	9659.9	7054.95	1052.139	12.445	17,779.43	1790	172,084	168,302	207,492.2

Table 5-9. Expected Annual National Per Capita Dollar Losses from Natural Hazard Exposures in the United States, by Type of Hazard (1970 and 2000 Exposures)

Hazard	Expected annual per capita loss for indicated year (1970$)	
	1970	2000
1. Rivering flooding	13.57	12.40
2. Tornado	8.12	20.38
3. Hurricane	5.20	13.77
4. Expansive soil	3.93	3.89
5. Earthquake	3.83	6.07
6. Storm surge	3.16	9.15
7. Landslide	1.82	3.40
8. Tsunami	.07	.16
9. Severe wind	.06	.19
All hazards	39.76	69.41

Mississippi (0.477%), Kansas (0.423%), Washington (0.404%), Texas (0.398%), Missouri (0.396%), Nebraska (0.391%), California (0.375%), Oklahoma (0.338%), Colorado (0.319%), South Dakota (0.311%), South Carolina (0.311%), North Carolina (0.310%), and Alabama (0.309%) (Table 5-15).

Although state-level aggregations of people and property certainly represent communities whose interests are affected by exposures to natural hazards, in most cases such aggregations do not provide a specific geographic picture of the

Table 5-10. Expected Annual U.S. Dollar Losses from Natural Hazard Exposures (1970 and 2000)

Hazard	Expected annual dollar loss[a] (millions)			% of increase
	1970	2000	Increase	
1. Water hazards:	3,414.5	5,558.6	2,144.1	63
Riverine flooding	2,758.3	3,175.3	417.0	15
Storm surge	641.2	2,342.9	1,701.7	265
Tsunami	15.0	40.4	25.4	169
2. Wind hazards:	2,730.0	8,798.8	6,068.8	222
Tornado	1,656.0	5,219.1	3,563.1	215
Hurricane	1,056.0	3,526.3	2,470.3	234
Other severe wind	18.0	53.4	35.4	197
3. Expansive soil	798.1	997.1	199.0	25
4. Earthquake	781.1	1,553.7	772.6	99
5. Landslide	370.3	871.2	500.9	135
Totals	8,094.0	17,779.4	9,685.4	120

[a]Losses calculated in 1970 dollars and are confined to dollar losses caused by damage to buildings and building contents as well as income and supplier losses.

Table 5-11. Relationship between State Damage Rates, Projected Increases in Building Wealth, and Estimated Increases in Natural Hazard Losses (1970–2000)

State category	Number and proportion in category		Building wealth increase, 1970–2000		Increase in total dollar losses, all hazards, 1970–2000	
	N	%	N	%	N	%
High damage rate states	13	25.5	902.3	31.54	5140.03	50.71
Medium damage rate states	25	49.0	1195.4	41.79	3295.98	35.07
Low damage rate states	13	25.5	763.0	26.67	1249.42	14.22
Totals[a]	51	100.0	2860.7	100.00	9685.43	100.00

[a]Includes Washington, D.C.

number of people and buildings directly exposed to a given hazard or combination of hazards. Accordingly, U.S. counties were tallied and rank-ordered by their 1970 building damage rates. For that year, 500 U.S. counties experienced building damage rates of 0.287% or higher as a result of their exposure to all hazards, except riverine flooding (Table 5-16). Containing 20.4% of the nation's population and 20.8% of its building wealth these counties accounted for 36% of the nation's annual expected dollar loss from building damage for 1970 exposures. In these counties the annual expected proportion of building value expected to be lost as a result of hazard-induced damages (excluding riverine flooding) ranged from a low of 0.287% to a high of 3.51% and the ratio of county building damage rates to the national rate ranged from 1.47 to 18.00. Table 5-17

Table 5-12. Hazard Exposure Areas Utilized in Analysis

Analysis area	Hazards
County-level agreggations of people and property	Tornado, hurricane winds, severe winds, earthquake
Proportions of counties	Expansive soils, landslide
Inundation areas (100-year and 500-year)	Tsunami
Storm-surge zones, by county segments (0–20′ mean sea level, adjusted)	Storm surge
Riverine floodplain zones, by multi state regions: 50–100 yr. (E); 25–50 yr. (D); 10–25 yr. (C); 5–10 yr. (B); 2–5 yr. (A).	Riverine floods

Table 5-13. Exposure of Building Values and Persons to Natural Hazards in the United States, by Type of Hazard for 1970

		Human population exposed		Value of building exposed to hazard	
Hazard	Type of area	N	% of U.S. total	N ($ × 10^6)	Percent of U.S. total
1. Severe wind	Counties	203,260,531	100.0	2,064,507.5	100.0
2. Tornado	Counties	181,198,749	89.1	1,788,989	89.1
3. Expansive soils:	Fractions of counties				
High zones		17,730,021[a]	8.7[b]	Not available	
Medium zones		23,710,910	11.7		
Low zones		98,320,710	48.4		
4. Earthquake any intensity	Counties	143,169,495	70.4	1,494,293	72.4
5. Earthquake	Fractions of states				
Zone 1		120,600,000	59.3	1,205,138	58.4
Zone 2		33,340,000	16.4	340,210	16.5
Zone 3		38,020,000	18.7	427,860	20.7
6. Hurricane wind	Counties	62,741,264	30.9	682,476	30.9
7. Landslide:	Counties				
High		44,068,071	21.7	Not available	
Moderate		39,426,341	19.4		

8. Storm surge	Counties	38,387,247	18.9	438,733	21.3
9. Storm surge	County segments 0–20′ mean sea level (adjusted)	9,940,101	4.89	104,670.25	5.1
10. Riverine flood	5,539 flood-prone cities:				
	All zones	24,112,000	11.86	731,977.34	35.5
	100 yr. floodplain	12,056,000	5.93	364,650.79	17.7
	50 yr. floodplain	8,776,416	4.32	265,471.01	12.9
	25 yr. floodplain	5,830,968	2.87	175,478.42	8.5
11. Tsunami	100 yr. inundation area	109,400	0.054		
	500 yr. inundation area	237,500	0.117		
	Counties containing above areas	18,200,851	8.95	217,327.63	10.5

[a] Number of people in single family residences.
[b] Percent of total population.

Table 5-14. Expected Hazard-Induced Building Damage Rates, by State, for Nine Natural Hazards in 1970

	All hazards			Flood		
State	Damage rate	Ratio to nat'l average	Rank	Damage rate	Ratio to nat'l average	Rank
FL	.825	2.96	1	.093	1.011	27
LA	.721	2.59	2	.125	1.355	12
MS	.477	1.71	3	.158	1.718	3
KS	.423	1.52	4	.134	1.460	6
WA	.404	1.45	5	.086	.930	32
TX	.398	1.43	6	.112	1.220	16
MO	.396	1.42	7	.133	1.447	8
NE	.391	1.40	8	.134	1.459	7
CA	.375	1.35	9	.078	.847	39
OK	.338	1.21	10	.115	1.251	15
CO	.319	1.15	11	.081	.880	36
SD	.311	1.12	12	.162	1.763	2
SC	.311	1.12	12	.111	1.202	17
NC	.310	1.11	14	.105	1.135	19
AL	.309	1.11	15	.108	1.173	18
AR	.299	1.07	16	.147	1.596	4
IA	.291	1.04	17	.141	1.532	5
UT	.286	1.03	18	.083	.901	35
RI	.284	1.02	19	.073	.794	40
CT	.284	1.02	19	.062	.676	47
MD	.277	.99	21	.167	1.810	1
IL	.274	.98	22	.104	1.133	20
MA	.272	.98	23	.064	.699	46
IN	.252	.90	24	.127	1.375	11
ME	.233	.84	25	.086	.929	33
MI	.230	.83	26	.116	1.263	14
GA	˙.223	.80	27	.093	1.007	28
VA	.220	.79	28	.078	.850	38
NH	.220	.79	28	.072	.778	41
MN	.215	.77	30	.129	1.405	9
WI	.213	.76	31	.128	1.387	10
TN	.210	.75	32	.100	1.089	21
MT	.202	.73	33	.091	.984	29
WY	.200	.72	34	.088	.950	30
OH	.194	.70	35	.119	1.300	13
KY	.192	.69	36	.099	1.077	22
DE	.184	.66	37	.079	.857	37
NM	.182	.65	38	.093	1.014	26
ID	.179	.64	39	.094	1.017	25
WV	.166	.60	40	.097	1.048	23
NY	.166	.59	40	.055	.598	50
AK	.164	.59	42	.062	.668	48
NJ	.159	.57	43	.059	.645	49
MD	.155	.56	44	.065	.710	45
NV	.153	.55	45	.071	.768	42
OR	.151	.54	46	.094	1.021	24
VT	.137	.49	47	.084	.915	34
PA	.136	.49	48	.070	.765	43
AZ	.130	.47	49	.087	.945	31
HI	.114	.41	50	.069	.749	44
DC	.087	.31	51	.050	.543	51
U.S.	.279			.092		

Table 5-14. (continued)

	Storm surge			Tsunami		
State	Damage rate	Ratio to nat'l average	Rank	Damage rate	Ratio to nat'l average	Rank
FL	.319	14.930	1	0		
LA	.279	13.053	2	0		
MS	.090	4.231	3	0		
KS	0					
WA	0			.002	4.622	4
TX	.015	.709	12	0		
MO	0			0		
NE	0					
CA	0			.002	4.692	3
OK	0			0		
CO	0			0		
SD	0			0		
SC	.036	1.671	5	0		
NC	.019	.901	10	0		
AL	.016	.749	11	0		
AR	0			0		
IA	0			0		
UT	0			0		
RI	.076	3.571	4	0		
CT	.028	1.310	6	0		
MD	0			0		
IL	0			0		
MA	.020	.922	9	0		
IN	0			0		
ME	.006	.282	17	0		
MI	0			0		
GA	.011	.515	14	0		
VA	.028	1.295	7	0		
NH	0			0		
MN	0			0		
WI	0			0		
TN	0			0		
MT	0			0		
WY	0			0		
OH	0			0		
KY	0			0		
DE	.009	.412	15	0		
NM	0			0		
ID	0			0		
WV	0			0		
NY	.022	1.019	8	0		
AK	0			.010	23.460	2
NJ	.014	.643	13	0		
MD	.008	.381	16	0		
NV	0			0		
OR	0			.0005	1.173	5
VT	0			0		
PA	0			0		
AZ	0			0		
HI	0			.041	96.187	1
DC	0			0		
U.S.	.0214			.0004		

Table 5-14. (continued)

State	Tornado			Hurricane		
	Damage rate	Ratio to nat'l average	Rank	Damage rate	Ratio to nat'l average	Rank
FL	.068	1.595	10	.312	9.398	1
LA	.044	1.030	20	.160	4.831	2
MS	.058	1.349	11	.060	1.795	9
KS	.150	3.523	2	0		
WA	.001	.026	49	0		
TX	.089	2.095	6	.030	.889	15
MO	.122	2.871	3	0		
NE	.086	2.017	7	0		
CA	.006	.141	43	0		
OK	.188	4.417	1	0		
CO	.041	.962	21	0		
SD	.055	1.283	13	0		
SC	.040	.945	22	.054	1.614	12
NC	.032	.745	26	.082	2.477	6
AL	.069	1.610	9	.045	1.355	13
AR	.054	1.266	14	0		
IA	.072	1.691	8	0		
UT	.005	.111	45	0		
RI	.005	.126	44	.112	3.368	4
CT	.016	.375	35	.125	3.765	3
MD	.046	1.074	16	0		
IL	.120	2.815	4	0		
MA	.056	1.306	12	.100	3.022	5
IN	.098	2.292	5	0		
ME	.012	.281	37	.076	2.286	8
MI	.044	1.037	18	0		
GA	.045	1.055	17	.020	.602	18
VA	.017	.408	33	.059	1.777	10
NH	.038	.887	24	.070	2.361	7
MN	.054	1.262	15	0		
WI	.044	1.036	19	0		
TN	.032	.762	25	0		
MT	.006	.147	42	0		
WY	.016	.366	36	0		
OH	.039	.924	23	0		
KY	.031	.733	28	0		
DE	.025	.577	29	.032	.952	14
NM	.024	.567	30	0		
ID	.006	.140	41	0		
WV	.007	.164	40	0		
NY	.005	.109	46	.056	1.698	11
AK	0			0		
NJ	.031	.734	27	.028	.831	16
MD	.018	.410	32	.026	.783	17
NV	.003	.081	47	0		
OR	.002	.047	48	0		
VT	.021	.494	31	.005	.158	21
PA	.017	.406	34	.014	.407	20
AZ	.011	.258	38	0		
HI	0			0		
DC	.009	.211	39	.015	.452	19
U.S.	.043			.033		

Table 5-14. (continued)

	Severe wind			Earthquake		
State	Damage rate	Ratio to nat'l average	Rank	Damage rate	Ratio to nat'l average	Rank
FL	.0003	.548	32	.002	.052	28
LA	.0003	.603	31	.006	.199	16
MS	0			.003	.086	21
KS	.001	1.699	20	.001	.029	34
WA	.0006	1.023	24	.272	8.577	1
TX	.001	1.762	17	.001	.0243	38
MO	.0004	.804	27	.034	1.069	8
NE	.001	2.558	10	.001	.044	29
CA	.0004	.639	29	.193	6.081	2
OK	.002	3.161	6	.002	.068	25
CO	.002	3.654	4	.089	2.804	5
SD	.002	3.453	5	.002	.060	27
SC	0			.010	.306	13
NC	.0005	.913	25	.0003	.008	41
AL	0			0		
AR	0			.024	.756	9
IA	.001	2.119	12	0		
UT	.001	1.717	19	.115	3.626	3
RI	.002	3.928	2	.001	.034	32
CT	.001	2.088	13	.002	.072	24
MD	.002	3.800	3	0		
IL	.0006	1.151	22	.0008	.025	37
MA	.001	2.302	11	.003	.084	23
IN	.0008	1.498	21	.0002	.006	42
ME	0			0		
MI	.0001	.201	35	.001	.031	33
GA	0			.001	.038	31
VA	.0006	1.114	23	.0008	.026	35
NH	0			.003	.085	22
MN	.001	1.937	15	0		
WI	.0005	.895	26	0		
TN	0			.049	1.544	6
MT	.002	2.850	9	.022	.689	11
WY	.003	5.709	1	0		
OH	.0003	.512	33	.001	.039	30
KY	0			.005	.160	17
DE	0			.002	.057	27
NM	.001	2.010	14	.013	.416	12
ID	.002	2.887	8	.024	.750	10
WV	0			.0007	.022	39
NY	.0004	.797	28	.009	.277	14
AK	0			.092	2.899	4
NJ	.001	1.746	18	.004	.128	18
MD	.0002	.365	34	.0002	.006	43
NV	.002	3.142	7	.047	1.467	7
OR	0			.009	.268	15
VT	0			0		
PA	.0003	.621	30	.0003	.011	40
AZ	0			.004	.126	19
HI	.001	1.827	16	.003	.095	20
DC	0			.001	.25	36
U.S.	.0005			.032		

Table 5-14. (continued)

	Landslide			Expansive soil		
State	Damage rate	Ratio to nat'l average	Rank	Damage rate	Ratio to nat'l average	Rank
FL	.0064	.357	49	.024	.621	27
LA	.0106	1.093	25	.087	2.250	5
MS	.0189	1.054	26	.090	2.328	4
KS	.0159	.806	34	.120	3.104	3
WA	.0292	1.628	4	.013	.336	39
TX	.0110	.613	42	.139	3.596	2
MO	.0238	1.327	13	.083	2.147	6
NE	.0185	1.031	27	.149	3.854	1
CA	.0162	.903	32	.080	2.069	7
OK	.0094	.524	44	.021	.543	30
CO	.0276	1.539	6	.078	2.018	8
SD	.0247	1.377	10	.066	1.707	10
SC	.0154	.914	31	.044	1.138	18
NC	.0244	1.360	11	.047	1.216	16
AL	.0172	.959	30	.054	1.397	13
AR	.0219	1.221	17	.052	1.345	14
IA	.0202	1.126	23	.057	1.474	12
UT	.0295	1.645	2	.052	1.345	14
RI	.0065	.362	46	.008	.207	47
CT	.0294	1.639	3	.020	.517	31
MD	.0235	1.310	14	.039	1.009	21
IL	.0262	1.461	7	.023	.595	26
MA	.0208	1.160	21	.008	.207	46
IN	.0139	.775	37	.013	.336	39
ME	.0255	1.422	8	.028	.724	24
MI	.0214	1.193	19	.047	1.216	16
GA	.0123	.686	41	.041	1.061	20
VA	.0183	1.020	29	.018	.466	33
NH	.0199	1.109	24	.009	.245	44
MN	.0146	.814	35	.015	.414	34
WI	.0253	1.411	9	.015	.380	35
TN	.0065	.362	46	.022	.569	29
MT	.0184	1.026	28	.063	1.630	11
WY	.0222	1.238	16	.072	1.862	9
OH	.0206	1.148	22	.014	.362	36
KY	.0233	1.299	15	.034	.879	23
DE	.0279	1.555	5	.011	.285	43
NM	.0090	.502	45	.042	1.086	19
ID	.0160	.892	33	.038	.983	22
WV	.0344	1.918	1	.027	.698	25
NY	.0128	.714	39	.006	.155	49
AK	0			0		
NJ	.0128	.714	39	.009	.230	45
MD	.0243	1.355	12	.014	.362	36
NV	.0110	.613	41	.019	.491	32
OR	.0211	1.176	20	.025	.647	26
VT	.0142	.792	36	.012	.310	42
PA	.0215	1.199	18	.013	.336	39
AZ	.0136	.758	38	.014	.362	36
HI	0			0		
DC	.0065	.362	46	.006	.155	48
U.S.	.018			.039		

presents the results of the analysis for a select subset of ten counties from the 500 expected to experience the highest damage rates for eight of the hazards. The geographic distribution of high damage rate counties and states is shown in Figures 5-2 to 5-8. Particularly striking in Figure 5-3 is the cluster of counties vulnerable to storm surge and hurricane winds along the shores of the Gulf Coast and Atlantic Ocean. Considering only the tornado hazards the unbroken belt of tornado-prone counties along the interior and middle sections of the United States as shown in Figure 5-8 is of particular significance. Counties expected to experience high level damage rates from earthquakes are shown in Figure 5-8.

Although the aggregate county data presented in Tables 5-16 and 5-17 suggests the steep gradient of natural hazard risk to which various population subsets are exposed, a clearer picture is revealed by examining specific hazard zones or areas, rather than the whole counties in which such areas are located. For example, as indicated in Table 5-18, building damage rates from exposure to storm surge in coastal county areas within the surge-prone zone (0 to 20 feet adjusted mean sea level) range from 0.025% per year in coastal counties of New Hampshire to 1.429% in the coastal counties of Mississippi, and annual per capita losses per *exposed* person range from \$2.98 in New Hampshire to \$121.37 in Georgia. Of course, in any given storm surge hazard area (where storm frequencies are held constant), these damage rates and annual per capita losses will vary markedly throughout the exposure zone, with persons and property at lower elevations experiencing much higher rates and per capita losses than those at higher elevations.

Similarly, riverine flood losses for 1970 level exposures also exhibited a steep gradient in annual expected building damage rates and in annual per capita losses (for all *types* of buildings) by type of exposure area. Table 5-19 identifies the exposed population, damage rate, and annual per capita expected loss by flood zone for both protected and unprotected communities. Comparable 1970 data for 500 year tsunami inundation areas are shown in Table 5-20.

The gradient in per capita losses arising from natural hazard exposures in 1970, together with estimates of the size of the population at risk of experiencing these losses, are presented in Table 5-21. Not only is there a large gradient in annual expected per capita losses, but there also is much variance in expected frequency of large losses.

The expected frequency of building damage occurrences of various magnitudes ranges over a broad continuum as a function of the type of hazard. For example, expansive soils may be viewed as producing frequent or near continuous damage of light magnitude, while major earthquakes, tsunamis, hurricanes, and floods may be classed as low probability but high consequence occurrences. Generally, risk producing events may be ordered into four classes corresponding to the simple typology presented below:

High Probability Events

Low Consequence Events ←→ High Consequence Events

Low Probability Events

Table 5-15. Damage to Buildings (in Millions of 1970 Dollars) by State (1970 Exposure)

1970 baseline % damage	State	Total structure value ($ bill.)	Population (millions)	Earth hazards			Water hazards			Wind hazards			All hazards
				EQ	ES[a]	LS	FL[b]	SS	TS	TO	HU	SW	
.309	AL	26.7	3.44	.0	14.45	4.6	28.9	4.3	0	18.4	11.9	0	82.6
.164	AK	3.9	0.27	3.6	0	0	2.4	0	0.4	.0	0	0	6.4
.130	AZ	16.9	1.77	0.7	2.3	2.29	14.7	0	0	1.9	0	0	21.9
.299	AR	13.6	1.92	3.3	7.01	2.98	20.0	0	0	7.3	0	0	40.6
.375	CA	228.0	19.96	439.6	182.59	36.83	178.5	0	3.5	13.8	0	0.8	855.6
.319	CO	22.7	2.21	20.2	17.77	6.27	18.4	0	0	9.4	0	0.4	72.3
.284	CT	35.0	3.03	0.8	7.08	10.30	21.8	9.7	0	5.7	43.6	0.4	99.4
.184	DE	5.7	0.55	0.1	.64	1.59	4.5	0.5	0	1.4	1.8	0	10.5
.087	DC	12.6	0.76	0.1	.79	.82	6.3	0	0	1.1	1.9	0	11.0
.825	FL	61.1	6.79	1.0	14.60	3.91	56.9	195.0	0	41.8	190.4	0.2	503.8
.223	GA	41.5	4.59	0.5	17.17	5.12	38.5	4.4	0	18.6	8.4	0	92.7
.114	HI	10.0	0.77	0.3	0	0	6.9	0	4.1	.0	0	0.1	11.4
.179	ID	6.3	0.71	1.5	2.42	1.01	5.9	0	0	0.4	0	0.1	11.3
.274	IL	126.8	11.12	1.0	28.69	33.16	132.2	0	0	152.2	0	0.8	348.0
.252	IN	48.8	5.20	0.1	6.31	6.80	61.8	0	0	47.7	0	0.4	123.1
.291	IA	25.8	2.83	.0	14.69	5.21	36.4	0	0	18.6	0	0.3	75.2
.423	KS	21.5	2.25	0.2	25.88	3.41	28.9	0	0	32.3	0	0.2	90.9
.192	KY	25.6	3.22	1.3	8.58	5.96	25.4	0	0	8.0	0	0	49.2
.721	LA	30.3	3.64	1.9	26.29	5.95	37.8	84.5	0	13.3	48.6	0.1	218.4
.233	ME	8.3	0.99	.0	2.30	2.12	7.1	0.5	0	1.0	6.3	0	19.3
.155	MD	49.2	3.92	0.1	6.66	11.97	32.2	4.0	0	8.6	12.8	0.1	76.4
.272	MA	63.4	5.69	1.7	4.83	13.21	40.8	12.5	0	35.3	63.6	0.8	172.7
.230	MI	90.8	8.88	0.9	42.31	19.47	105.6	0	0	40.1	0	0.1	208.5
.215	MN	37.7	3.80	.0	5.87	5.52	48.8	0	0	20.3	0	0.4	80.9

.477	MS	14.6	2.22	0.4	13.19	2.76	23.1	13.2	0	8.4	8.	0	69.7
.396	MO	45.1	4.68	15.3	37.22	10.75	60.1	0	0	55.2	0	0.2	178.8
.202	MT	6.4	0.69	1.4	4.02	1.18	5.8	0	0	0.4	0	0.1	12.9
.391	NE	14.3	1.49	0.2	21.33	2.64	19.2	0	0	12.3	0	0.2	55.9
.153	NV	5.8	0.49	2.7	1.12	.64	4.1	0	0	0.2	0	0.1	8.9
.220	NH	7.4	0.74	0.2	.70	1.47	5.3	0	0	2.8	5.8	0	16.3
.159	NJ	83.7	7.17	3.4	7.47	10.70	49.7	11.5	0	26.2	23.1	0.8	132.9
.182	NM	9.1	1.02	1.2	3.81	.82	8.5	0	0	2.2	0	0.1	16.6
.166	NY	229.6	18.24	20.2	14.76	29.36	126.5	50.0	0	10.7	129.4	1.0	381.9
.310	NC	40.0	5.08	0.1	18.92	9.76	41.8	7.7	0	12.7	32.	0.2	124.1
.277	ND	4.8	0.62	.0	1.88	1.13	8.0	0	0	2.2	0	0.1	13.3
.194	OH	106.3	10.66	1.0	14.57	21.87	126.8	0	0	41.9	0	0.3	206.4
.338	OK	23.1	2.56	0.5	4.86	2.16	26.6	0	0	43.5	0	0.4	78.0
.151	OR	19.9	2.09	1.7	4.97	4.19	18.7	0	0.1	0.4	0	0	30.1
.136	PA	116.1	11.80	0.4	14.84	24.91	81.8	0	0	20.1	15.7	0.4	158.2
.284	RI	9.3	0.95	0.1	.74	.60	6.8	7.1	0	0.5	10.4	0.2	26.4
.311	SC	19.6	2.59	1.9	8.66	3.21	21.7	7.0	0	7.9	10.5	0	60.9
.311	SD	5.3	0.67	0.1	3.49	1.31	8.6	0	0	2.9	0	0.1	16.5
.210	TN	30.8	3.92	15.1	6.67	2.00	30.9	0	0	10.0	0	0	64.7
.398	TX	103.7	11.20	0.8	143.69	11.36	116.5	15.7	0	92.6	30.6	1.0	412.3
.286	UT	10.6	1.06	12.2	5.53	3.13	8.8	0	0	0.5	0	0.1	30.3
.137	VT	3.8	0.44	.0	.44	.54	3.2	0	0	.8	0.2	0	5.2
.220	VA	48.8	4.65	0.4	8.87	8.94	38.2	13.5	0	8.5	28.8	0.3	107.5
.404	WA	35.6	3.41	96.9	4.71	10.41	30.5	0	0.7	.4	0	0.2	143.8
.166	WV	14.2	1.74	0.1	3.87	4.88	13.7	0	0	1.0	0	0	23.6
.213	WI	41.2	4.42	.0	6.28	10.43	52.6	0	0	18.2	0	0.2	87.7
.200	WY	3.2	0.33	.0	2.27	.71	2.8	0	0	0.5	0	0.1	6.4
.279	U.S.	2064.5	203.24	655.2	798.1	370.3	1901.0	441.1	8.8	880.2	685.4	11.3	5751.4

[a]Only residential expansive soil losses.
[b]Flood is estimated utilizing a regional model.

Table 5-16. Distribution of Most Damage-Prone Counties of the United States Classified by Expected Annual Losses and by Annual Proportion of Building Value Expected To Be Lost through Exposure to Eight Natural Hazards[a] (1970 Exposure)

Category, by annual percent of building value expected to be lost (%)[b]	Number of counties in category		Building wealth at risk (billions 1970$)		Expected annual building losses (millions 1970$)		Population at risk (millions)		Expected annual per capita losses (1970$)
	N	CUM	N	CUM	N	CUM	N	CUM	
3.00+	2	2	.308	.308	$ 10.72	$ 10.72	.049	.049	$218.37
2.50–2.99	0	2	–	.308	–	10.72	–	.049	–
2.00–2.49	1	3	.422	.730	9.92	20.64	.053	.102	188.54
1.75–1.99	3	6	1.247	1.977	23.42	44.06	.189	.290	124.24
1.50–1.749	2	8	.094	2.071	1.54	45.60	.012	.302	128.98
1.25–1.49	11	19	9.332	11.403	126.13	171.73	.958	1.260	131.65
1.00–1.249	9	28	22.557	33.960	267.93	439.66	2.133	3.393	125.62
0.75–0.999	29	57	26.040	60.000	222.00	661.66	2.508	5.901	88.53
0.50–0.749	39	96	46.029	106.029	270.75	932.41	4.816	10.717	56.22
0.40–0.499	64	160	84.346	190.375	371.37	1303.78	7.673	18.389	48.40
0.30–0.399	251	411	206.371	396.747	670.31	1974.09	19.418	37.807	34.52
<0.30	89	500	33.518	430.264	98.26	2072.35	3.623	41.430	27.12

[a] All study hazards, except riverine flooding.
[b] Range is 0.287 to 3.51%.

Table 5-17. Expected Total Annual and Annual Per Capita Building Losses from Exposure to Eight Natural Hazards[a] in a Selected Subset of the 500 Most Damage-Prone Counties in the United States (1970 Exposure)

County	Damage rate	Rank	Index[b]	Value at risk (millions, 1970$)	Expected annual loss (1970$)	Population (1000s)	Expected annual per capita loss (1970$)
1. Stoddard, MO	.0351	1	18.00	161.0	5,651,100	25.7	219.89
2. Iberia, LA	.0186	5	9.53	380.0	7,068,000	57.4	123.14
3. Collier, FL	.0144	10	7.38	306.0	4,406,400	38.0	115.96
4. Washington, NC	.0109	25	5.59	88.8	967,920	14.0	69.14
5. St. Lucie, FL	.00784	50	4.02	408.0	3,198,720	50.8	62.97
6. Manatee, FL	.00483	100	2.48	576.0	2,782,080	97.1	28.65
7. Refugio, TX	.00368	200	1.87	61.9	227,792	9.5	24.00
8. Murray, OK	.00332	300	1.70	79.0	262,280	10.7	24.51
9. Sangamon, IL	.00303	400	1.55	2060.0	6,241,800	161.0	38.77
10. Grady, OK	.00287	500	1.47	217.0	622,790	29.4	21.18

[a]Does not include riverine flooding.
[b]County damage rate divided by national average building damage rate for 1970 for all hazards except rivering flooding (.00195).

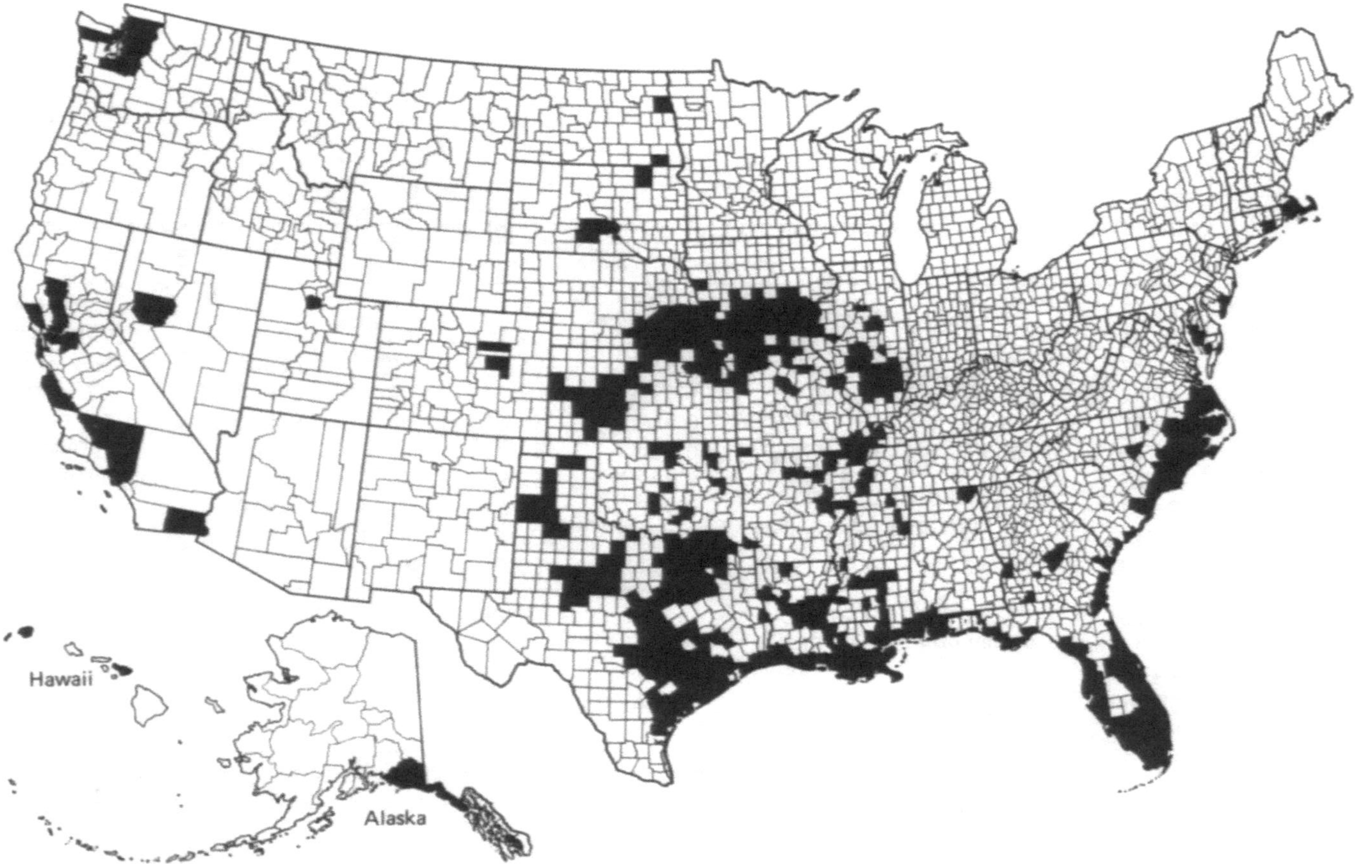

Figure 5-2. Five hundred high damage rate counties for eight natural hazards (excludes riverine floods) (1970 exposure).

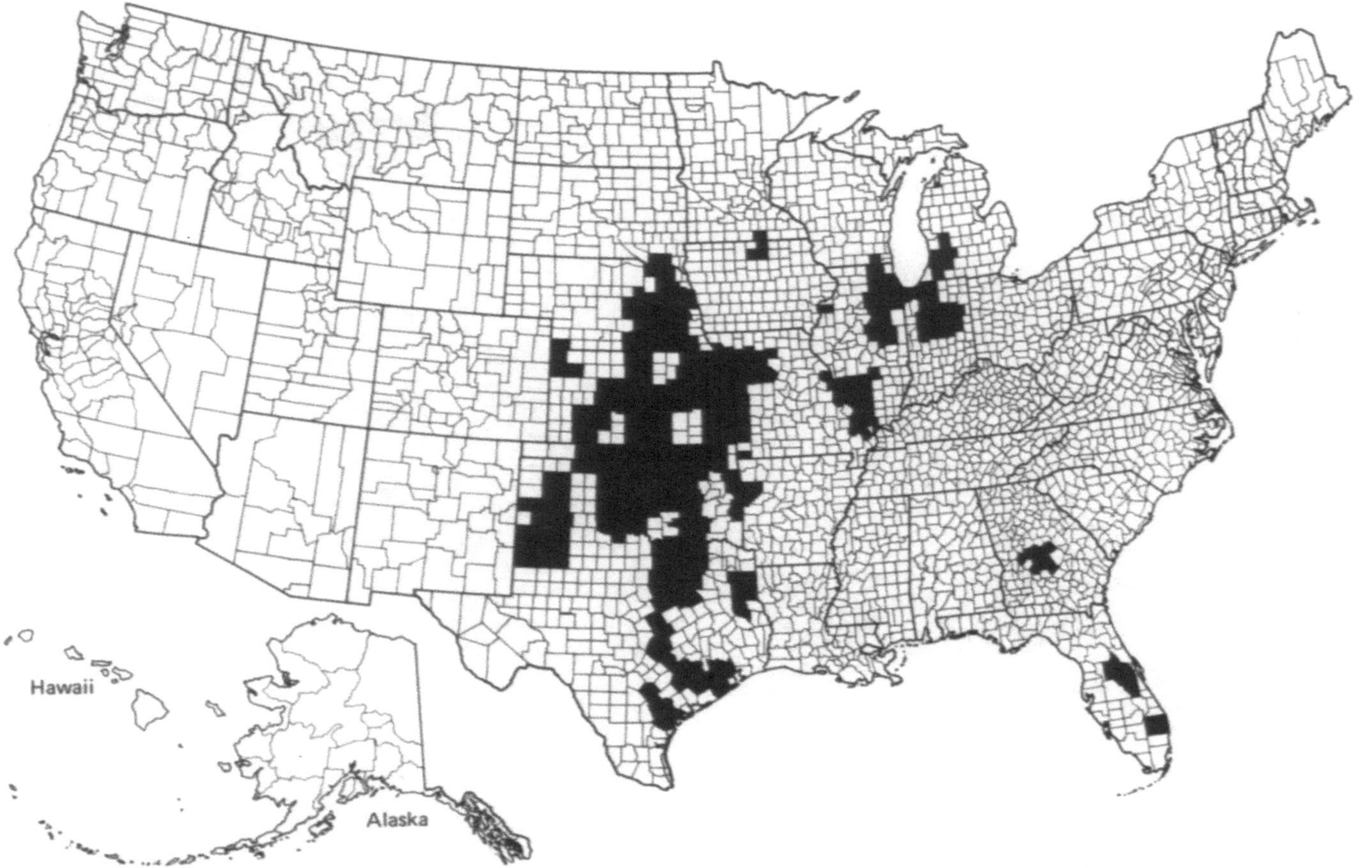

Figure 5-3. Counties with expected damage rate two or more times the NBDR for tornado (1970 exposure).

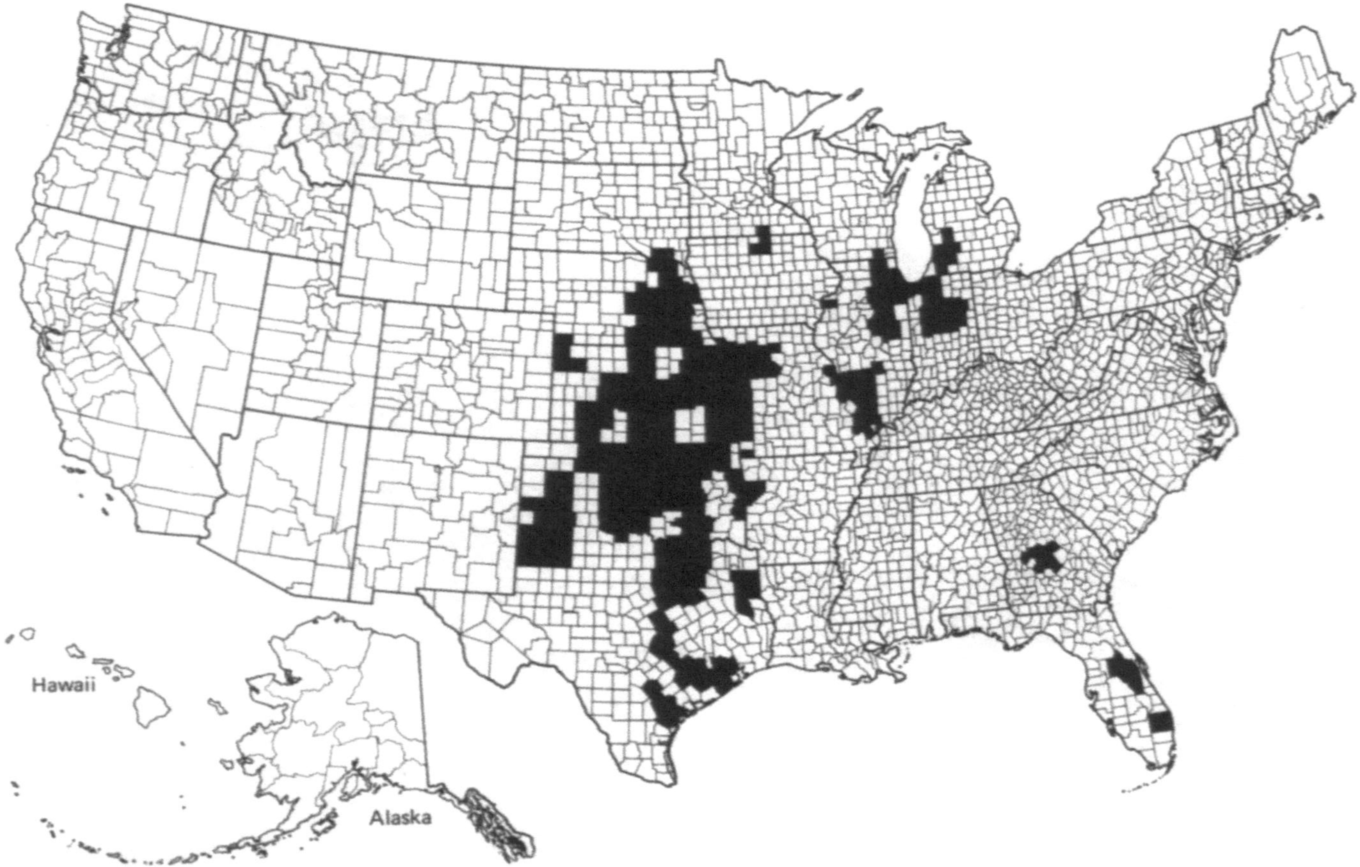

Figure 5-4. Counties with expected damage rate two or more times the NBDR for earthquake (1970 exposure).

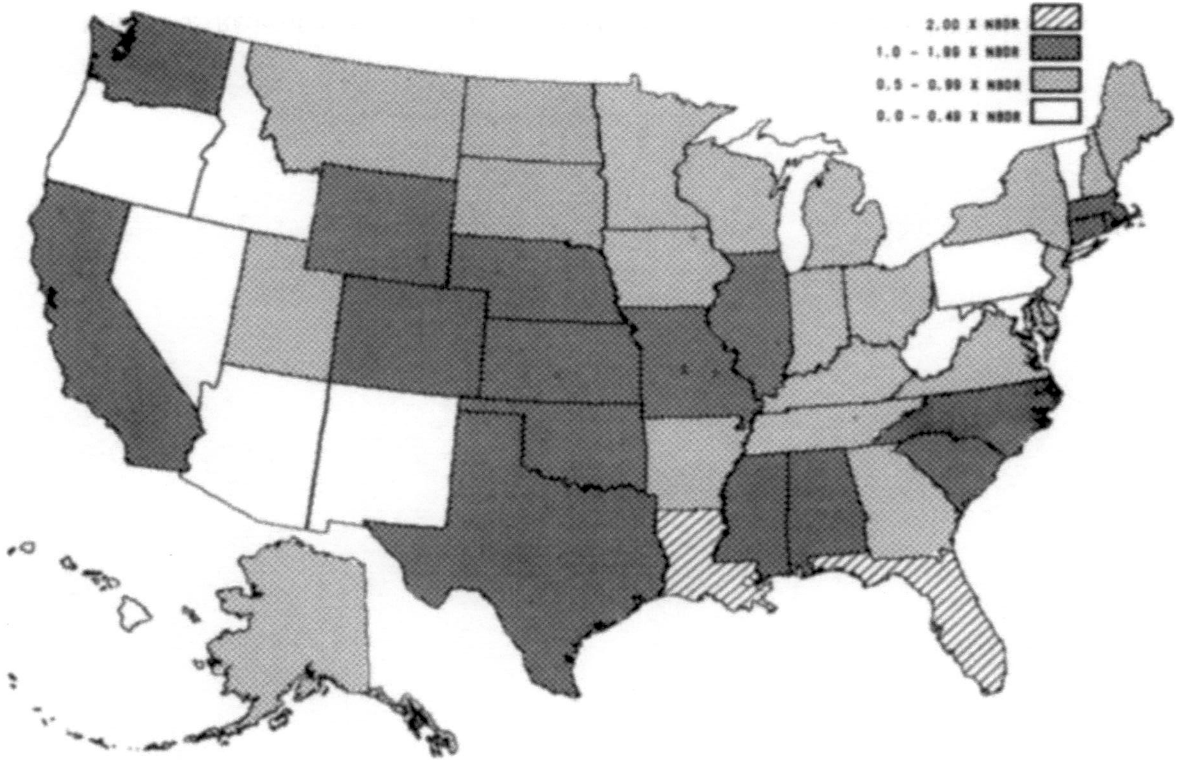

Figure 5-6. Ratio of state building damage rate to the NBDR for all hazards (2000 exposure).

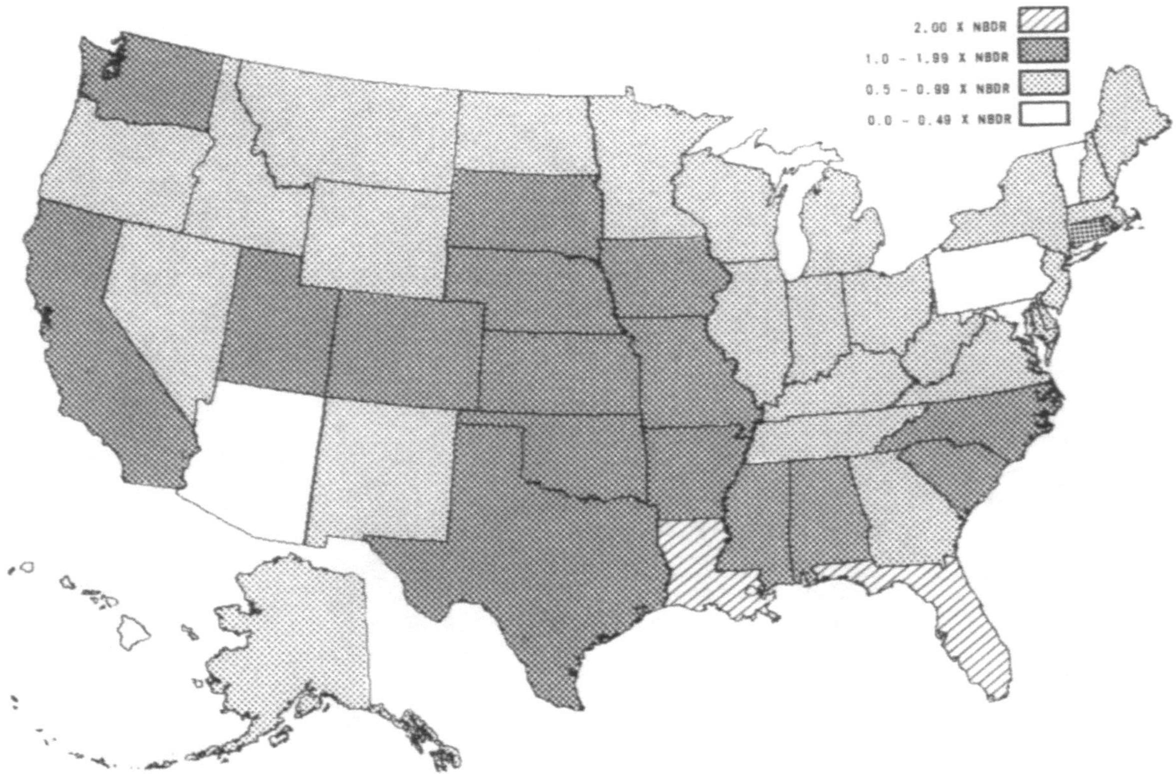

Figure 5-5. Ratio of state building damage rate to the NBDR for all hazards (1970 exposure).

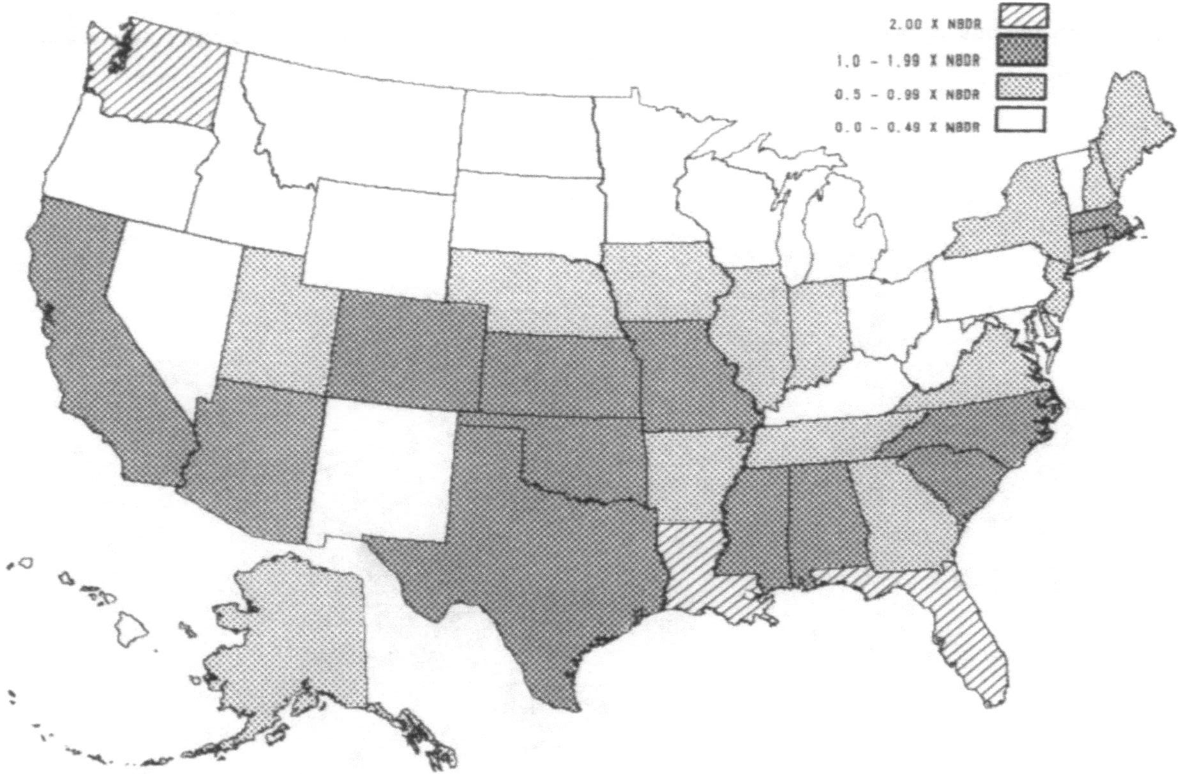

Figure 5-7. Ratio of state building damage rate for earthquake, storm surge, tornado, and hurricane to the NBDR (1970 exposure).

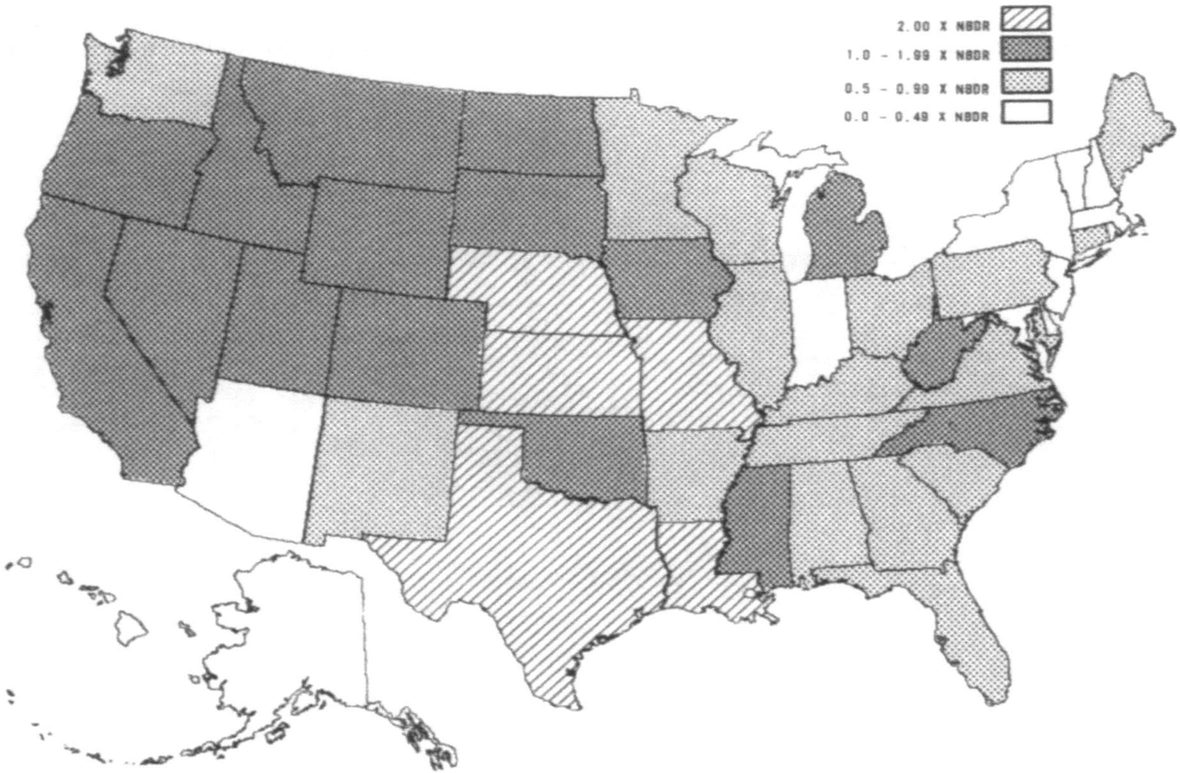

Figure 5-8. Ratio of state building damage rate for earthquake, storm surge, tornado, and hurricane to the NBDR (2000 exposure).

Table 5-18. Population Levels, Building Exposures, Building Damage Rates, and Building Losses from Storm Surge in Areas of 0–20 Feet AMSL (1970 Exposure)

State	Population	Building wealth ($ million)	Annual[a] expected loss ($ million)	Building damage rate (%)	Annual[a] expected per capita loss ($)
1. Mississippi	116,924	922.18	13.178	1.4290	112.71
2. Georgia	35,883	312.68	4.355	1.3928	121.37
3. Florida	2,467,506	22,720.49	195.271	0.8595	79.14
4. Alabama	78,774	593.82	4.323	0.7280	54.88
5. Louisiana	1,664,001	15,000.57	84.482	0.5632	50.77
6. Rhode Island	135,288	1,321.94	7.075	0.5352	52.30
7. N. Carolina	269,991	1.895.59	7.723	0.4074	28.61
8. Texas	408,987	4,142.19	15.725	0.3797	38.45
9. Massachusetts	297,700	3,490.19	12.635	0.3620	42.44
10. Connecticut	257,364	2,854.71	9.684	0.3392	37.63
11. So. Carolina	336,588	2,828.94	6.999	0.2474	20.79
12. Maine	26,370	236.64	0.527	0.2227	19.98
13. Delaware	25,770	268.91	0.536	0.1993	20.80
14. New York	1,926,873	27,456.65	50.049	0.1823	25.97
15. Virginia	877,110	8,763.71	13.547	0.1546	15.45
16. New Jersey	689,514	8,271.66	11.406	0.1379	16.54
17. Maryland	314,715	3,462.94	4.044	0.1168	12.85
18. N. Hampshire	10,743	126.44	0.032	0.0253	2.98
Totals	9,940,101	104,670.25	441.591	0.4219	44.43

[a] 1970 dollars.

Table 5-19. Riverine Flood Exposure Areas, Building Damage Rate, Exposed Population, and Per Capita Annual Loss (1970 Exposure)

Riverine flood exposure area	Building damage rate (%)	Exposed population	Per capita annual loss (1970$)
1. Flood Zone A– Unprotected communities	2.78	805,730	799.75
2. Flood Zone B– Unprotected communities	1.20	895,241	349.59
3. Flood Zone C– Unprotected communities	0.58	1,193,654	173.20
4. Flood Zone A– Protected communities	0.46	860,378	132.42
5. Flood Zone B– Protected communities	0.29	955,960	83.42
6. Flood Zone C– Protected communities	0.21	1,274,614	41.93
7. All flood-prone communities	0.11	3,312,216	33.55

Table 5-20. Tsunami Inundation Areas, Building Damage Rate, Exposed Population, and Per Capita Annual Loss

Variable	Counties in inundation areas	Inundation areas, only
Population	18,200,851	237,500
Building wealth[a] ($ million)	217,324.4	2,802.219
Building damage loss[a] ($ million)	8.7	8.7
Building damage rate (%)	0.0040	0.3105
Per capital building[a] damage loss	0.48	36.63

[a] 1970 dollars.

Table 5-21. Expected Annual Per Capita Building Damage Loss from Natural Hazards for Selected Populations at Risk (1970 Exposure)

Population at risk of exposure to indicated hazard	1970 Population (millions of persons)	1970 Annual expected loss (millions of 1970$)	1970 Annual expected per capita loss (1970$)
1. Severe wind:			
a. State with highest annual expected damage rate for hazard (Wyoming)	0.33	0.1	0.30
b. National population of all counties at risk of exposure to hazard	203.23	11.4	0.06
c. National population of states with building damage rates (BDR) equal to or greater than 2 X the national building damage rate (NBDR)	23.29	3.4	0.15
d. National population of all states with BDR = 1 X NBDR	73.92	7.7	0.10
2. Tornado:			
a. Three counties with highest BDR—and over 100 tornado strikes annually (Oklahoma)	.066	1.58	23.97
b. State with highest annual expected damage rate for hazard (Oklahoma)	2.56	43.5	16.99

Table 5-21. (continued)

Population at risk of exposure to indicated hazard	1970 Population (millions of persons)	1970 Annual expected loss (millions of 1970$)	1970 Annual expected per capita loss (1970$)
c. National population of all counties at risk of exposure to hazard	181.199	879.8	4.86
d. National population of states with BDR equal to or greater than 2 X the NBDR	38.5	435.8	11.32
e. National population of all states with BDR ⩾ 1 X NBDR	88.01	681.2	7.74
f. National population of all states with BDR less than 0.50 NBDR	78.07	76.5	0.98
3. Expansive soils:			
a. Intracounty populations in high hazard zones	17.73	–	26.00
b. Intracounty populations in moderate hazard zones	23.71	–	8.00
c. Intracounty populations in low hazard zones	98.32	–	1.50
4. Earthquake:			
a. State with highest annual expected damage rate for hazard (Washington)	3.41	96.9	28.42
b. National population of states with BDR ⩾ 2 X NBDR	26.91	572.5	21.27
c. National population of all counties at risk of exposure to hazard	143.169	655.2	4.58
d. National population of states with BDR less than 0.5 X NBDR	163.92	43.4	0.26
5. Hurricane:			
a. State with highest annual expected damage rate for hazard (Florida)	6.79	190.4	28.04
b. National population of states with BDR 2 X NBDR	26.91	401.6	14.92
c. National population of all counties at risk of exposure to hazard	62.741	685.4	10.92

Table 5-21. (continued)

Population at risk of exposure to indicated hazard	1970 Population (millions of persons)	1970 Annual expected loss (millions of 1970$)	1970 Annual expected per capita loss (1970$)
d. National population of states with BDR less than 0.5 X NBDR	117.76	17.8	0.15
6. Landslides:			
a. National population of all counties in high risk landslide areas	44.07	.6	4.25
b. National population in medium risk areas	39.43	–	0.80
c. State with highest annual expected damage rate for hazard (West Virginia)	1.74	4.88	2.80
d. State with highest per capita loss (Connecticut)	3.03	10.30	3.40
e. National population of states with BDR ⩾ 1 X NBDR	104.53	23.87	2.29
f. National population of states with BDR less than 0.5 X NBDR	13.46	7.2	.53
7. Storm surge:			
a. Surge-prone areas of Georgia	0.036	4.355	121.37
b. Surge-prone areas of Florida	2.468	195.271	79.14
c. Surge-prone areas of Gulf Coast (not including Florida)	2.269	117.708	51.88
d. Surge-prone areas of South Atlantic (Virginia, North Carolina, South Carolina, Georgia)	1.520	32.624	21.46
e. Surge-prone areas of Middle and North Atlantic	3.683	95.988	26.06
f. All surge-prone areas of United States	9.940	441.591	44.43
g. Surge-prone areas of New Hampshire	0.011	0.032	2.98
8. Riverine flooding:			
a. Flood Zones A–C in communities not protected from 50-year floods	2.895	1,164.09	402.16
b. Flood Zones A–C in communities protected from 50-year floods	3.091	247.09	79.94

Table 5-21. (continued)

Population at risk of exposure to indicated hazard	1970 Population (millions of persons)	1970 Annual expected loss (millions of 1970$)	1970 Annual expected per capita loss (1970$)
c. Flood Zone E (50–100 yr. t.p.) in all flood-prone communities	3.332	111.8	33.55
d. Flood Zones A–E, all flood-prone communities	12.341	1,710.4	138.59
e. Flood Zones A–F, all flood-prone communities	24.683	1,901	77.02
9. Tsunami:			
a. 500-year tsunami	.23	8.7	36.63
b. State with highest damage rate (Hawaii)	.77	4.1	5.32
c. Counties in which inundation areas are located	18.201	8.7	0.47

At the scale of whole communities, states, regions, and the nation, low probability but high consequence hazard events may produce catastrophic or near catastrophic losses in the single year or month of their occurrence. At the time of their occurence they may be more highly significant than high probability but low consequence events which may produce larger cumulative losses over any given time period. If the real community and familial impacts of a loss are those related to the loss experiencer's perceptions of economic and other burdens *at the time of loss*, then the probabilistically derived annualized estimates of natural hazard losses may be misleading. That is, they may tend to obscure the loss experiencer's real burden and perceptions of consequences at the time a severe, but infrequently occurring, hazardous event takes place. Thus, it is important to understand the differences between anticipated and unanticipated events and their impact on individuals or communities.

Several of the hazards examined in this analysis qualify as potentially severe but infrequently occurring events. Interestingly, these potentially catastrophic hazards account for the major fraction of the expected annual losses for both 1970 and 2000: expected annual per capita losses for earthquakes, hurricanes, riverine flooding, storm surge, tornadoes, and tsunamis total $33.95 for 1970, or 85.39% of all expected natural hazard per capita losses for that year. For the year 2000, expected per capita annual losses for the same catastrophic events total $61.93 or 89.22% of the expected per capita hazard losses. If only one half of the annual expected per capita hazard losses for a community were due to infrequently occurring but high damage producing events, per capita losses in the single year of the event could well press community recovery capacities and create substantial social costs as a result of the socioeconomic dislocations and lengthy recovery times associated with community rehabilitation responses.

For example, if an earthquake of Richter magnitude 8.3 at its epicenter were to occur in the 39-county San Francisco area in the year 2000, it could be expected to produce damage intensities reaching 9.5 on the MMI scale within the area, to produce building value losses at $18.2 billion (1970 dollars), and a total loss of $39.9 billion (see Chapter 7). Such a loss is 124% greater than the national, annual average losses of $17.7 billion for nine hazards in the same year. The latter, of course, represents a value derived from averaging into each year's expected losses, the cumulative, long-term losses expected to result from low probability, high consequence events. In one recent 10 year period, 13 discrete events of this type produced approximately $6.4 billion in losses, and a 10 year average annual loss equal to 7.9% of the total, all-hazard, annual expected loss estimated through this analysis for the year 1970.

It seems likely that the most troublesome impacts, at time of occurrence, are *unanticipated* and involve a comparatively high drain on annual income and/or the loss of a large fraction of one's capital accumulation. The least troublesome probably are anticipated and involve either comparatively low annual demands on income or only a trivial proportion of one's capital assets, such as the annual costs of fire insurance and of tax burdens for fire protection abatement services. Annual costs of this sort probably are more willingly borne by the public and are viewed as less burdensome than the comparatively large costs associated with unanticipated, low probability, high consequence hazard events. However, if these events are of such low probability that the public does not choose to engage in appropriate mitigation, then a social anomaly is presented. If the public is not highly motivated to cope with the threats posed by low probability, high consequence occurrences, nonetheless, when such phenomenon do occur, they probably are perceived as being far more burdensome and distressing than other types of events. In short, human reaction to such events may be inconsistent with the logic and the calculus of the actuarial sciences.

Referent Effects and Problems

The possible significance of projected natural hazard losses may be examined contextually by comparison with other social costs that constitute problems within U.S. society. One suggested set of comparisons is presented in Table 5-22. As shown, the 30 year total annual expected increase in natural hazard losses of $9685 million is greater than the sum of the 1970 aggregated costs represented by all crimes against property, all annual investments in water pollution control facilities, all business losses due to 6 major types of criminal activities, and all building losses due to fire. Similarly, that increase approaches the value of all health insurance premium payments in 1970 and is greater than other costs such as those resulting from auto liability insurance premiums, accidents at work, losses from air pollution related morbidity and mortality, air pollution effects on the value of property, and all expenditures by state and local police departments. The total annual expected loss from natural hazard exposures in

Table 5-22. Annual Expected Losses from Nine Natural Hazards in 1970, Compared with Annual Value of other Types of Losses and Events

Type of loss or event	Cost (millions 1970 $)
1. All property tax collections by state and local governments	34,054
2. All accidents	27,000
3. Expected annual natural hazard losses (2000 exposure)	17,779
4. All traffic accidents	16,200
5. Total economic effects of air pollution	16,000
6. Increase in annual expected losses from natural hazards, 1970–2000	12,960
7. Health insurance premiums	11,546
8. Pollution control costs (air, water, solid wastes)	9,300
9. Auto liability insurance premiums	8,958
10. Expected annual natural hazard losses (1970 exposure)	8,258
11. Losses from accidents at work	8,000
12. Losses from air pollution-related morbidity and mortality	6,000
13. Air pollution effects on value of property	5,200
14. Air pollution effects on materials and vegetation	4,900
15. Expenditures by all state and local police departments	4,494
16. All crimes against property	4,264
17. Investments in water pollution control facilities	3,100
18. Business losses due to six types of criminal activities	3,049
19. Building losses due to fires	2,209

Sources: U.S. Bureau of the Census (1976); National Safety Council (1970); Insurance Information Institute (1971); Council on Environmental Quality (1971).

2000 is greater than the total 1970 economic effects of air pollution, greater than the losses caused by all traffic accidents in 1970, and equal to about 50% of all 1970 state and local property tax collections.

The annual expected losses from natural hazards in 1970 were nearly equal to the total sums expended by owner/occupants of one unit properties in the United States for the improvement, maintenance, and repair of their dwelling units. In the group of 13 highest damage rate states, in 1970, total annual expected household losses from natural hazards equaled $196 as compared to the national average owner/occupant expenditures of $265 for the improvement, maintenance, and repair of one unit housing structures (U.S. Bureau of the Census, 1975). Similarly, the loss analysis revealed that 113,884 housing units could be expected to be lost from 1970 exposures to natural hazards. This number is equal to 7.75% of all the new housing units produced in the United States in that year (U.S. Bureau of the Census, 1975).

Still another referent useful in judging the real impacts of natural hazard losses is the following comparison of annual natural hazard losses per household as contrasted to the annual costs involved in a 1% shift in the interest rates for home mortgages. As shown in Table 5-23, a 1% shift in the interest demand for a $25,000, 25 year home mortgage produces a change in annual household costs of $202.08. In contrast, 1970 annual expected household losses due to natural hazards in three member households totaled $119.28. The ratio of annual interest rate costs (assuming a 1% upward movement in interest rates) to annual natural hazard losses varies from 1.69:1 to 3.05:1, depending on the size of the mortgage.

Natural hazard losses also may be compared with losses from fires and expenditures for fire abatement services. According to the National Household Fire Survey (U.S. Department of Commerce, 1976) annual national fire losses for single detached structures, garages, sheds, apartments, and vacation homes total $1213 million annually. In 1974, all state and local expenditures for fire protection payrolls totaled $2700 million. Thus, in 1970 natural hazard losses in the United States from 9 natural hazards ($8094 million) are substantially in excess of the sum of comparatively recent annual losses from fires to structures and the cost of support of all state and local fire abatement and fire prevention personnel.

However, since there are approximately 2.4 million loss-producing fires annually, the threat of fires to structures may be viewed as comparatively high probability events, as compared with most natural hazards. Since such events seem to produce more acute responses from the general public than do low probability events, inspite of the possible higher long-term accumulated losses arising from the latter, there is substantially less than full certainty that the public will respond to natural hazard losses with the same vigor that they have entered into payment of fire insurance premiums and into support of local fire prevention and abatement services.

Nonetheless, the referent information presented above suggests that natural hazard losses, both in the aggregate and in terms of their per capita and per hosuehold values, are of a magnitude approximating other expenditures or losses generally viewed as extremely important in our society.

Table 5-23. Annual Expected Lossses per Household as Compared with Household Annual Costs Associated with a 1% Shift in Mortgage Interest Rates

Size of mortgage $	Annual payment at 8%	9%	Annual difference in interest rate	Annual hazard costs[a]	Ratio of increased annual interest rate costs to natural hazard losses
25,000	2315.52	2517.6	202.08	119.28	1.69:1
30,000	2778.6	3021.12	242.52	119.28	2.03:1
35,000	3241.68	3524.64	282.96	119.28	2.37:1
40,000	3704.75	4028.16	323.41	119.28	2.71:1
45,000	4167.68	4531.68	363.84	119.28	3.05:1

[a] Based on 1970 exposure and a three-person household.

Cost Escalations in Catastrophic Losses

Since many hazardous natural events occur only infrequently, but are capable of producing extraordinarily large losses, the use of estimates of annual expected losses can be misleading to the unwary. Arithmetically, high consequence events that occur only infrequently can produce annual average losses that appear—if not comparatively small—easily manageable. However, at the time of occurrence, the loss may be of such magnitude as to produce staggering immediate burdens on those affected by the disaster, create immediate demands for capital which press the capacity of the financial community, and impose the cost of borrowing on loss-experiencing parties and on those to whom some fraction of the burden of the losses is transferred. This can be demonstrated through examination of Table 5-24.

Using this table, imagine that a high consequence event is expected in a state like California once every 200 years, and at the time of its occurrence is expected to produce a loss of $80 billion. Averaged over the 200-year period, this event produces an annual expected loss of $400 million, or approximately $20 per capita, per year (assuming a state population of 20 million persons). However, if appropriate plans have not been laid to prepare for the event, if no reserves have been accumulated to meet the sudden capital and other financial demands that it creates, then the state's recovery from the effects of the disaster may require the negotiation of a loan from other parties. Assume that the state financed the loss through a 50 year loan for the full amount of the catastrophe at a true interest rate of four percent (viz., actual interest minus impacts of anticipated future inflation). To amortize a loan of this size and interest rate over the ensuing 50 year period would require annual repayment costs totaling $3724 million, or an average of $182.20 per capita, and an amount equal to 9.4% of all federal individual income tax collections within the state in 1979. Assume further that in this same state four lesser catastrophes can be expected once each century, each capable of producing $10 billion in losses, or an annual average loss of $20 per capita.

Table 5-24. Fifty-Year Amortized Cost of Probable Maximum Loss Events in Relationship to Selected Tax Yields in Selected Areas of the United States

(1) Value of event ($)	(2) 50-year annual repayment cost (ARC) @ 4%	(3) ARC as % of total federal individual income & employment tax receipts, 1979 — Total national	(3) California	(4) ARC as % of all state/local tax collections in California, 1979
80 billion	$3724 million	1.02%	9.35%	13.61%
40 billion	$1862 million	0.51%	4.70%	6.81%
20 billion	$931 million	0.26%	2.35%	3.40%
10 billion	$466 million	0.13%	1.18%	1.70%
5 billion	$233 million	0.07%	0.6%	0.84%
2.5 billion	$116 million	0.03%	0.3%	0.43%
1.00 billion	$47 million	0.02%	0.15%	0.21%

Having successfully negotiated a favorable loan to meet the costs of the major catastrophe (the $80 billion event), the residents of the state decide to employ the same strategy to meet the burden of the 4 lesser events. Within a short while, the first of these events occurs and, as expected, a loss of $10 billion is experienced, and is financed over a 25 year period at a true interest rate of 4%. On this basis, the annual average amortized cost of the loan is $640.1 million. The citizens proceed with the loan and, on the year of its full retirement, experience another catastrophe of the same magnitude and therefore are required to continue this annual payment for an additional quarter century. Thus, over the 50 year period following the first major catastrophe (the $80 billion event), the citizens of this state experience loan repayment burdens totaling $4364 million a year, or an average of $218 per capita, as the result of three catastrophic events whose *annual expected losses* over a hundred year period totaled only $40 per capita. The combined annual average half-century repayment costs of these hazard-induced loans would be equal to approximately 16% of all the federal individual income taxes collected within the State of California in 1979.

Federal Liability for Natural Hazard Losses

A substantial part of the direct financial losses sustained by the owners and occupants of hazard-exposed properties, as well as a portion of the related borrowing costs, are shifted to the federal government and thence to taxpayers in areas not exposed to the loss-producing hazards. These shifts in loss burdens are produced through a variety of federal policies, 3 of the more powerful of which are embedded in the Internal Revenue Code, the Disaster Relief Act of 1974, and in the Congressionally-authorized SBA Disaster Loan Program.

Policy Assumptions

The provisions of these 3 major policies are as follows:

1. Internal Revenue Code Provision. Current Internal Revenue Service regulations permit taxpayers to deduct casualty losses from earned income. Conventionally, such losses are defined as unexpected or unusual property losses attributable to fire, storm, or other causes.

Both individual taxpayers and business may deduct casualty losses from their reported incomes and, when such losses are in excess of current income, they may be carried back three years or carried forward five years to offset previous or future tax liabilities. Since IRS regulations also permit deduction of interest payments on loans from reported income, individuals sustaining property loss as the result of hazardous natural occurrences may receive double relief from unreimbursed losses: first, for the casualty loss itself, and additionally, from interest paid on any relief loans.

Rettger and Boisvert (1979) report that the casualty loss which may be deducted:

> is limited to actual physical damages to the taxpayer's own property and may not include reductions in property values that do not arise in actual physical damages.
>
> The amount that may be deducted for personal casualty loss is the sustained loss or the adjusted basis of the property, whichever is less (sustained loss equals market value immediately before the loss, minus the value immediately afterward). The adjusted basis equals the original cost plus improvement costs, minus depreciation and obsolescence. Excluded from this deduction must be any insurance reimbursement, any payments received as relief to restore the property, any other compensation for loss of property, and $100 (if non-business property). . . . Clean-up costs may be included in figuring the loss deduction.

Because of these policies the Comptroller General of the United States (1980) has observed, "Tax transfers alone account for reductions of losses borne by disaster victims to 64.7 percent and 19.2 percent of total losses for residential and commercial property, respectively . . . (assuming) average tax brackets of 21 percent for residential property owners and 48 percent for commercial property owners." The estimate is based on replacement of loss with nonsubsidized loans at 10%, reduction in tax liability (at appropriate tax rate) by amount of loss and annual 20 year interest payments, and appropriate discounting to present value of payments in future years.

In addition to the "casualty loss" provisions of the Internal Revenue Code, other expenses arising from natural hazard exposures are deductible in determining tax liability. Thus, for businesses, insurance premium payments are deductible as a business operating expense, but are not deductible for individuals.

2. Disaster Relief Act of 1974. This Act authorizes the extension of a variety of financial aids to individuals, governments, and non-profit organizations located in Presidentially-defined disaster areas. Under the Act, federal funds are provided to state and local governments and qualifying private non-profit organizations for debris clearance, emergency protective measures, and the repair or replacement of public and private non-profit facilities, buildings, and utilities. Current policies provide for 75% federal cost sharing for this type of assistance. In addition, Section 408 of the Act authorizes grants to individuals and families to meet disaster related necessary expenses, when the individuals concerned are unable to meet such needs through other programs or other means. Grants of up to $5000 are authorized, and in fiscal year 1979, alone, 66,207 grants were approved under this program. Under the Act, temporary housing assistance can be provided by federal authorities, with no rental charge during the first 12 months of occupancy. Funds also are made available for limited home repairs to owner-occupied residences so as to permit the return of families to their homes in lieu of utilizing temporary housing resources. For individuals facing the loss of their residences because of a disaster-induced inability to meet mortgage or rental payments, the Act authorizes the award of federal grants.

3. SBA Disaster Loan Program. Since 1953 the Small Business Administration has been authorized to provide low cost loans to persons, businesses, and nonprofit organizations suffering physical or economic loss from a natural disaster. Physical disaster loans may be used to finance repairs or replacement of damaged structures and personal property, economic injury loans provide working capital for the payment of debts and other expenses, and product disaster loans provide an opportunity for businesses to recover from the burdens faced when a product is no longer marketable because of contamination or toxicity. Because of the low interest associated with SBA disaster loans, approximately 30% of the typical loan amounts to little more than a federal grant. Moreover, approximately 22% of all current disaster assistance loans are past due, delinquent, deferred, or in liquidation (Boucher and Cardenas, 1981).

Exposure Assumptions

To determine the probable impacts of these three classes of policies on federal revenue and expenditures, an appropriate natural hazard exposure model is required. Such a model must classify both exposed structural wealth and estimated hazard losses so that the impacts of each class of policy may be separately examined. This is accomplished, below:

1. National Structure Exposure Model. Utilizing the Statistical Abstract of the United States (U.S. Bureau of the Census, 1976), Table 5-25 was prepared. Assuming that the current mix of U.S. structures is representative of the mix of new structures added to the U.S. inventory between 1960 and 1975, the data reported in Table 5-25 may be used as the exposure model for all subsequent calculations in this analysis.

2. Policy Impact Model. The structural classes depicted in Table 5-25 must be regrouped to represent the categories of losses to which the 3 classes of federal policies relate. This is accomplished in Table 5-26. As shown in Column #2, policy impacts vary as a function of the circumstances in which the loss occurred, some policies being applicable only to losses sustained in zones in which federally-declared disasters have occurred.

For Class "A" structures, this analysis assumes that 100% of the hazard losses are subject to being "federalized" as a result of the combination of IRS "casualty loss" and "operating expense" policies. Although the "casualty loss" policies presumably would apply only to the depreciated book value of structures in the class, the actual cost of insurance premiums would be fully deductible. If one assumes that insurance premium rates are fixed at twice the annual expected loss rate (in order to provide for administrative, profit, and "risk" factors), then even a 50% insurance coverage of the structures in the class would result in 100% of the structural value of the class being subject to IRS operating expense policies.

For Class "B," 100% of the value of the structures is deemed eligible for reimbursement. Thus, all of the hazard-induced losses initially sustained are deductible at the income tax rate of the structure owner (after a minimum

Table 5-25. Estimated Proportion of U.S. Structural Wealth, by Category of Structure.

Structural category	Estimated % of: Total national structural value [a]	Total national building value	Total national value, other structures
1. Private residential buildings	36.4	55.0	–
2. Public utilities	10.8	–	31.7
3. Commercial buildings	9.4	14.3	–
4. Highways and streets	9.4	–	27.6
5. Educational buildings	6.4	9.7	–
6. Industrial buildings	6.3	9.6	–
7. Water supply and sewer systems	3.6	–	10.7
8. Hospital buildings	3.1	4.8	–
9. Miscellaneous public buildings	2.8	4.2	–
10. Conservation and development projects	2.4	–	6.9
11. Nonresidential farm buildings	1.6	2.4	–
12. Military facilities	1.2	–	3.6
13. Other	6.6	–	19.5
Total	100.0	100.0	100.0

Source: U.S. Bureau of the Census (1976).
[a] Assumes that current mix of U.S. structural wealth conforms to the value of new construction in censal and mid-censal years over the period, 1960–1975.

deductible), and all of the interest charges on any repair/reconstruction loans are similarly deductible.

For Classes "C" and "D," the financial impacts of federal policies must consider the fact that class-relevant policies apply only to losses sustained within zones in which "disasters" have been declared by federal authorities. Accordingly, the annual expected hazard losses which are later presented for these classes must be reduced so as to count only that fraction of the losses which occurs during declared disasters.

Loss Assumptions

Based on the exposure model presented in Table 5-25, and the loss estimates presented in Table 5-3, the data presented in Table 5-27 were calculated. Therefore, adoption of assumptions other than those listed on Table 5-27 will influence all subsequent calculations.

Table 5-26. Classification of U.S. Structural Wealth by Class of Federal Disaster Relief Policy

(1)	(2)	(3)	(4) % of U.S. structural wealth	
Policy category	Operative circumstances	Structural class	Buildings	All structures
A. IRS policies toward insurance premium payments and casualty losses of business/industry	Any hazard losses	1. Commercial bldgs.	14.3	9.4
		2. Industrial bldgs.	9.6	6.3
		3. Nonresidential farm bldgs.	2.4	1.6
		4. Public utilities (privately owned)	—	8.1
		5. Other structures	—	6.6
		Subtotal	26.3	49.3
B. IRS policies toward casualty losses to private residences	Any hazard losses	Private residential buildings	55.0	36.4
C. Disaster relief policies concerning repair/replacement of buildings, building contents, utilities and other facilities (Disaster Relief Act of 1974)	Losses in designated disaster zones	1. Educational bldgs.	9.7	6.4
		2. Hospital bldgs.	4.8	3.1
		3. Misc. public bldgs.	4.2	2.8
		4. Public utilities (owned by public entities)	—	—
		5. Highways and streets	—	9.4
		6. Water supply & sewer systems	—	3.6
		7. Conservation and development projects	—	6.9
		Subtotal	18.7	34.9
D. SBA interest-subsidized physical disaster loans	Losses in designated disaster zones	1. Private residential bldgs.	55.0	36.4
		2. Nonresidential farm bldgs.	2.4	1.6
		3. Other	not calculated	not calculated
		Subtotal	57.4	38.0

Table 5-27. Estimated Annual Expected Losses in the U.S. from Seven Potentially Catastrophic Hazardous Occurrences in 1970[a] (1970 Dollars)

Type of loss	Annual expected loss ($ million)	As % of total	As % of building damage
1. Building damage			
Private residential	2718.16	–	55.0
Commercial	706.72	–	14.3
Educational	479.38	–	9.7
Industrial	474.44	–	9.6
Hospital	237.22	–	4.8
Misc. public	207.57	–	4.2
Nonresidential farm	118.61	–	2.4
Subtotal	4942.1[b]	52.21–56.79	100.0
2. Damage to other structures			
Public utilities	403.53– 645.65		
Highways and streets	351.34– 562.14		
Water supply and sewer systems	136.21– 217.93		
Conservation and development projects	87.84– 140.54		
Military facilities	45.83– 73.32		
Other	248.23– 397.16		
Subtotal	1272.97–2036.74[c]	14.63–21.52	
3. Building content losses			
Private residential	1133.47	–	–
Commercial	294.70	–	–
Educational	199.90	–	–
Industrial	197.84	–	–
Hospital	98.92	–	–
Misc. public	86.56	–	–
Nonresidential farm	49.46	–	–
Subtotal	2060.86[d]	21.77–23.69	
4. Income and supplier losses	425.3[e]	4.49– 4.89	
Total	8701.23–9465.0	176.7–192.2	100.0

[a] Earthquake, hurricane wind, landslide, riverine flooding, storm surge, tsunami, tornado.
[b] Annual expected losses to all buildings derived from Table 5-3 and allocated to the several building categories based on the proportionate contribution of each to total national building value, as depicted in Table 5-25.
[c] Annual expected losses to *other structures* is based on the proportionate contribution of these categories to total national structure value, as depicted in Table 5-25. Also assumes that the damage rate (as % of value) to other structures is equal to 50–80% of the damage rate to buildings.
[d] Assumes that building content losses are equal to 41.7% of building damage (see Table 5-3). Contents losses then have been distributed to buildings on the basis of the contribution of each building category to total building damage losses, as shown in item 1.
[e] Derived from Table 5-3.

Estimated Federal Liability for Annual Expected Catastrophic Natural Hazard Losses

Estimates of the federal "share" of annual expected catastrophic natural hazard losses are sensitive to several major assumptions including (1) the proportion of annual expected losses which are chargeable to hazardous events in which federal aid-qualifying "disasters" or "emergencies" have been declared; (2) the cost and market coverage of insurance against losses sustained during hazardous occurrences of all types; (3) the tax bracket of the party experiencing loss and the proportion of the predisaster value of the damaged property which is represented by its "book value" for tax purposes; and (4) the proportion of federal aid, during disasters, which is represented by direct assistance versus tax relief, and whether these types of aid are funded out of current revenues or borrowed at the federal government's effective borrowing interest rate.

Within the scope of this inquiry, definitive answers to these and other questions could not be developed, and thus an attempt was made to place the federal share of natural hazard losses within a crudely acceptable range for the purpose of judging the gross policy significance of the subject.

This is accomplished in the data which are presented in Table 5-28 and in the following discussion. The "Estimate 1" values in Table 5-28 are the result of calculations based on the assumptions specified in Table 5-29. Because of their highly conservative nature, these estimates may be viewed as comprising the "lower bounds" of the federal liability for natural hazard exposures. The assumptions underlying these values were modified when "Estimate 2 values" were calculated.

Specifically, the assumptions were altered as follows:

1. Losses during declared disasters equal 50% of all annual expected losses, rather than 25%.
2. Federal borrowing (10 years at 8.75%) will be necessary to fund liabilities during declared disasters, and this requirement will escalate estimated federal costs by 1.533 for: (a) casualty losses during disasters (estimated at 50% of all losses) sustained by farm buildings and contents, commercial and industrial buildings and their contents, residences, privately-owned public utilities; (b) reimbursement to government and non-profit entities for damages sustained by educational, hospital, and public buildings and their contents, government-owned public utilities, highways, streets, water supply and sewer systems, conservation and development projects; and (c) direct federal grants for temporary housing, emergency repairs, and similar services.

As noted in Table 5-28, "Estimate 1" assumptions yield federal loss liabilities which range from approximately 56–63% of all building damage losses. In contrast, "Estimate 2" values produce federal liabilities which range from approximately 100.8 to 117.2% of all annual expected building damage losses. Since the latter estimates do not include the full range of costs resulting from federal disaster relief and loan policies, and since the estimate of the proportion of

Table 5-28. Estimated Federal Liability for 1970 Annual Expected Natural Hazard Losses

(1) Type of loss	(2) Annual expected loss ($ million)	(3) Federal share of loss			
		Estimate 1		Estimate 2	
		%	Total ($ million)	%	Total ($ million)
1. Damage to residential buildings and building contents	3851.63	35–39.5	1348–1521	52–61	2011–2359
2. Damage to nonresidential farm buildings and building contents	168.07	25	42	27	45
3. Damage to commercial and industrial buildings and their contents	1922.13	42	807	60	1153
4. Damage to educational, hospital and public buildings and their contents	1309.55	18.8	246	76.6	1003
5. Damage to military facilities	45.83–73.32	100	46–73	153.3	69.5–112.4
6. Damage to privately owned public utilities (75% of all PUs)	302.65–484.24	42	127–203	60	181.6–290.5
7. Damage to governmentally owned public utilities (25% of all PUs)	100.88–161.41	18.8	19–30	76.6	77.3–123.6
8. Damage to highways and streets, water supply and sewer systems, conservation and development projects, other structures	575.4–920.61	18.8	108–173	76.6	440.8–705.2
9. Income and supplies losses	425.3	–	–	–	–
Total			2743–3095		4980–5791.7
As % of building damages (see Table 5-27)			55.5–62.6		100.8–117.2

Table 5-29. Estimate 1—Assumptions Governing Calculation of Federal Liabilities for Losses from Occurrence of Catastrophic Hazardous Events

		Estimated federal liability	
Building loss	Assumptions	$ (millions)	%
Residential buildings and their contents	1. 90% of all losses deductible as casualty losses at effective tax rate of 25%.	866.62	35–39.5
	2. 25% of all losses occur during declared disasters. 90% of the losses which so occur are covered by SBA subsidized loans under conditions where the federal share of losses is equal to 50–70% (see Comptroller General of the United States, 1980, Table 1).	433–607	
	3. Direct federal grants for temporary housing, emergency repairs, and similar services equal 5% of losses during declared disasters.	47.77	
	Total	1347.39–1521.39	
Residential farm buildings and their contents	1. An amount equal to 50% of all losses is fully covered by insurance whose premium payments are 1.5 times the annual expected building damage rate. All such payments are tax deductible at effective tax rate of 25%.	31.53	25
	2. An amount equal to 50% of all losses is tax deductible as casualty loss (at effective tax rate of 25%) but at depreciated value which cuts losses by one-half.	10.50	
	Total	42.03	

Damage to: (1) commercial and industrial buildings and their contents; (2) other structures; (3) privately owned public utilities	1. An amount equal to 50% of all losses is fully covered by insurance whose premium payments are 1.5 times the annual expected building damage rate. All such payments are tax deductible at effective tax rate of 48%.	800.92–866.29	42
	2. An amount equal to 50% of all losses is tax deductible as casualty loss (at effective tax rate of 48%) but at depreciated value which cuts losses by 75%.	133.49–144.38	
	Total	934.41–1010.67	
Damage to: (1) educational, hospital, and public buildings and their contents; (2) governmentally owned public utilities; (3) highways, streets, water supply and sewer systems, conservation and development projects	Assumes that 25% of all losses occur during declared disasters and are federally reimbursed at 75% level.	372.34–448.42	18.8
Military facilities	100% federal liability	45.83–73.32	100.0

annual expected losses from the "catastrophic" hazards which occur during declared disasters (50%) are probably still within a conservative range, the "Estimate 2" values probably do not represent the "upper bounds" of federal liability. Accordingly, the "federal share" of annual expected natural hazard losses was fixed at a value equal to 100% of the building damage losses for the purpose of the detailed tax-transfer calculations which appear in the following section. Should the reader choose to adopt "Estimate 1" assumptions, then all values entered on the tables should be reduced by approximately one-half.

Interstate Tax Transfers Produced by Federal Liability for Natural Hazard Losses

Since the estimated federal liability for natural hazard losses is approximately equal to annual expected building damage losses alone, the latter data can be utilized for each state to estimate the annual expected federal dollar flow into each state. If one postulates that the federal dollar flow is derived from a "federal fund for natural hazard losses" which, in turn, is derived from each state's ability to pay (as measured by federal individual income tax collections within each state), then the development of the federal fund would have required an annual expected amount equal to 5.9 percent of total federal income tax collections in 1970. Using the latter figure, the "contributions" of each state to the federal fund can be calculated, as well as the amount due each state as a result of natural hazard losses. Such data can be utilized to crudely identify which states receive less than they pay into the fund and which receive more. These data (Tables 5-30 and 5-31) indicate the amounts theoretically credited to "beneficiary states" and contributed by "donor states" in 1970 based on natural hazard losses and individual income tax collections in that year. Thus, the total value of 1970 interstate tax transfers from "low risk" to "higher risk" states exceeded $1 billion (1970 constant dollars). When the value of this interstate transfer is escalated on the basis of changes in the national consumer price index between 1970 and 1980, the current value of the identified interstate tax transfer is $2.112 billion (1980 constant dollars).

In terms of the magnitude of the received transfers, the principal beneficiary states are Florida, California, Mississippi, and Louisiana (see Table 5-32). Collectively, these states would be recipients of more than 75% of the interstate transfers, with California alone accounting for nearly half (47.92%) of the total interstate transfers. Similarly, six states account for approximately 75% of this donation, i.e., New York, Michigan, Pennsylvania, New Jersey, Ohio, and West Virginia.

The value of these interstate transfers to beneficiary states may be measured in terms of the impact of the received funds on the individual capacities of the recipient states to bear the financial burdens associated with their annual expected natural hazard losses. If interstate transfers are therefore viewed as an offset against a "hazard loss" surcharge that would otherwise have to be imposed on family income as reported on federal income tax returns, the leading beneficiaries of the transfers are Mississippi and Louisiana where the interstate tax transfers are equal to 31.4% and 13.54% of all federal income tax collections within those states, respectively. However, Washington and Oklahoma also received transfers equal to or greater than 3% of the federal income tax collections within those states. The burden of interstate tax transfers on donor states may be comparably measured. In these terms, only four states contribute sums equal to 3% or more of their federal income tax collections, including Michigan, the District of Columbia, Pennsylvania, and Nevada (see Table 5-33).

The data suggest that few states receive interstate tax transfers of sufficient magnitude and importance to be of high value to them, and that few, if any, states contribute sums in sufficiently large quantities and at sufficiently burdensome levels to be of much concern to them.

If net natural hazard loss burdens, by state, are viewed as the sum of the natural hazard losses which they experience, modified by their donation and receipt of interstate tax transfers, then typical burdens among the states are as indicated in Table 5-34. In this table, the annual expected natural hazard loss for each state is assumed to be twice the building damage loss reported in Tables 5-30 and 5-31. The net burden is the annual expected loss plus the value of the interstate transfer.

Interstate Tax Transfers Produced by Other Federal Policies

Currently, federal revenue losses (e.g., "tax expenditures") due to the tax deductibility of casualty losses totals $590 million per year, or 0.25% of all federal individual income tax collections (U.S. Bureau of the Census, 1980). Annual expected losses from catastrophic hazardous occurrences can be expected to push that value to a substantially higher proportion. Considered as a whole, the current federal share of annual expected natural hazard losses may be placed at a level approximately equal to 5.9% of all federal individual income tax collections, and the total mix of federal disaster relief policies (including those embedded in the Internal Revenue Code) result in net interstate natural hazard losses totaling approximately $2112 million (1980 constant dollars). Of course, interstate tax transfers are also produced by other federal policies directed toward other subjects and problems.

For example, federal aid to state and local governments for all purposes except transportation and highway aids (which may be viewed as being funded from excise taxes on sales of gasoline) totaled approximately $80,000 million in 1979 ($91,021 million in 1980 constant dollars), of which approximately $9186 million ($10,348 million in 1980 constant dollars) represented a net interstate transfer. Selected data on the net transfer are presented in Table 5-35. Thus, interstate tax transfers (in 1980 constant dollars) produced by natural hazard losses are equal to 20.4% of the total value (in 1980 dollars) of

Table 5-30. Estimated 1970 Interstate Tax Transfers from Beneficiary States under Existing Federal Policies for Mitigating the Financial Impacts of Natural Hazard Losses (Millions of 1970 Constant Dollars)

State	Annual expected building damage losses		
	All hazards	Non-catastrophic	Catastrophic
1. Mississippi	208.5	42.41	166.09
2. Louisiana	218.4	26.39	192.01
3. Washington	143.8	4.71	139.09
4. Oklahoma	78.0	5.26	72.74
5. S. Dakota	16.5	3.59	12.91
6. Florida	503.8	14.80	489.00
7. S. Carolina	60.9	8.66	52.24
8. Kansas	90.9	26.08	64.82
9. Utah	30.3	5.63	24.67
10. Alabama	82.6	14.45	68.15
11. Missouri	178.8	37.42	141.38
12. N. Dakota	13.3	1.98	11.32
13. California	855.6	183.39	672.21
14. Arkansas	40.6	7.01	33.59
15. N. Carolina	124.1	19.12	104.98
16. Rhode Island	26.4	0.94	25.46
17. Nebraska	55.9	21.53	34.37
18. Texas	412.3	144.69	267.61
19. Colorado	72.3	18.17	54.13
20. Iowa	75.2	14.99	60.21
21. Massachusetts	172.7	5.63	167.07
Totals	3426.7	599.84	2826.86

[a]Federal individual income tax collections (FIITC). Source: 1970 data on federal individual income tax collections, by state, derived from U.S. Bureau of Census (1980).

FIITC[a] 1970 ($ million)	Payment to federal fund for natural hazard losses at 5.913% of FIITC	Net interstate transfer	As % of FIITC
445	26.31	139.78	31.40
987	58.36	133.65	13.54
1380	81.60	57.49	4.17
799	47.24	25.50	3.19
156	9.22	3.69	2.37
2680	158.47	330.53	2.33
644	38.08	14.16	2.20
803	47.48	17.34	2.16
307	18.15	6.52	2.12
856	50.62	17.53	2.05
1798	106.32	35.06	1.95
149	8.81	2.51	1.69
8844	522.95	149.26	1.69
466	27.55	6.04	1.30
1518	89.76	15.22	1.00
373	22.06	3.40	0.91
510	30.16	4.21	0.83
4098	242.31	25.30	0.62
852	50.38	3.75	0.44
969	57.30	2.91	0.30
2700	159.65	7.42	0.27
31334	1852.78	1001.27	

Table 5-31. Estimated 1970 Interstate Tax Transfers from Donor States under Existing Federal Policies for Mitigating the Financial Impacts of Natural Hazard Losses (Millions of 1970 Constant Dollars)

	Annual expected building damage losses		
State	All hazards	Non-catastrophic	Catastrophic
1. Michigan	80.9	6.27	74.63
2. D.C.	11.0	0.79	10.21
3. Pennsylvania	158.2	15.24	142.96
4. Nevada	8.9	1.22	7.68
5. Arizona	21.9	2.3	19.6
6. Hawaii	11.4	0.1	11.3
7. Oregon	30.1	4.97	25.13
8. Vermont	5.2	0.44	4.76
9. New Jersey	132.9	8.27	124.63
10. Wyoming	6.4	2.37	4.03
11. Delaware	10.5	0.64	9.86
12. Maryland	76.4	6.76	69.64
13. West Virginia	23.6	3.87	19.73
14. New York	381.9	15.76	366.14
15. Ohio	206.4	14.87	199.53
16. Minnesota	69.7	13.19	56.51
17. Montana	12.9	4.12	8.78
18. Alaska	6.4	–0–	6.4
19. New Mexico	16.6	3.91	12.69
20. Kentucky	49.2	8.58	40.62
21. Idaho	11.3	2.52	8.78
22. Tennessee	64.7	6.67	58.03
23. Wisconsin	87.7	6.48	81.22
24. Connecticut	99.4	7.48	91.92
25. Georgia	92.7	17.17	75.53
26. New Hampshire	16.3	0.70	15.6
27. Virginia	107.5	9.17	98.33
28. Illinois	348.0	29.49	318.51
29. Indiana	123.1	6.71	116.39
30. Maine	19.3	2.30	17
Totals	2824.7	209.37	2115.33

[a]Federal individual income tax collections (FIITC). Source: 1970 data on federal individual income tax collections, by state, derived from U.S. Bureau of Census (1980, Table 456, p. 273).

FIITC[a] 1970 ($ million)	Payments to fed. fund from nat. hazard losses at 5.913% of FIITC	Net interstate transfer	As % of FIITC
4008	236.99	162.36	4.06
386	22.82	12.61	3.27
5129	303.28	160.32	3.13
269	15.91	8.23	3.06
669	39.56	19.96	2.98
379	22.41	11.11	2.93
791	46.77	21.64	2.74
149	8.81	4.05	2.72
3863	228.42	103.79	2.69
124	7.33	3.30	2.66
299	17.68	7.82	2.62
1961	115.95	46.31	2.37
539	31.87	12.14	2.25
9334	551.92	185.78	1.99
4711	278.56	87.03	1.84
1382	81.72	25.21	1.82
213	12.59	3.81	1.79
150	8.87	2.47	1.65
296	17.50	4.81	1.63
920	54.40	13.78	1.50
197	11.65	2.87	1.46
1232	72.85	14.82	1.20
1644	97.21	15.99	0.97
1844	109.04	17.12	0.93
1481	87.57	12.04	0.81
303	17.92	2.32	0.77
1863	110.16	11.83	0.64
5761	340.65	22.14	0.38
2061	121.87	5.48	0.27
292	17.27	.27	0.09
52250	3089.55	1001.41	

Table 5-32. Major Beneficiary and Donor States, by Absolute Magnitude of Interstate Tax Transfer for Natural Hazard Losses, 1970

	Beneficiary states				Donor states		
	$ received (millions)		% of total interstate transfer		$ donated (millions)		% of total interstate transfer
State	N	Cum			N	Cum	
1. Florida	330.53	330.53	33.01	1. New York	185.78	185.78	18.55
2. California	149.26	479.79	47.92	2. Michigan	162.36	348.14	34.77
3. Mississippi	139.78	619.57	61.88	3. Pennsylvania	160.32	508.46	50.78
4. Louisiana	133.65	753.22	75.23	4. New Jersey	103.79	612.25	61.15
5. Washington	57.49	810.71	80.97	5. Ohio	87.03	699.28	69.84
6. Missouri	35.06	845.77	84.47	6. W. Virginia	46.31	745.59	74.41
7. Oklahoma	25.50	871.27	87.02	7. Minnesota	25.21	770.80	76.98
8. Texas	25.30	896.57	89.54	8. Illinois	22.14	792.94	79.19
9. Alabama	17.53	914.1	91.29	9. Oregon	21.64	814.58	81.35
10. Kansas	17.34	931.44	93.03	10. Arizona	19.96	834.54	83.35
11. Others (11)	69.83	1001.27	100.0	11. Others (20)	166.46	1001.27	100.0
Total		1001.27					

Table 5-33. Major Beneficiary and Donor States, by *Comparative* Impact of Interstate Tax Transfers on Natural Hazard Loss Burdens, 1970

	Beneficiary states					Donor states			
	Interstate transfer as % of FIITC in state	$ received (millions)				Interstate transfer as % of FIITC in state	$ donated (millions)		
State		N	Cum	Cum %	State		N	Cum	Cum %
1. Mississippi	31.4	139.78	139.78	13.96	1. Michigan	4.06	162.36	162.36	16.2
2. Louisiana	13.54	133.65	273.43	27.3	2. D.C.	3.27	12.61	174.97	17.48
3. Washington	4.17	57.49	330.92	33.05	3. Pennsylvania	3.13	160.32	335.29	33.49
4. Oklahoma	3.19	25.50	356.42	35.6	4. Nevada	3.06	8.23	343.52	34.31
5. S. Dakota	2.37	3.69	360.11	35.97	5. Arizona	2.98	19.96	363.48	36.30
6. Florida	2.33	330.53	690.64	68.98	6. Hawaii	2.93	11.11	374.59	37.41
7. S. Carolina	2.20	14.16	704.8	70.39	7. Oregon	2.74	21.64	396.23	39.57
8. Kansas	21.6	17.34	722.14	72.12	8. Vermont	2.72	4.05	400.28	39.98
9. Utah	2.12	6.52	728.66	72.77	9. New Jersey	2.69	103.79	504.07	50.34
10. Alabama	2.05	17.53	746.19	74.52	10. Wyoming	2.66	3.30	507.37	50.67
11. Others (11)		255.08			11. Others (20)	–	493.9	1001.27	100.0

Table 5-34. Impact of Interstate Transfers on Catastrophic Natural Hazard Loss Burdens[a] of Selected States (Millions of 1970 Constant Dollars)

States	Type	Annual expected natural hazard loss	Interstate transfer	Net burden N	Net burden As % of FIITC
Mississippi	Beneficiary	332.18	−139.78	192.4	43.24
Louisiana	"	384.02	−133.65	250.37	25.37
Florida	"	978	−330.53	647.47	25.16
Washington	"	278.18	−57.49	220.69	15.99
Kansas	"	129.64	−17.34	112.3	13.99
Alabama	"	136.3	−17.53	118.77	13.88
Missouri	"	282.16	−35.06	247.1	13.74
California	"	1344.42	−149.26	1195.16	13.52
Texas	"	535.22	−25.30	509.92	12.44
All beneficiary states	"	5653.72	−1001.41	4652.32	14.85
All donor states	Donor	4230.66	+1001.41	5232.07	10.01

[a]Natural hazard loss burden is defined as the annual expected natural hazard losses (catastrophic hazards) ± interstate tax transfer.

interstate tax transfers produced by such federal programs as public assistance, Medicaid, general revenue sharing, aids to elementary and secondary education, grants for construction of waste treatment facilities, low-rent public housing, CETA programs, and block grants for community development.

By combining the federal aid data with the data presented in Tables 5-30 and 5-31 (converted to 1979 dollars), the results shown in Table 5-35 are produced. As indicated, when hazard-induced tax transfers are considered, several states experience substantial reductions in the net impact of aid to state and local governments (Minnesota, New York, Pennsylvania, Tennessee, Wisconsin) and one state (Pennsylvania) moves completely out of the set of states which are net beneficiaries of these aid policies. Considered as a group, however, this set of beneficiary states receives only a modest (2.2%) negative impact from hazard-induced interstate tax transfers. On the whole, the general impact of hazard-induced tax transfers is to slightly reduce the net cost otherwise borne by the set of states which contributes net dollar flows to the states which are the beneficiaries of tax transfers produced by federal policies concerning aids to state and local governments.

Conclusion

The data presented in the loss analysis seem to support 6 major problem-focused conclusions:

Table 5-35. Interstate Tax Transfers Resulting from Natural Hazard Losses and other Policies Concerning Aids to State and Local Government (all Federal Aids Except Transportation and Highways) (Millions of 1979 Constant Dollars)

Beneficiary states	(1) All federal aids[a]	(2) Net hazard tax transfer[b]	(3) Net federal aids
1. Alabama	324	32.41	356.4
2. Alaska	89	(−4.616)	84.4
3. Arkansas	320	11.29	331.29
4. D.C.	841	(−23.57)	817.43
5. Georgia	577	(−22.50)	554.50
6. Hawaii	38	(−20.77)	17.23
7. Idaho	75	(−5.36)	69.64
8. Kentucky	274	(−25.75)	248.25
9. Louisiana	243	249.79	492.79
10. Maine	216	(−.51)	215.50
11. Massachusetts	734	13.87	747.87
12. Minnesota	126	(−47.12)	78.88
13. Mississippi	521	261.25	782.25
14. Montana	155	(−7.12)	147.89
15. New Mexico	222	(−8.99)	213.0
16. New York	2473	(−347.41)	2125.59
17. North Carolina	262	28.45	290.45
18. North Dakota	83	4.69	87.69
19. Oklahoma	35	47.66	82.66
20. Oregon	149	40.45	108.56
21. Pennsylvania	183	(−299.64)	(−116.64)
22. Rhode Island	111	6.36	117.36
23. South Carolina	255	26.47	281.47
24. South Dakota	127	6.90	133.90
25. Tennessee	215	(−27.70)	187.30
26. Utah	65	12.19	77.19
27. Vermont	107	(−7.57)	99.43
28. West Virginia	172	22.69	194.69
29. Wisconsin	165	(−29.89)	135.12
30. Wyoming	29	(−6.17)	22.83
Totals	9186	(−200.97)	8985

[a]Source: U.S. Bureau of the Census (1980).
[b]Derived from Tables 5-30 and 5-32 with adjustment to 1979 dollars.

1. Natural Hazard Dollar Losses Are of a Magnitude Equal to or Approching the Costs of Other Phenomena Viewed as Problems in Our Society. The 9 natural hazards examined in this study produced annual expected dollar losses totaling $8094 million in 1970 and are expected to produce annual losses totaling $17,779 million in the year 2000. Representing economic losses sustained by wage earners, suppliers, and the owners of buildings and building contents, these losses exceed the 1970 value of such generally disvalued phenomena as all building losses due to fires, all crimes against property, all expenditures by state and local police departments, and the value of all losses resulting from work accidents. The 30 year expected increase in annual expected natural hazard losses ($9685 million) approaches the 1970 value of all annual premium payments for health insurance policies. For the nation, the nine natural hazards considered in this analysis produce annual expected per capita losses of $39.76 in 1970 and will produce expected annual per capita losses of $69.41 in 2000. In both years, 4 of the 9 natural hazards account for 77.52% to 72.67% of these losses–riverine flooding, tornado, expansive soil, and hurricane.

2. Risk Takers and Their Related Communities at Interest Currently Experience Much Unevenness in the Actual Natural Hazard Losses Sustained Each Year. Many of the hazards examined in this analysis, and for which annual expected damage losses were calculated, may be viewed as comparatively low probability, high consequence occurrences. Thus, hurricane, earthquake, storm surge, and tsunami accounted for nearly 31% of the annual expected national per capita hazard losses for 1970 exposure. In any particular heavily populated area, the probability in any year that a major high-intensity event of these types will occur is comparatively small. However, when such events do occur, they are capable of producing losses representing a substantial fraction of the annual income and building value of the impacted area. For example, as shown in Chapter 7, if a Richter magnitude 8.3 earthquake were to occur in the 39-county San Francisco Bay Area in the year 2000, it could be expected to produce building value losses of $18 billion and a total dollar loss of $39.9 billion. Such a loss would be more than double the national, annual, all-hazard expected losses for the same year. Sudden, large magnitude dollar losses of this type may severely impair the recovery capacity of a community, state, or region because of the large short-term demands on available capital and annual incomes. If actuarily sound financial reserves were developed to finance recovery from low probability, high consequence events, then the problems produced by unevenness in annual economic losses might be corrected. In the absence of such systems, however, the uneven temporal distribution of losses must itself be viewed as a potential problem of considerable magnitude.

3. Substantial Unevenness in the Distribution of Annual Per Capita Loss Burdens Is Exhibited by the Several Hazard-Specific and Geographically Identifiable Risk Taker and Community at Interest Groups. On a total national basis, the annual expected per capita losses resulting from natural hazard exposures in 1970 and 2000 are not particularly large. For example, the 1970 annual expected per capita loss is $39.76 and is equal to only 1% of per capita national income in that year. Assuming 3.2 persons per household, average national household burdens are $127 and $222 for 1970 and 2000, respectively. Such annual losses per household do not appear to be very burdensome, as compared with other major household expenditures.

However, per capita losses vary widely among the several populations exposed to hazards or exposed to the risk of bearing hazard-induced community costs. For example, if all 1970 riverine flood losses are distributed among the population that actually reside in U.S. riverine floodplains, this loss burden would produce an annual expected per capita loss of $114.40, or an annual expected household loss of $366.08. Similarly, several million persons in the United States reside in intracounty areas in which they are exposed to a high level of expansive soil hazard and where they will experience annual per capita losses from this hazard totaling $26.00. The nearly 27 million people residing in states exposed to a high level of earthquake hazard are at risk of experiencing an annual per capita loss of $21.27 from this hazard exposure. Residents in the State of Florida, where a comparatively high frequency of hurricane episodes can be expected, were at risk of experiencing 1970 annual expected per capita losses, totaling $28.04 for hurricanes and $28.72 for storm surge. In contrast, the 90 million persons residing in states with the lowest riverine flood losses were expected to experience an annual per capita loss of only $7.76 for this hazard in 1970. In the United States, total annual expected losses for all hazards varied from a low of $16.86 in Vermont to a high of $110.32 in Florida.

When this unevenness in annual expected hazard losses between the several populations at risk is combined with the year by year unevenness in actual loss experience, it is clear that personally catastrophic losses from natural hazards may be experienced by numerically significant population subsets in the United States in any given year.

4. Improvident Patterns of Interstate Migration and Capital Investment Are the Major Causes of the Large Projected Increase in Natural Hazards Between 1970 and 2000. The 13 states that exhibited the highest 1970 building damage rate for exposure of property to natural hazards will account for nearly 32% of the projected increase in building wealth between 1970 and 2000. In contrast, the 13 states with the lowest building damage rates will account for only 27%. When combined with the higher probability that the population, buildings, and other wealth drawn to those states will be at higher risk of exposure to more frequent and higher magnitude hazards, the higher rates of growth in the high damage rate states result in a situation where the high damage rate states will account for nearly 51% of the increase in annual expected hazard-induced dollar losses between 1970 and 2000.

These improvident interstate patterns of migration unquestionably are matched by equally improvident intrastate and intracounty patterns of community development that further exacerbate the economic and other effects of natural hazard occurrences. Continued development of such hazardous areas as those characterized by high soil expansivity, by their location in or proximity to 50 and 100 year floodplains, and by their proximity to areas in which hurricane wind, storm surge, and tsunami damages will be experienced increases the magnitude of projected hazard-induced losses. Large as 1970 annual expected hazard losses may have been, they are dwarfed by the 30 year increase expected between 1970 and 2000. Almost the entirety of that projected increase is unnecessary and avoidable in the sense that hazard-avoidance strategies would, alone, substantially reduce, if not eliminate, those increases.

5. Annual Expected Per Capita Losses from Natural Hazards Generally Are Insufficiently High to Provide Incentives for Individuals to Make Appropriate Hazard-Avoiding or Hazard-Mitigating Decisions. In spite of the large estimated aggregated annual losses from natural hazard exposure, and in spite of the fact that some populations at risk may experience burdensome annual per capita losses from such exposures, these per capita burdens generally are insufficiently high to stimulate migrants and investors in property to avoid high hazard areas. The natural, climatologic, and other amenities associated with migrant receptor areas apparently outweigh, in the minds of migrant and investment risk takers, the burdens associated with high hazard exposures in those areas. Although low probability, high consequence events may be most disvalued by loss experiencers at the time of occurrence, these same events apparently do not motivate potential risk takers to retreat from situations in which those events may occur. Instead, the largest fraction of the U.S. population apparently draws its motivations from its knowledge of comparatively short-term cycles of events. Thus, such comparatively high probability but low consequence events as structural fires act as more powerful motivators for personal use of risk mitigating strategies than do such low probability, high consequence events as earthquakes and tsunamis.

6. As Compared with Other Significant Phenomena, Natural Hazard Exposures Do Not Produce High Annual Expected Losses of Lives, Jobs, and Homes. For 1970, the analysis revealed that an average of 784 deaths per year could be expected from the exposure of populations to earthquake, hurricane, storm surge, tornado, and tsunami hazards; and for the year 2000, an average of 1620 deaths per year can be expected from population exposures to the same hazards. Even if these annual expected mortality levels were doubled or tripled in order to take account of potential life loss from riverine flooding, the resulting numbers of death per year are not impressive when compared with other causes of death within the United States. Thus, for the 15- to 24-year age segment of the U.S. population, alone, annual deaths from all accidents total 27,175. Suicide within this same age group results in nearly 4200 deaths. Motor vehicle accidents in 1970 caused nearly 55,000 deaths among the total U.S. population and all other accidents were responsible for approximately 60,000 deaths. Suicide produced 23,500 deaths while homocides produced another 16,800. Diseases of early infancy were responsible for 43,200 deaths in the same year, while cirrhosis of the liver was responsible for 31,400 deaths (U.S. Bureau of the Census, 1975).

If a large dollar investment is necessary to substantially reduce expected life loss from natural hazard exposures, then it is more than conceivable that a diversion of those funds into other public life-saving strategies would secure even greater annual reductions in life loss. Thus, measured by number of human lives saved per year, there may be substantial "opportunity costs" associated with investment of national resources in hazard-mitigating policies in order to reduce expected life loss from population exposures to natural hazards.

7. Current Federal Policies Result in a Shift of a Significant Portion of Natural Hazard Loss Burdens to Nonexposed Populations. As a result of current federal income tax and disaster relief policies, a significant fraction of the annual expected losses from hazardous natural occurrences are shifted to nonexposed tax payers. In 1980 constant dollars, the value of the interstate tax transfers produced by these policies exceeded $2 billion in 1970 and was equal to approximately 20% of the net interstate tax transfers produced by all federal aid to state and local governments, except those concerning transportation and highways. These net interstate tax transfers occurred between a set of 30 "donor" and 21 "beneficiary" states. However, considered as a whole, the states which are the beneficiaries of federal policies concerning aid to state and local governments received only a slightly negative impact from hazard-induced interstate tax transfers. The latter reduced their net receipt of aid-related funds by approximately 2.2%

References

Bates, F. L. 1963. The Social and Psychological Consequences of a Natural Disaster: A Longitudinal Study of Hurricane Audrey. National Research Council Study No. 18. National Academy of Sciences, Washington, D.C.

Boucher, P. R., and M. Cardenas. 1981. Statement by SBA Inspector General. Emergency Preparedness News, May 21, 1981, pp. 77–78.

Cochrane, H. C. 1975. Natural Hazards and Their Distributive Effects. Program on Technology, Environment and Man. Monograph No. NSF-RA-E-75-003. Institute of Behavioral Science, University of Colorado, Boulder.

Comptroller General of the United States. 1980. Federal Disaster Assistance: What Should the Policy Be? U.S. General Accounting Office, Washington, D.C.

Council on Environmental Quality. 1971. Second Annual Report. U.S. Govt. Printing Office, Washington, D.C.

Dacy, D. C., and H. Kunreuther. 1969. The Economics of Natural Disasters, Implications for Federal Policy. Free Press, New York.

Haas, J. E., R. W. Kates, and M. J. Bowden. 1977. Reconstruction Following Disaster. MIT Press, Cambridge, Massachusetts.

Hart, G. C. 1976. Natural Hazards: Tornado, Hurricane, Severe Wind Loss Models. NTIS No. PB294594/AS. J.H. Wiggins Company, Redondo Beach, California.

Insurance Information Institute. 1971. Insurance Facts 1971. III, New York.

Jones, D. E., and W. G. Holtz. 1973. Expansive soils, the hidden disaster. Civ. Eng. 43(8):49–57.

National Safety Council. 1971. Accident Facts 1971. NSC, Chicago, Illinois.

Rettger, M. J., and R. Boisvert. 1979. Economics of federal flood insurance and loan programs. Search 9(2):

U.S. Bureau of the Census. 1975. Statistical Abstract of the United States. U.S. Govt. Printing Office, Washington, D.C.

U.S. Bureau of the Census. 1976. Statistical Abstract of the United States. U.S. Govt. Printing Office, Washington, D.C.

U.S. Bureau of the Census. 1980. Statistical Abstract of the United States, U.S. Govt. Printing Office, Washington, D.C.

U.S. Department of Commerce. 1976. Highlights of the National Household Fire Survey. U.S. Department of Commerce, Washington, D.C.

U.S. Department of Commerce, National Oceanic and Atmospheric Administration. 1976. Natural Hazard Management in Coastal Area. USDC/NOAA, Washington, D.C.

White, G. F., and E. J. Haas. 1975. Assessment of Research on Natural Hazards. MIT Press, Cambridge Massachusetts.

Wright, J., P. H. Rossi, S. R. Wright, and E. Weber-Burdin. 1979. After the Clean-up, Long-range Effects of Natural Disasters. Sage Publications, Beverly Hills, California.

6
Natural Hazard Mitigation Costs and Impacts

Introduction

In theory, 5 major classes of loss-reducing technologies may be employed to reduce the natural hazard losses projected for the year 2000. As shown in Figure 6-1 these include: *Avoidance, Area Structural Protection, Building Strengthening, Site Preparation,* and *Building Removal Strategies,* each of which is capable of being implemented through adoption or amendment of appropriate public investment, land use, disaster assistance, building code, and subdivision policies. However, since none of these strategies is completely free of cost to someone, somewhere, under some circumstances, a *loss reduction* (benefit) and *cost analysis* was performed for each identified strategy. The analysis revealed that these loss-reducing strategies could be grouped into three subsets: those whose use will result in (1) the largest estimated reductions in hazard exposure losses in year 2000 (Most Effective Group); (2) hazard exposure losses at least equal to the annual amortized cost of applying the mitigation (Cost Feasible Group); (3) the largest net savings (Highest Net Savings Group). The latter group was shown by the analysis to produce annual net savings equal to 13.7% of the losses expected for the year 2000, while the former were found to produce amortized mitigation costs substantially in excess of the annual loss reduction they secured.

Estimate of Losses for Year 2000

The analysis of the loss reduction effectiveness of each of the specific strategies required the preparation of 6 sets of annual expected building damage loss estimates. The first of these estimates for 1970 exposures are presented in Table 5-3. The remaining 5 sets of loss estimates are for the year 2000 and are based on the following different sets of assumptions.

Technology by class and title	Hazard to which applicable								
	Riverine flooding	Storm surge	Tsunami	Hurricane	Tornado	Severe wind	Earthquake	Landslide	Expansive soil
1.0 Hazard Avoidance Strategies and Technologies									
1.1 Zero growth on fifty-year flood plains after 1980		●	●						
1.2 Zero growth on 100-year flood plains after 1980		●	●						
1.3 Zero growth on fifty-year riverine flood plains in specified additional numbers of flood-prone cities each year, to 2000	●								
1.4 Zero growth in counties exhibiting high Tornado Strike Risk (greater than 10^{-4} tornado strikes per year per square mile).					●				
2.0 Area Structural Protection Strategies									
2.1 Structural protection (dams, levees, etc.) of cities with riverine flood problems.	●								
2.2 Construction of sea-walls to protect four additional counties per year from 100-year storm surge heights. Construct in order of decreasing damages in affected counties.		●							
3.0 Building Strengthening Strategies									
3.1 Require tie-downs on all mobile homes.				●	●	●			
3.2 Increase designed wind resistance capability of new buildings to level equalling 1.5 X the level specified in the Uniform Building Code (1.5 X UBC)				●	●	●			
3.3 Increase designed wind resistance capability of new buildings to level equalling 3.0 X the level specified in the Uniform Building Code (3.0 X UBC).				●	●	●			
3.4 Increase strength of new buildings to level required in UBC Earthquake Zone #3. (UBC 3).							●		
3.5 Floodproof 2% annually of all structures in fifty-year riverine flood plains to provide zero damage to height of four feet.	●								
3.6 Floodproof 2% annually, of all structures in 100-year riverine flood plains to provide zero damage to height of four feet.	●								
3.7 After 1980, floodproof all new buildings in storm surge areas to height of four feet.		●							
3.8 Modify and retrofit existing buildings in high seismic risk areas to meet seismic safety standards.							●		
4.0 Site Preparation Strategies									
4.1 Require soils testing and improved site grading standards in landslide-prone areas.								●	
4.2 Require soils testing and pre-construction moisture control and/or soil stabilization on construction sites.									●
5.0 Building Removal Strategies									
5.1 Purchase and/or condemn and accelerate removal of high vulnerability structures in high hazard areas.	●	●	●				●	●	

Figure 6-1. Hazard-mitigating technologies, by type and applicability to nine natural hazards.

Level One Estimates

These estimates provide a baseline against which the impact and cost–payoff relationships for selected mitigations may be calculated. For only 2 hazards, earthquake and riverine flooding, do level 1 and level 2 estimates differ. For these 2 hazards, level 1 estimates assume that mitigations *now being applied* are suspended in the period 1980 through 2000. In the case of riverine flooding, the estimate assumes that *current* levels of construction of area flood protection works will be halted in 1980, and that, beginning in the same year, no further progress will be made to limit building construction in the nation's 50 year riverine floodplains.

Level Two Estimates

These estimates represent the level of annual expected losses that will be experienced in 2000 if current population trends, migration patterns, building practices, and zoning policies are continued. In the case of wind resistance standards, the basic assumption is that all areas of the country are now observing equivalent UBC standards. Since many local areas may not be conforming to such standards, these loss estimates probably are somewhat lower than would be generated by a more detailed analysis.

Level Three Estimates

These estimates depict annual expected building damage losses in the year 2000 if the "most effective" mitigations examined in this study are applied to the management of each hazard. The "most effective" mitigation for any hazard produces the largest loss reduction in the year 2000, regardless of economic and/or political feasibility. These estimates should be considered as representing the *lowest attainable loss levels.*

Level Four Estimates

Utilizing gross methods for assessing the cost feasibility of alternative loss-reducing strategies, these estimates represent the loss levels that will be experienced in 2000 if the nation adopts those natural hazard management policies that will both reduce annual expected losses by the largest amount and, at the same time, accomplish this end without imposing annual mitigation costs that are higher than the annual loss reductions expected.

Level Five Estimates

These estimates are intended to represent the loss levels that will be experienced in 2000 if the nation adopts those natural hazard management policies that will produce the highest annual *net* loss reduction. For any specific hazard such a reduction is defined as the largest difference between the actual annual loss reduction achievable through use of any mitigation and the annualized cost of implementing that mitigation.

The 5 sets of building damage loss and mitigation cost estimates for the year 2000 are the major subjects of this chapter.

Alternative Mitigation Strategies

The five sets of strategies for which costs and benefits were calculated in this analysis are presented below.

Avoidance Strategies

As noted in Chapter 5, patterns of interstate population migration and capital investment are the major reasons for the projected $4 billion increase in natural hazard building damage losses between 1970 and 2000. The large projected increases in building wealth additions in high damage rate states and counties, principally along the Gulf Coast and South Atlantic, account for the major fraction of the large increases in annual expected natural hazard losses between 1970 and 2000. Loss-accelerating patterns of interstate migration are matched by similar movements of population and building investments to high-hazard areas internal to states and counties. If past patterns of development continue, there will be substantial increases in natural hazard losses. For example, if community growth in riverine flood plains continues as it has in the past decades, building damage losses from riverine floods can be expected to jump from an annual expected value of $1502 million in 1980 to $2634 million in 2000, assuming that no mitigations are applied during that period. Specifically, this loss estimate assumes that the construction of area flood control works is halted in 1980, that no regulation of growth in floodplains occurs thereafter, and that no building floodproofing mitigations are applied. Similarly, if continued development occurs in areas subject to storm surge and hurricane wind hazards, annual expected building damage losses from these phenomena can be expected to jump from $1127.5 million in 1970 to $2918 million in 2000.

Accordingly, this analysis sought to determine the natural hazard loss reductions that could be achieved by 2000 through the use of hazard avoidance policies. The impact of 4 such policies on annual expected hazard losses in the year 2000 are presented in Table 6-1. As shown, avoidance policies alone could produce about a 24% reduction in estimated building damage losses from tornado, tsunami, storm surge, and riverine flooding. Totaling $1370.6 million this annual loss reduction is equal to 13% of the estimated level 1 building damage loss estimates for all hazards in that year.

Loss reductions associated with avoidance mitigations for tornado, tsunami, and storm surge were derived directly from the computer loss models used by Hart (1976) and Lee et al. (1976). For riverine flooding, the loss reductions were calculated using computer outputs, as follows: (1) building wealth values and building damage losses in all six flood zones were determined for 1970 and 1980 (Tables 6-2 and 6-3); (2) building damage rates for each flood zone were then calculated from these data; and (3) the building damage rate for Zone F (area outside 100 year floodplain) was applied to the projected 1980 to 2000 increase in building wealth to flood-prone communities. The latter was determined from the average value of buildings shown in Tables 6-2 and 6-3 and from the building population data and from the 1980 baseline conditions presented in Tables 6-26 and 6-27.

Table 6-1. Estimated Payoffs in Year 2000 from Application of Selected Avoidance Strategies

		Annual expected building damage losses (millions 1970$)			
				Loss reduction	
Avoidance strategy	Hazard	2000 baseline loss	2000 loss with use of strategy	Total	%
1. Zero growth in counties at risk of more than 10^{-4} tornado strikes per year per square mile	Tornado	2058.0	1890.0	168	8.2
2. Zero residential growth on tsunami 100-year floodplains after 1980	Tsunami	19.8	19.5	0.3	1.5
3. Zero growth of all types on 100-year storm surge floodplains after 1980	Storm surge	1176.0	913.0	263.0	22.4
4. Zero net growth in 100-year riverine floodplains between 1980–2000	Riverine flooding	2634.0[a]	1694.2	939.8	35.7
Totals	—	5887.0	4516.4	1370.6	23.3

[a]Year 2000 expected losses for riverine flooding as shown in Table 5-8 are based upon the assumption of a continuation of current practice of providing structural protection and land use control in cities through the year 2000. The estimate of $2634 million loss is based upon the assumptions that current measures are discontinued in 1980 (Lee et al., 1978).

Table 6-2. Building Exposures, Losses, and Building Damage Rates, by Flood Zone in Flood-Prone Communities (1970 Exposures)

Variable	Zone A 2–5 yr. floodplain	Zone B 5–10 yr. floodplain	Zone C 10–25 yr. floodplain	Zone D 25–50 yr. floodplain
Average value, residential units	11,640	13,296	15,860	15,860
Average value, commercial-industrial gov't units	184,000	184,000	184,000	184,000
Residential units as percent of all units	61.5	61.5	61.5	61.5
Exposed units, all types (1000s)	614.8	683.1	910.8	1,115.7
Average value, all units	78,061.9	79,018.5	80,660.2	80,660.2
Total building value ($ millions)	47,992.5	54,020.6	73,465.3	89,922.5
Building losses, ($ millions)	758.3	392.6	287.9	159.8
Building damage rate (%)	1.58	0.727	0.392	0.178

Source: Lee et al. (1978).

As shown in Table 6-4, the annual expected building damage losses from riverine floods in 2000 may be held to $1694 million if new construction in 100 year floodplains is avoided between 1980 and 2000. This will occur even if the construction of new area flood control works is suspended during that period, and even if no other mitigations are employed to reduce losses within the 100 year floodplain. The total economic and noneconomic losses associated with use of the "most effective" mitigation for riverine floods and tsunami are presented in Tables 6-5 and 6-6.

Area Structural Protection Strategies

All other things being equal, it seems clear that the hazard-proneness of any given area may be reduced through construction of protective facilities to reduce the vulnerability to one or more specific hazards. Flood control dams, levees,

Zone E 50–100 yr. floodplain	Zone F 100 yr. + floodplain	Total	Zones A–D 50 yr. floodplain
15,860	15,860	–	–
184,000	184,000	–	–
61.5	61.5	61.5	61.5
1,229.6	4,554	9,108	2,324.4
80,660.2	80,660.2	80,366.4	79,855.3
99,179.7	367,326.5	731,977.3	265,471.0
111.8	190.6	1,901.0	1,598.6
0.113	0.052	0.259	0.602

channel improvements, seawalls, revetments, and similar constructed systems are exemplary of this approach.

In this analysis the flood control model revealed that annual expected losses for 1970 were approximately $1 billion below the value estimated for the modeled cities, had no flood control facilities been provided in prior years to protect the communities included in the model. The same riverine flood model showed that the annual expected building damage loss of $2634 million in the year 2000 could be reduced to $1653 million by equipping all flood-prone cities with 50 year area flood protection facilities.

Similarly, if counties subject to storm surge losses are protected through construction of seawalls capable of handling 100 year surge heights, annual expected losses from this hazard in the year 2000 can be reduced substantially. If four counties per year are so protected, in decreasing order of their annual expected storm surge damage losses, then annual expected storm surge losses in 2000 may

Table 6-3. Building Exposures, Losses and Building Damage Rates, by Flood Zone, in Flood-Prone Communities, 1980

Variable	Zone A 2–5 yr. floodplain	Zone B 5–10 yr. floodplain	Zone C 10–25 yr. floodplain	Zone D 25–50 yr. floodplain
1. Total buildings exposed[a] (1000s)	728.3	809.3	1,079.0	1,321.8
2. Average value	78,061.95	79,081.55	80,660.20	80,660.20
3. Total building value ($ million)	56,852.52	64,000.70	87,032.36	106,616.65
4. Building losses ($ million)	475.06	277.49	238.07	163.31
5. Building damage rate (%)	0.835	0.433	0.273	0.153

Source: Lee et al. (1978).

[a]These data assume no population growth diversion from floodplains between 1970–1980, but assumed extension of 50-year area flood protection to 200 communities per year, 1970–1980.

be reduced from approximately $1177 million to 335 million, *even if no other mitigations are applied* to deal with this hazard.

Building Strengthening Strategies

Substantial reductions in damage rates for selected mixes of buildings can be achieved through use of building strengthening technologies. Accordingly, 4 possible technology-forcing building code modifications were considered in this inquiry. The first involves modification of all U.S. Building Codes so as to impose 1973 Uniform Building Code, Zone 3 Earthquake standards to all structures constructed after 1980 in the United States in the seismically active areas mapped in that code (Wiggins, et al, 1978). The second involves the imposition of new nationwide construction standards producing 1.5 times the wind resistance level now required in the UBC. The third involves the imposition of new nationwide construction standards producing 3.0 times the wind resistance level now specified in the UBC (Hart, 1976). The fourth involves the requirement that all buildings in riverine floodplains be subject to floodproofing to a height of four feet (Lee, et al, 1978).

In the year 2000, enforcement of the most rigorous code modifications could reduce annual expected dollar losses of all types from $10,353 to $8027 million for earthquake, tornado, hurricane, and severe wind (Tables 6-7 and 6-8).

Zone E 50–100 yr. floodplain	Zone F 100 yr. + floodplain	Total	Zones A–D 50 yr. floodplain
1,456.6	5,395	10,790	3,938.4
80,660.20	80,660.20	80,336.42	79,855.32
117,489.65	435,161.78	867,153.66	314,502.23
129.88	218.61	1,502.50	1,154.00
0.110	0.050	0.173	0.366

The use of these "most effective" building strengthening mitigations could reduce annual losses in 2000 by approximately 8% for earthquake (Table 6-9), 24% for tornado (Table 6-10), 27% for hurricane (Table 6-11), and 23% for severe winds (Table 6-12).

Since the loss reduction analysis assumed that enforcement of these mitigations would be limited to structures constructed after 1980, the estimated loss reduction cited above fails to indicate the real scope of the improvement possible through the use of building-strengthening technologies. For example, had these same technologies been applied at the time of their construction to all buildings which were in place in 1970, expected annual losses in that year from earthquakes, tornadoes, hurricanes, and severe winds could have been reduced by increments ranging from 41% (hurricane) to 51% (severe winds) (Tables 6-13 and 6-14).

Substantial loss reductions also can be achieved through application of floodproofing standards to buildings located in 100 year riverine and storm surge floodplains. As shown in Table 6-15, a four foot floodproofing mitigation reduces storm surge losses in the year 2000 by nearly 59%. Similarly, the riverine flood model suggests that between 1980 and 2000, four foot floodproofing of 40% of the structures located in structurally protected 100 year riverine floodplains can reduce annual expected building damage losses in 2000 by approximately 17%.

Table 6-4. Impact of 100 Year Floodplain Avoidance Policy on Annual Expected Building Damage Losses in 2000

	1980			2000		
	Zones A–E	Zone F	Total	Zones A–E	Zone F	Total
Total structures exposed (1000s)	5331.0	5459.0	10,790	5331	10,398	15,729
Total value of structures ($ millions)	426,809.31	440,324.03	–	426,809.31	838,704.76	–
Average value of structures	79,855.32	80,660.2	–	79,855.32	80,660.2	–
Building damage rate (%)	0.297	0.051	–	0.297	0.051	–
Estimated loss ($ millions)	1266.05	224.76	1490.81	1266.05	428.16	1694.21

Table 6-5. Annual Expected Flood Losses and Loss Reductions, without Mitigation and with Most Effective Mitigation, Year 2000

	Annual expected loss, 2000		Loss reduction	
Variable	No mitigation	Most effective mitigation	$	%
1. Building damage[a]	2634	953.4	1680.6	63.8
2. Building contents[a] value	2559.13	940.78	1658.35	63.8
3. Income[a]	14.35	5.19	9.16	63.8
4. Supplier costs[a]	.17	.06	.11	64.7
Subtotal[a]	5247.65	1899.2	3348.22	63.8
5. Human lives	263	95	168	63.9
6. Housing units	–	–	–	–
7. Home use (person-years)	–	–	–	–
8. Employment (person-years)	2280.8	825.6	1455.2	63.8

[a]Millions 1970 dollars.

Building Removal Strategies

An additional strategy considered for reducing natural hazard losses involves the public purchase and removal of buildings from high-hazard areas. Riverine floodplains were selected for examination of the possible cost feasibility of this strategy.

As shown in Table 6-16, more than 60% of all 1970 annual expected building damage losses from riverine floods was produced by buildings located in the 2–10 year segments of the nation's riverine floodplains. Not including the value of the land on which these structures were located, the aggregate worth of these structures was equal to slightly less than the sum of 89 years of the annual losses expected to be sustained by these same structures. Public purchase and removal of these structures to flood Zone F in 1970 would have reduced total flood losses in that year by $1099 million and would have liberated much "open space" and potential park land for urban use. Even if land values were equal to 50% of the value of the structures in these flood zones, the expected annual reduction in building damage flood losses when combined with the recreational and aesthetic benefits associated with public ownership of these lands might well make this strategy attractive in some areas.

As an alternative to the aforementioned building removal strategy, consideration also might be given to application of this strategy to selected flood-insured structures that experience damage from any given flood event. Structures that

Table 6-6. Annual Expected Tsunami Losses and Loss Reductions, 1970–2000, Based on Applications of Zero Residential Growth Policy in Tsunami Zones

Variable	National value at risk		Annual expected losses				Loss reduction (Most effective mitigation)	
			No mitigation		Mitigation			
	1970	2000	1970	2000	50 year Floodplain	100 year Floodplain	Total	%
1. Building value[a]	217,327	508,852	8.7	19.8	19.7	19.5	.3	1.52
2. Building contents[a] value	117,907	608,870	5.54	19.1	19.0	18.8	.3	1.57
3. Income[a]	106,500	268,700	.727	1.479	1.453	1.433	.046	3.10
4. Supplier costs[a]	–	–	.0124	.2073	.0266	.0263	.0010	3.66
Subtotal[a]	–	–	14.98	40.41	40.18	39.76	.647	1.60
5. Human lives	18,200,851	23,606,782	20	44	44	43	1	2.27
6. Housing units	6,340,000	9,410,000	234	335	333	329	6	1.79
7. Home use (person-years)	–	–	345	389	380	376	13	3.34
8. Employment (person-years)	–	–	97.5	195.9	193.1	189.8	6.1	3.11

[a]Millions 1970 dollars.

Table 6-7. Annual Expected Building Damage Losses and Loss Reductions in 2000 as a Function of Specified Mitigations for Four Hazards

	Annual expected losses (millions 1970 $)				Reduction with most effective mitigation	
Hazard	No mitigation	UBC Zone 3 earthquake standards	UBC X 1.5 wind resistance standards	UBC X 3.0 wind resistance standards	Total	%
1. Earthquake	1177	1084	Not applicable	Not applicable	93.0	7.90
2. Tornado	2058	Not applicable	1746	1564	494.0	24.00
3. Hurricane wind	1742	Not applicable	1453	1280	462.0	26.52
4. Severe wind	24.8	Not applicable	21.2	19.1	5.7	23.00
Totals	5001.8	1084	3220.2	2863.1	1054.7	21.09

Table 6-8. Annual Expected Losses of All Types under Various Conditions of Exposure to Earthquake, Tornado, Hurricane, and Severe Wind Hazards, Year 2000

Variable	Annual expected loss, 2000: No mitigation	Annual expected loss, 2000: Most effective mitigation	Loss reduction: Total	Loss reduction: %
1. Building damage[a]	5,001.8	3,947.1	1,054.7	21.09
2. Building contents[a] value	4,302.6	3,294.9	1,007.7	23.42
3. Income[a]	1,035.8	777.2	258.6	24.97
4. Supplier costs[a]	12.24	7.76	4.48	36.60
Subtotal[a]	10,352.7	8,027.0	2,325.7	22.46
5. Human lives	1,484	1,185	299	20.15
6. Housing units	127,992	98.173	29.819	23.30
7. Home use (person-years)	157,583	117,773	39,810	25.26
8. Employment (person-years)	206,279	155,259	51,020	24.73

[a]Millions 1970 dollars.

might be targeted for such removal could be those meeting specific criteria based on the size of the insurance claim and the annual expected future damage rate for the repaired structure in relation to the removal cost minus the amount of the insurance claim. In this analysis, no estimates were made to determine the possible impact of this strategy.

Site Preparation Strategies

To deal with landslide and expansive soil hazards, 2 site preparation mitigations were considered (Wiggins, et al., 1978). For landslide, the mitigation involved the enforcement of improved site grading and soils testing ordinances between 1980 and 2000. For expansive soils, the mitigation involved chemical stabilization of soils on new single family dwelling unit sites between 1980 and 2000. Use of these mitigations result in the following loss estimates.

Hazard	2000 baseline (million $)	2000 with mitigation (million $)	Loss reduction: (million $)	Loss reduction: %
Landslide	871.2	349.4	521.8	59.9
Expansive soils	997.1	760.1	237.0	23.8
Totals	1868.3	1109.5	758.8	40.6

Table 6-9. Annual Expected Earthquake Losses and Loss Reductions, 1970–2000, Based on Application of UBC Zone 3 Construction Standards to All Structures Built after 1979

	National value at risk		Annual expected losses			Loss reduction	
Variable	1970	2000	1970	2000 (without mitigation)	2000 (with mitigation)	Total	%
1. Building value[a]	1,494,293.0	3,577,016.0	655.2	1,177.0	1,084	93	7.90
2. Building contents[a] value	858,053.0	4,466,392.0	123.23	372.8	341	32	8.58
3. Income[a]	757,428.0	1,953,000.0	2.65	3.91	3.62	.283	7.25
4. Supplier costs[a]	–	–	.030	.048	.043	.005	11.37
Subtotal[a]	–	–	781.1	1,553.7	1,428.7	125	8.05
5. Human lives	143,169,495.0	182,491,865.0	272	400	370	30	7.50
6. Housing units	47,790,000.0	69,960,000.0	20,485	22,888	20,585	2,303	10.06
7. Home use (person-years)	–	–	736	648	597	51	7.87
8. Employment (person-years)	–	–	414	635	585	50	7.88

[a]Millions 1970 dollars.

Table 6-10. Annual Expected Tornado Losses and Loss Reductions, 1970–2000, Based on Application of More Stringent UBC Wind Resistance Construction Standards to All Structures Built after 1979

	National value at risk		Annual expected losses				Loss reduction (Most effective mitigation)	
			No mitigation		With bldg. code mitigation			
Variable	1970	2000	1970	2000	2000 (1.5 X UBC)	2000 (3.0 X UBC)	Total	%
1. Building value[a]	1,788,989	4,300,491	879.8	2,058	1,746	1,564	494	24.00
2. Building contents[a] value	1,065,396	5,539,277	469.9	2,401	2,043	1,828	573	23.87
3. Income[a]	917,200	2,421,000	302.8	751	637	569	182	24.23
4. Supplier costs[a]	–	–	3.5	9.0	6.9	5.7	3.3	36.41
Subtotal[a]	–	–	1,656.0	5,219.0	4,432.9	3,966.7	1,252.3	24.00
5. Human lives	181,198.749	228,720,498	392	920	781	694	226	24.57
6. Housing units	60,010,000	86,990,000	36,212	52,119	44,081	39,303	12,816	24.59
7. Home use (person-years)	–	–	86,122	107,650	91,120.1	81,542	26,108	24.25
8. Employment (person-years)	–	–	57,541	146,569	124,610	110,774	35,795	24.42

[a]Millions 1970 dollars.

Table 6-11. Annual Expected Hurricane Losses and Loss Reductions, 1970–2000, Based on Application of More Stringent UBC Wind Resistance Construction Standards to All Structures Built after 1979

	National value at risk		Annual expected losses				Loss reduction (Most effective mitigation)	
			No mitigation		With bldg. code mitigation			
Variable	1970	2000	1970	2000	2000 (1.5 X UBC)	2000 (3.0 X UBC)	Total	%
1. Building value[a]	682,476	1,680,245	685.5	1,742	1,453	1,280	462	26.52
2. Building contents[a] value	379,800	2,038,041	267.6	1,505	1,256	1,105	400	26.58
3. Income[a]	343,400	913,000	101.8	276.2	229	201	75.2	27.23
4. Supplier costs[a]	–	–	1.0	3.1	2.3	1.9	1.1	37.74
Subtotal[a]	–	–	1,056.0	3,526.3	2,988.1	2,587.9	938.4	26.61
5. Human lives	62,741,264	83,615,242	62	153	127	112	41	26.80
6. Housing units	20,720,000	31,890,000	31,885	52,237	43,130	37,713	14,524	27.80
7. Home use (person-years)	–	–	34,505	48,271	39,857	34,860	13,411	27.78
8. Employment (person-years)	–	–	21,004	58,224	48,839	43,242	14,982	25.73

[a] Millions 1970 dollars.

Table 6-12. Annual Expected Losses and Loss Reductions from Severe Winds, 1970–2000, Based on Application of More Stringent UBC Wind Resistance Construction Standards to All Structures Built after 1979

	National value at risk		Annual expected losses				Loss reduction (Most effective mitigation)	
			No mitigation		With bldg. code mitigation			
Variable	1970	2000	1970	2000	2000 (1.5 X UBC)	2000 (3.0 X UBC)	Total	%
1. Building value[a]	2,064,507	4,924,075	11.4	24.8	21.2	19.1	5.7	23.0
2. Building contents[a] value	1,207,498	6,252,952	4.5	23.8	22.0	20.9	2.9	12.2
3. Income[a]	1,053,000	2,753,000	2.09	4.70	4.02	3.61	1.09	23.2
4. Supplier costs[a]	–	–	.02	.05	.04	.03	.01	33.3
Subtotal[a]	–	–	18.01	53.35	47.26	43.64	9.7	18.2
5. Human lives	203,260,531	256,179,160	5	11	10	9	2	18.2
6. Housing units	67,560,000	97,770,000	547	748	640	572	176	23.5
7. Home use (person-years)	–	–	852	1,014	865	774	240	23.7
8. Employment (person-years)	–	–	373	851	727	658	193	22.7

[a]Millions 1970 dollars.

Table 6-13. Building Damage Rates from Selected Mixes of Buildings Exposed to Earthquakes, Tornadoes, Hurricane Winds, and Severe Winds

Building Mix	Building damage rate for specified hazard (Annual expected damage ÷ building value)			
	Earthquake (Rate $\times 10^{-2}$)	Tornado (Rate $\times 10^{-3}$)	Hurricane (Rate $\times 10^{-2}$)	Severe wind (Rate $\times 10^{-3}$)
1. 1970 mix of buildings	0.438	0.492	0.100	.006
2. 2000 mix of buildings	0.329	0.47855	0.104	.005
3. New buildings constructed to UBC Zone 3 earthquake standards	0.251	Not applicable	Not applicable	Not applicable
4. New buildings constructed to UBC X 1.5 wind resistance standards	Not applicable	0.345	0.077	.003
5. New buildings constructed to UBC X 3.0 wind resistance standards	Not applicable	0.272	0.059	.003

Table 6-14. Annual National Expected Building Damage Losses from Exposure to Earthquake, Tornado, Hurricane, and Severe Wind under Various Conditions of Building Strength, 1970

	Annual expected loss (millions of dollars) in indicated year for specified hazard under indicated condition			
Conditions	Earthquake	Tornado	Hurricane	Severe wind
1. Actual conditions, 1970	655.2	879.8	686	11.4
2. UBC Zone 3 earthquake standards in all areas of the United States	374	Not applicable	Not applicable	Not applicable
3. UBC X 1.5 in all areas of the United States	Not applicable	617	525	7.1
4. UBC X 3.0 in all areas of the United States	Not applicable	487	406	5.6
Maximum possible savings under most effective mitigation $	281.2	392.8	280	5.8
Maximum possible savings under most effective mitigation %	42.92	44.65	40.82	50.88

Loss Reduction Analysis

A *most effective* mitigation strategy or combination of strategies was devised for each of the 9 hazards considered. Without regard to political or economic feasibility, they were the strategies that produced the largest reductions in annual expected losses for the year 2000. Each mitigation is summarized in Table 6-17. For earthquake, hurricane winds, severe winds, and tornado, they were the mitigations shown in Tables 6-8 to 6-12. For landslide and expansive soil, the most effective mitigations were those discussed under *Site Preparation Strategies* above. Similarly, for tsunami, the most effective strategy was that discussed under *Avoidance Strategies*.

For storm surge, the most effective mitigation determined through the loss reduction analysis was one which would involve the construction of seawalls capable of providing protection of surge-prone counties up to the 100 year surge height. The specific mitigation utilized in the analysis assumed that, between 1980 and the year 2000, four counties per year could be so protected, and that construction of such walls would proceed in order of decreasing volume of damages within each county. This mitigation provided a reduction in annual expected losses from storm surge in the year 2000, from $1177 million to a level of $335 million. However, 4-foot floodproofing of essentially all the buildings in the storm surge zone produced nearly equal loss reductions. In the

Table 6-15. Annual Expected Storm Surge Losses and Loss Reductions, 1970–2000, Based on Application of Four-Foot Floodproofing Standards to New Buildings Constructed after 1979

	National value at risk		Annual expected losses			Loss reduction (Most effective mitigation)	
			No mitigation		2000 Floodproofing mitigation		
Variable	1970	2000	1970	2000		Total	%
1. Building value[a]	438,733	1,061,972	442	1,176	479	697	59.27
2. Building contents[a] value	238,241	1,275,127	197	1,160	479	681	58.71
3. Income[a]	222,700	581,300	2.367	6.407	2.660	3.747	58.48
4. Supplier costs[a]	–	–	.028	.077	.049	.028	36.19
Subtotal[a]	–	–	641.4	2,342.5	960.7	1,381.8	58.99
5. Human lives	38,387,247	51,328,807	37	103	42	61	59.22
6. Housing units	12,890,000	19,890,000	24,521	43,757	16,917	26,882	61.43
7. Home use (person-years)	–	–	7,290	10,330	3,989	6,341	61.38
8. Employment (person-years)	–	–	370	1,018	511	507	49.80

[a]Millions 1970 dollars.

Table 6-16. Building Wealth and Building Damage Losses, 1970, by Riverine Floodplain Zone

	Building wealth			Annual expected building damage loss		
Floodplain zone	N ($ millions)	%	Cum %	N ($ millions)	%	Cum %
Zone A (2–5 year return)	47,992.50	6.56	6.56	758.3	39.9	39.9
Zone B (5–10 year return)	54,020.61	7.38	13.94	392.6	20.7	60.6
Zone C (10–25 year return)	73,465.31	10.04	23.98	287.9	15.2	75.8
Zone D (25–50 year return)	89,992.59	12.29	36.27	159.8	8.4	84.2
Zone E (50–100 year return)	99,179.78	13.55	49.82	111.8	5.9	90.1
Zone F (outside 100 year floodplain)	367,326.55	50.80	100.00	190.6	9.9	100.0
All zones	731,977.34	100.00	100.00	1,901.0	100.0	100.0

latter mitigation, the analysis revealed that storm surge losses could be reduced from $1176 to $479 million per year (Table 6-15).

The most effective mitigation for riverine floods involved the following mix of strategies: (1) provision of area flood protection facilities to 200 additional cities per year between 1970 and 1980; (2) application of "zero" net growth land use policies to the 100 year floodplains of all flood-prone cities between 1980 and 2000; and (3) removal of all structures from the 2–10 year floodplain by the year 2000. This combination of strategies produced an annual estimated riverine flood loss from damage to buildings totaling $954 million for the year 2000. Annual expected riverine flood losses of all types in 2000 are as shown in Table 6-18. As indicated, year 2000 baseline losses are reduced by 64% through the application of the most effective mitigations.

When these most effective mitigations are applied to all hazards in the year 2000, the annual expected building damage losses are as shown in Table 6-34. Annual expected total losses (i.e., building damage, contents damage, income loss, and supplier loss) are shown in Tables 6-19 and 6-20.

The reduction of loss is tied directly to the reduction of building damage. The most effective mitigations discussed above are but a few strategies available. Building damage reductions are achievable in the year 2000 through the use of other hazard specific mitigations. The following discussion covers the various mitigations considered for each hazard. In all cases the loss reductions were calculated using the year 2000 loss estimate as a baseline.

Table 6-17. Most Effective Mitigations by Hazard

Hazard	Mitigation
1. Hurricane winds 2. Tornado 3. Severe winds	Increase designed wind resistance capability of new buildings to level equaling 3 X the level specified in the UBC. Apply the mitigation to all new buildings constructed after 1979.
4. Earthquake	Increase strength of new buildings to level required in 1973 UBC for Earthquake Zone 3. Apply mitigation to all new buildings constructed in seismically active zones after 1979.
5. Expansive soils	Require application of chemical soil stabilization techniques to all residential construction sites, *or* improved foundation design. Apply mitigation to all new residential construction in low, moderate, and high expansive soil zones after 1979.
6. Landslide	Require soils testing and improved site-grading standards in all landslide-prone areas. Apply mitigation to new construction after 1979.
7. Tsunami	Permit zero net residential growth in tsunami-prone areas after 1979.
8. Storm surge	Construct sea walls to protect 4 additional counties each year from 100 year storm surge heights. Begin construction in 1980, in order of decreasing damages in affected counties.
9. Riverine flooding	Between 1970 and 1980, construct area flood control facilities to protect 200 additional cities each year from 50 year flood levels. Thereafter suspend all construction of flood control facilities and apply the following mitigations: (1) prohibit all net new growth in 100 year floodplain; and (2) purchase and remove all structures from 2–10 year floodplains.

Earthquake

The loss reduction estimates for earthquake are a product of a 2-part mitigation assumption: (1) that all new buildings constructed in California will continue to comply with the lateral force standards specified for 1973 UBC Earthquake Zone 3 structures; and (2) that, after 1980, all new buildings constructed in all *other* areas of the U.S. which are mapped as seismically active in the same 1973 UBC will also comply with Zone 3 standards. The loss estimates resulting from these assumptions are

	2000 baseline (million $)	2000 with most effective mitigation (million $)	Loss reduction	
			(million $)	(%)
California counties	952.1	748.7	203.4	21.36
All other	426.4	335.3	91.1	21.36
Total	1378.5	1084.0	294.5	21.36

Expansive Soils

All expansive soils loss estimates for 2000 were calculated utilizing the following procedure: (1) county-level losses in 1970 were aggregated, by state, and then increased by the estimated 1970-2000 growth rate for the state; (2) the sum of the resulting state-level loss estimates then was used as the baseline for the 2000 loss reduction calculation; (3) the national 1970 loss estimate was reduced by

Table 6-18. Total Annual Expected Riverine Flood Losses in 2000 under Three Selected Mitigations

	Annual expected losses) (millions 1970 $)			
			Reduction from baseline	
Condition	A Building damage	B Total (A X 1.99)[a]	N	%
1. Baseline conditions	2634	5241.66	–	–
2. Losses if all projected growth in Flood Zones A–E, 1980–2000, is shifted to Zone F	1694	3371.06	1870.6	35.7
3. Losses if Condition 2 exists and all 1980 buildings in Flood Zones A and B of unprotected communities are removed and replaced by structures in Zone F	1329.75	2646.21	2595.45	49.5
4. Losses if Condition 2 exists and all 1980 buildings in *all* Zones A and B are removed and replaced by structures in Zone F	953.62	1897.7	3343.96	63.8

[a]1.99 is a factor applied to building damage in order to account for contents losses, income losses, and supplier costs.

Table 6-19. Annual Expected Total Dollar Losses of All Types of Nine Natural Hazards under Two Conditions of Exposure, 2000[a]

Hazard	1970 dollar loss (all types)	2000 dollar losses (all types)		Achievable loss reduction	
		No mitigation (Level 1)	Most effective mitigation (Level 3)	$	%
Earthquake	781.1	1,553.7	1,428.7	125	8.05
Expansive soils	798.1	997.1	760.1	237	23.77
Landslide	370.3	871.2	349.4	521.8	59.89
Hurricane	1,056.0	3,526.3	2,587.9	938.4	26.61
Severe wind	18.0	53.4	43.64	9.7	18.2
Tornado	1,656.0	5,219.1	3,966.7	1,252.3	24.00
Riverine flood	2,758.3	5,241.7	1,897.3	3,344.4	63.8
Storm surge	641.2	2,342.9	960.7	1,381.8	58.99
Tsunami	15.00	40.4	39.73	.65	1.6

[a]Millions 1970 dollars.

10% to account for building retirements between 1980 and 2000; (4) the resulting value was subtracted from the 2000 baseline projection and the difference was assumed to be the loss chargeable to new residential construction between 1980 and 2000; and (5) this value was reduced 85% to account for loss reductions achievable through use of soil stabilization techniques and by 10% to

Table 6-20. Annual Expected Losses of all Types for Nine Natural Hazards under Two Conditions of Exposure, 2000

Variable	Annual expected loss, 2000		Loss reduction	
	No mitigation	Most effective mitigation	Total	%
1. Building damage[a]	10,699.9	6,364.3	4,335.6	40.5
2. Building contents[a] value	8,080.7	4,589.3	3,491.4	43.2
3. Income[a]	1,058.0	785.7	272.3	25.7
4. Supplier costs[a]	12.5	7.9	4.6	37.1
Subtotal[a]	19,851.4	11,747.2	8,103.9	40.8
5. Housing units	172,084	110,326	61,758	35.9
6. Home use (person-years)	168,302	120,937	47,365	28.1
7. Employment (person-years)	209,773.7	156,631.4	53,142.3	25.3

[a]Millions 1970 dollars.

account for loss reduction achievable through use of construction-site moisture control techniques. The latter reduction assumed that moisture control techniques can reduce losses by 30% in areas where the building season is marked by wide variability in rainfall levels and that approximately 30% of all expansive soil-affected construction will be in such areas. The loss estimates resulting from these calculations follow.

Mitigation 1. Application of soil stabilization techniques are required for all sites on which single family dwelling units are constructed after 1980. Loss estimates are

2000 baseline (million $)	2000 with most effective mitigation (million $)	Loss reduction	
		(million $)	(%)
997.1	760.1	237.0	23.77

Mitigation 2. Application of moisture control techniques to soils are required on 30% of all sites on which new single family dwelling units are constructed after 1980. Loss estimates are

2000 baseline (million $)	2000 with mitigation (million $)	Loss reduction	
		(million $)	(%)
997.1	969.2	27.9	2.8

Landslide

The experience of the City of Los Angeles suggests that adequate grading and soils analysis ordinances can reduce landslide losses by 97%. Accordingly, the loss reduction estimates presented below assume that 2000 losses associated with new construction between 1980 and 2000 can be reduced 97%. The annual new construction baseline losses for that year were calculated using the procedure described, above, for expansive soils.

2000 baseline (million $)	2000 with most effective mitigation (million $)	Loss reduction	
		(million $)	(%)
871.2	349.4	521.8	59.9

Riverine Flood

Mitigation 1. This *baseline* mitigation assumes that flood control facilities are constructed to protect, from 50 year flood heights, 200 additional cities each year between 1970 and 1980. Thereafter, no mitigations of any type are applied. Building damage losses are as follows:

1970 losses	$1901 million
1980 losses	$1502 million
2000 losses	$2634 million

Mitigation 2. Apply Mitigation 1 through 1980, then shift all net additional community growth to areas outside the 100 year floodplain and engage in no further construction of flood control projects. The resulting building damage losses appear below. The calculations for this mitigation assumed that (1) the building damage rate for Flood Zone F (the area outside the 100 year floodplain) is 0.05%; and (2) the total building growth in flood-prone cities between 1980 and 2000 will be 4,939,000 net new units.

The loss estimates are:

1. 1970 expected losses	$1901 million
2. 2000 baseline losses	$2634 million
3. 1980 expected losses	$1490.81 to $1502 million
4. 2000 expected losses from new growth in Zone F	$203.4 million
5. Total expected losses, 2000 (Line 3 + Line 4)	$1694.21 to $1706 million
6. Expected loss reduction (Line 2 minus Line 5)	$939.79 to $928.0 million
7. Percentage reduction	35.23 to 35.68%

Mitigation 3. Apply Mitigation 1 through 1980 and Mitigation 2 thereafter. Also, provide for removal of all structures in 2-10 year floodplains by 2000. This mitigation results in an annual expected building damage loss of $953.4 million in 2000, as shown by the following:

1. 1970 loss	$1901 million
2. 2000 baseline loss	$2634 million
3. 2000 loss under Mitigations 1 and 2	$1706 million
4. Loss chargeable to Flood Zones A and B in 1980	$740.4 million
5. 2000 loss, this Mitigation	$953.4 million (most effective)
6. Loss reduction from 2000 baseline	$1680.4 million
7. Percentage reduction	63.8%

Mitigation 4. Apply Mitigation 1 and provide flood control facilities to protect from 50 year flood heights 167 additional cities per year between 1980 and 1990; and 133 additional cities per year thereafter, until all cities are so protected.

2000 baseline (million $)	2000 with mitigation (million $)	Loss reduction (million $)	Loss reduction (%)
2634	1653	981	37.2

Mitigation 5. Apply Mitigations 1 and 4, and assume that, from 1970 to 2000, 25 cities per year provide for no net building additions in the 50 year floodplain.

2000 baseline (million $)	2000 with mitigation (million $)	Loss reduction (million $)	Loss reduction (%)
2634	1594	1040	39.5

Mitigation 6. Apply Mitigations 1 and 5, and provide flood control facilities to protect from 50 year flood heights 300 additional cities per year until all cities are so protected.

2000 baseline (million $)	2000 with mitigation (million $)	Loss reduction (million $)	Loss reduction (%)
2634	1330	1304	49.5

Mitigation 7. Apply Mitigations 1, 4, and 5, and beginning in 1980, floodproof to a height of four feet, 2% each year of all structures in the 50 year floodplain.

2000 baseline (million $)	2000 with mitigation (million $)	Loss reduction (million $)	Loss reduction (%)
2634	1406	1228	46.6

Mitigation 8. Apply Mitigations 1, 4, and 5, and provide that, beginning in 1980, 75 additional cities per year provide for no net building additions in the 50 year floodplain.

2000 baseline (million $)	2000 with mitigation (million $)	Loss reduction (million $)	Loss reduction (%)
2634	1570	1064	40.4

Mitigation 9. Apply Mitigations 1, 4, and 5, and beginning in 1980, floodproof to a height of four feet, 2% each year of all structures in the 100 year floodplain.

2000 baseline (million $)	2000 with mitigation (million $)	Loss reduction	
		(million $)	(%)
2634	1329	1305	49.5

Storm Surge

Mitigation 1. In all 100 year storm-surge floodplains, require all new buildings (1980–2000), all building replacements (1980–2000), and essentially all others to be floodproofed after 1980 to a level of 4 feet. Loss estimates are

2000 baseline (million $)	2000 with mitigation (million $)	Loss reduction	
		(million $)	(%)
1177.1	477.5	699.6	59.43

Mitigation 2. In decreasing order of the annual storm surge damages they sustain, protect 4 surge-vulnerable counties per year, after 1980, with seawalls constructed up to the 100 year surge height. Loss estimates are

2000 baseline (million $)	2000 with most effective mitigation (million $)	Loss reduction	
		(million $)	(%)
1177.1	334.8	842.3	71.6

Mitigation 3. After 1980, prohibit net new building construction on the 50 year floodplain (permits replacement of existing structures but no net growth in building stock). Loss estimates are

2000 baseline (million $)	2000 with mitigation (million $)	Loss reduction	
		(million $)	(%)
1177.1	912.8	264.3	22.5

Mitigation 4. After 1980, prohibit net new building construction on the 100 year floodplain (i.e., includes replacement of existing structures but no new growth in building stock). Loss estimates are

2000 baseline (million $)	2000 with mitigation (million $)	Loss reduction	
		(million $)	(%)
1177.1	912.3	264.8	22.5

Mitigation 5. After 1980, prohibit construction of buildings *with basements* on the 100 year floodplain. Loss estimates are

2000 baseline (million $)	2000 with mitigation (million $)	Loss reduction	
		(million $)	(%)
1177.1	922.7	254.4	21.6

Tsunami

Since building strengthening strategies offer little prospect of reducing tsunami-related damages to conventional buildings, the only mitigation tested is one that provides for no net increase in the residential building stock of tsunami-prone zones between 1980 and 2000. The large 1970-2000 increase in losses under this mitigation results from 2 factors: (1) continued increase in total building stock between 1970 and 1980; and (2) replacement of low-valued structures by high-valued structures between 1980 and 2000. Loss estimates are

2000 baseline (million $)	2000 with most effective mitigation (million $)	Loss reduction	
		(million $)	(%)
19.8	19.4	0.4	2.0

Wind Hazards

Mitigation 1. Increase wind-resistance capacity of all buildings constructed after 1980 to a level representing 1.5 X current UBC requirements. The estimated losses under this mitigation are

Hazard	2000 baseline (million $)	2000 with mitigation (million $)	Loss reduction	
			(million $)	(%)
Hurricane	1742	1453	289	16.6
Tornado	2058	1746	312	15.2
Severe wind	24.8	21.1	3.6	14.5
Total	3824.8	3220.2	604.6	15.8

Mitigation 2. Increase wind-resistance capacity of all new buildings constructed after 1980 to a level representing 3.0 × current UBC requirements. The estimated losses under this mitigation are

Hazard	2000 baseline (million $)	2000 with most effective mitigation (million $)	Loss reduction (million $)	Loss reduction (%)
Hurricane	1742	1280	462	26.5
Tornado	2058	1564	494	24.0
Severe wind	24.8	19.1	5.7	22.9
Total	3824.8	2863.1	961.7	25.1

Mitigation 3. Population growth is halted, after 1980, in all counties where tornado strike probability is greater than 10^{-4} per year per square mile:

2000 baseline (million $)	2000 with mitigation (million $)	Loss reduction (million $)	Loss reduction (%)
2058	1813	245	11.9

Cost Analysis

As noted earlier, projected building damage losses from all hazards for the year 2000 may be reduced by 42% from the year 2000 baseline (level 1) estimates through use of the "most effective" mitigations. This level of mitigation would hold annual expected building damage losses in the year 2000 to a value exceeding that of 1970 by only 11%, in spite of the substantial projected population increases between 1980 and 2000. As noted, however, the "most effective" mitigations were selected without regard for either their political or economic feasibility. Accordingly, a cost analysis was performed to determine the gross relationships between the estimated costs of selected loss-reducing strategies and the annual loss reductions expected through their application. The analysis focused on the cost and cost to loss reduction ratios of selected mitigations applicable to each hazard. Table 6-21 summarizes the *most effective cost feasible* mitigations determined by this analysis.

Avoidance Strategies

No direct dollar costs were assigned to avoidance strategies in this analysis on the assumption that, at the national scale, the bulk of the costs of such mitigations are simply "redistributional" in character (i.e., the "costs" to property owners in one area are offset by "benefits" received by property owners in other areas). Accordingly, for the purposes of this analysis, all avoidance strategies are considered cost effective by definition.

Table 6-21. Most Effective Cost Feasible Mitigations by Hazard

Hazard	Mitigation
1. Hurricane winds 2. Tornado 3. Severe winds	For new buildings constructed after 1979 in 229 cost-feasible counties (Table 6-25), increase designed wind resistance capability to level equaling 1.5 X the level specified in the UBC. In the cost-feasible subset numbering 72 of these same counties (Table 6-25), increase designated wind resistance capability to level equaling 3.0 X the level specified in the UBC.
4. Earthquake	Continue application of UBC Earthquake Zone 3 lateral force requirements in all California counties and extend requirements to the two non-California counties in which this mitigation is cost feasible (Table 6-25).
5. Expansive soils	Apply construction site moisture control techniques in areas where the building season is marked by wide variability in rainfall levels.
6. Landslide	Require soils testing and improved site-grading techniques in all landslide-prone areas. Apply mitigation to new construction after 1979.
7. Tsunami	Permit zero net residential growth in tsunami-prone areas after 1979.
8. Storm surge	Require 4 foot floodproofing of all new structures added to surge-prone zones (20′ or less, adjusted mean sea level) after 1979. Confine the use of the mitigation to the one-half of the storm surge area in which this mitigation is estimated to be cost feasible.
9. Riverine flooding	Provide area flood control facilities sufficient to protect all flood-prone communities from 50 year flood levels.

Site Preparation

The expansive soils analysis assumed the following values as representing the annual per capita losses associated with foundation failures, *only*, in areas with the indicated expansive soils rating:

High expansivity areas	\$26.00 per capita per year
Moderate expansivity areas	\$8.00 per capita per year
Zero and low expansivity areas	\$1.50 per capita per year

At an average of 3 persons per dwelling unit, the annual expected costs of expansive soils damage per structure in "high" expansive soils zones therefore is \$78.00. Assuming a cost of \$0.65 per square foot for application of soil stabilization measures and an average application of 2000 square feet per structure site, the mitigation cost per structure is \$1300. Since such mitigation measures are estimated to yield an 85% reduction in annual expected losses, the cost effectiveness of this mitigation is as follows:

Expected annual costs without mitigation	\$78.00
Expected annual costs with mitigation	\$11.70
Expected annual loss reduction	\$66.30
One time mitigation cost	\$1300.00
Payout period required, in years (0.0% interest)	22.2
Fifty year annual mitigation cost at 0.0% interest	\$26.00
Ratio of annual loss reduction to annual cost	2.55:1
Fifty year annual mitigation cost at 6% interest	\$82.11
Ratio of annual loss reduction to annual cost	0.81:1

In appropriate areas, control of the moisture content of soils during construction is estimated to yield a 30% reduction in annual expected losses from expansive soils while creating only negligible additional construction costs. Thus, if the interest value of investment capital for the 50 year repayment period is to be considered, then only site moisture control methods are cost effective. Thus, approximately 30% of that fraction of the year 2000 annual expected losses from expansive soils that are attributable to new construction between 1980 and 2000 should be considered cost feasible, and the total cost feasible loss reduction is only \$2.8 million in the latter year.

Building Strengthening Strategies

Building strengthening strategies are defined as including (1) floodproofing of structures in both storm surge zones and in riverine floodplains; (2) increasing the wind-resistance capacity of new buildings to a level representing either 1.5 or 3.0 × the level specified in the UBC; and (3) increasing the capacity of new structures in seismically active areas to withstand ground shaking to the level specified for Earthquake Zone 3 in the UBC.

As a first step in this analysis, it was necessary to prepare a set of tables expressing the proportionate increase in building costs expected with the enforcement of selected building code mitigations. The cost factors, which were developed either through examination of relevant literature or through use of expert panels, are presented in Chapter 4. Using the cost factors, the total national costs were calculated for implementation of these mitigations on the several mixes of buildings to be constructed between 1980 and 2000 (Table 6-22).

A 50 year period was selected as the effective owner-life of each building mix. By treating all buildings constructed in each category between 1980 and 2000 as though they constituted a single cohort built at a single time, it was possible to calculate the annual amortized cost for each mitigation, as well as the ratio of annual loss reduction to annual 50 year repayment cost for each mitigation at three specified interest rates (Table 6-23).

Excluding riverine flood mitigations, considered elsewhere, and assuming a cost-feasible mitigation is defined as exhibiting an annual loss reduction to 50 year repayment ratio of 1/1 or greater, then only one of the mitigations con-

Table 6-22. Estimated Building Value Increases and Hazard Mitigation Costs, 1980–2000[a]

Building location	Total building value, 1980	Total building value, 2000	Building wealth increase, 1980–2000	Mitigation applicable to new buildings	Mitigation cost as % of building value[b]	Estimated cost of mitigation
1. 100 year riverine floodplain	Not applicable	Not applicable	Not applicable	2 ft. protection	13.2	Not applicable
				4 ft. protection	16.6	Not applicable
2. Hurricane affected areas[c]	926,502	1,680,245	753,743	1.5 X UBC	2.1	15,828.6
				3 X UBC	5.1	38,440.9
3. Storm surge affected areas all areas of exposed counties[d]	593,278	1,061,972	468,694	2 ft. protection	13.2	61,867.6
				4 ft. protection	16.6	77,803.2
4. Earthquake affected areas[e]	1,692,350	3,037,664	1,345,314	2 X UBC Zone 3	6.5	87,714.6
				Zone 3 UBC STIS.	2.1	27,699.3
5. Tornado affected areas[c]	2,383,114	4,300,491	1,917,377	1.5 X UBC	2.2	42,813.2
				3 X UBC	5.4	103,974.8
6. Severe wind areas[c]	2,747,291	4,924,075	2,176,784	1.5 X UBC	2.1	45,712.5
				3 X UBC	5.1	111,015.9

[a]Millions of 1970 dollars.
[b]See Tables 4-12, 4-13, and 4-14.
[c]Hart, 1976.
[d]Lee et al., 1978.
[e]Wiggins et al., 1978 (excludes California).

sidered can be judged to be cost effective at a 0.0% interest level for cost repayment. No mitigation can be scored as cost feasible when 50 year repayment of mitigation costs were calculated at 6.0% and 9.0% interest, respectively (Table 6-23).

The annual expected building damage rates were calculated for buildings constructed to "normal" standards, in order to yield, at the "mitigated" level of construction, a loss reduction to cost repayment ratio of 0.75/1 or greater for a building life of 50 years and an interest rate of 6.1% (see Chapter 4). The calculations were based on three parameters: (1) the reduction in building damage rate expected to be experienced by buildings constructed in compliance with a specific mitigation in a specified risk category (as indicated by the baseline building damage rate); (2) the time period (50 years) over which reduced building damage rates are expected to be experienced and loss reductions (savings) therefore accumulated; (3) the rate (6.1%) at which future loss reductions (savings) are to be discounted to account for the value at interest of current investments.

The break-even damage rates presented in Table 6-23 include loss reductions chargeable to nonbuilding damage, and are defined as the baseline rate that must be experienced for a specified hazard by a county level aggregation of buildings in order for total loss reduction in that county to equal total mitigation costs. A single damage rate was calculated for all wind hazards considered as a group. The rates are therefore not a function of the region in which the buildings are located, but of hazard intensity, structure type, discount rate, and estimated life of building. These "break-even" damage rates were then used to identify the set of counties in the United States in which application of each specified type of mitigation to newly constructed buildings could be considered to be cost feasible (i.e., exhibit loss reduction to cost ratios of 0.75/1, or higher). The resulting set therefore consists of those counties in which appliction of the specified mitigation to new construction between 1980 and 2000 will result in projected 50 year building life accumulated savings in 2000 that, at a 6.1% discount rate, are equal to the estimated initial cost of the mitigation at the time of building construction (Table 6-24).

Application of this procedure resulted in identification of a comparatively small number of U.S. counties where cost-feasible building strengthening mitigations may be employed to reduce annual expected losses from earthquakes, storm surge, and all wind hazards. At the "most effective" levels of mitigation, and loss reduction to cost ratios of 0.75:1 or greater, 104 counties qualify as candidates for selected wind resisting mitigations, 24 for storm surge and 7 for California counties for earthquake. In counties where the loss reduction to cost ratios were 1:1 or greater for the "most effective" mitigation, annual building damage losses in 1970 totaled 11% of all national losses for all wind hazards, 5% for all non-California earthquake hazards, and 13% for all storm surge losses (Table 6-25). These values may be used as a gross estimate to determine the proportion of the "most effective" loss reductions judged to be "cost feasible."

Table 6-23. Relationship between National Building Strengthening Mitigation Costs and Annual Estimated Natural Hazard Loss Reductions, 2000

Building location	Initial mitigation cost[a]	Annual estimated loss reduction: Building losses[a]	Annual estimated loss reduction: All losses[a]	Required payout period at 0.0% interest (in years)
1. Hurricane wind zones	38,440.9	462	933.24	41.19
2. Storm surge zones	15,828.6	699.6	1292.2	55.86
3. Earthquake zones	77,803.2	294.5	389.0	71.21
	61,867.6			
4. Tornado zones	103,974.8	494.0	1252.8	82.99
	42,813.2			
5. Severe winds	111,015.9	5.7	9.2	12066.9

[a]Millions of 1970 dollars.

Riverine Flood Mitigations

The cost feasibility of several major policy alternatives for reducing riverine flood losses in the year 2000 were examined. In all cases, 1980 was selected as the date for initiation of each policy, and the conditions shown in Table 6-26 were assumed to exist in that year. Exposed building values, damage rates, and loss for each flood zone are given in Table 6-27.

These baseline conditions were developed as follows:

1. The proportion of U.S. structures, building wealth, and population located in flood-prone cities in the period 1970 to 2000 was as shown in Table 6-28. The number of cities with area structural protection against 50 year floods represent basic assumptions used in developing the flood model that establishes the 1980 base values against which the impacts of future alternative loss-reducing strategies might be tested. The percentage of *all* buildings located in cities provided with area protection against 50 year floods were determined using the data presented in Tables 6-2, 6-3 and 6-28.
2. As shown in Table 6-29 building damage rates for Zones F, E, and the portion of the floodplain represented by Zones A through D were determined for 1970, 1980, 2000 and 1970 assuming no flood protection. Computer model values assume only one mitigation, provision of area facilities to protect communities against 50 year floods with year 2000 representing complete protection. The 1970 no flood protection assumption represents a condition in which no 50 year protection has been extended to any city.

Annual 50 year amortized cost (12 payments/yr.)			Ratio of annual loss reduction to annual amortized 50 yr. cost			
0.0%	6.0%	9.0%	0.0%	6.0%	9.0%	Mitigation
768.8	2428.26	3499.20	1.21/1	0.38/1	0.27/1	3 X UBC
						1 X UBC
1556.1	4914.67	7082.27	0.90/1	0.28/1	0.20/1	4 ft. floodproofing
						2 ft. floodproofing
554.0	1749.71	2521.41	0.70/1	0.22/1	0.15/1	UBC Zone 3 STDS.
2079.5	6567.83	9464.62	0.60/1	0.19/1	0.13/1	3 X UBC
						1.5 X UBC
2220.3	7012.60	10105.56	0.004/1	0.001/1	0.0009/1	3 X UBC
						1.5 X UBC

3. To determine the costs and payoffs associated with alternative mitigations, building damage rates for Zones A through D were needed for 2 conditions: (1) no area protection provided against 50 year floods; and (2), area protection provided against 50 year floods. The rates for each of these conditions (Tables 6-27 and 6-29) were determined as follows: (a) for both conditions rates for Flood Zones F and E were set at the average of the rates shown for these zones in Tables 6-2 and 6-3, or 0.051% and 0.112% respectively; (b) the 1970–1980 reduction in damage rates in each of Zones A through D for each 1% of additional structures provided with 50 year flood protection was determined; (c) these values then were applied to the percent increase in structure protection between 1980 and 2000 to determine individual flood zone rates.
   These rates are defined as the building damage rates for Zones A–F under a condition where all structures have been provided with area protection against 50 year floods. Using these rates, the percent increase in buildings with area protection against 50 year floods for 1970 (Table 6-28), and the data shown in 6-2, the building damage rates were determined for flood Zones A–D where no area protection is provided against 50 year floods. The full range of these rates is displayed in Table 6-29.
4. For baseline purposes, 1980 riverine flood exposures, losses and building damage rates in floodprone cities were assumed to be as shown in Table 6-26. Values for all flood-prone cities were derived using the following procedure: (a) the total number of exposed structures shown in Table 6-28 for 1980 was distributed among the several zones in the proportions shown in Table 4-3;

Table 6-24. U.S. Counties, Classified by Cost-Loss Reduction Ratio Associated with Selected Natural Hazard Technical Mitigations, 1980–2000

		Number of counties in specified category of cost-loss reduction ratios					
Hazard	Technical mitigation	1–0.75 or greater	1–1.00 or greater	1–1.25 or greater	1–1.50 or greater	1–1.75 or greater	1–2.00 or greater
1. Earthquake[a]	Imposition of Earthquake Zone 3 building code standards	7	2	2	2	2	2
2. Storm surge	4 ft. elevation of structure, above floodplain level	24	17	11	8	6	2
3. Storm surge	2 ft. elevation of structures above floodplain level	30	18	17	12	9	7
4. Storm surge	Floodproofing of structures to 2 ft. elevation above level of floodplain	36	25	18	13	10	8
5. Winds	Escalation of building code standards to level equaling 1.5 X UBC requirements	250	229	102	76	52	28
6. Winds	Escalation of building code standards to level equaling 3.0 X UBC requirements	104	72	28	20	16	12

[a]Does not include California counties, since these standards already in effect.

(b) the number of structures allocated to Zones A–D then was depressed by 64,000 to account for the anticipated 1970–1980 diversion of population growth away from the fifty-year floodplains as a result of on-going floodplain zoning activities during that period (Lee et al., 1978). The 64,000 was then added to Zone F; (c) the resulting building counts then were distributed to "protected" and "unprotected" communities in accordance with the percentages presented in Table 6-28, Row 7. Building damage rates and average structure values were taken from Tables 6-29 and 6-2, respectively.

Cost factors for floodproofing and elevating structures were based on the storm surge flood proofing cost factors (Table 4-14). Property purchase costs were calculated from the building values presented in Table 6-26 to which were applied a land purchase cost factor of 6.95% (Lee et al., 1978) and a factor of 21.6% to account for conveyance, demolition, and site restoration costs (Johnson, 1978). Costs of area flood protection were derived by first setting 1936–1976 U.S. expenditures for area flood control works at $10.5 billion (Miller, 1978). The $10.5 billion was then converted into 1970 "constant" dollars by assuming that it was equally divided among three decades (1940–1950, 1950–1960, and 1960–1970) and that the complete investment was made at the midpoint of each decade. Based on these assumptions, the accumulated 1930 to 1969 investment in U.S. flood control facilities was found to total $13,523.4 million 1970 dollars. Assuming that 2037 cities had been protected from 50 year floods in 1970 as a result of this investment (Table 6-28), the cost per city protected was set at $6.638 million.

Using the data discussed above, the cost feasibility of alternative riverine flood mitigations was found to be as shown below. All annual repayment costs were calculated on the basis of equal monthly payments, and annual estimated *total* loss reductions were based on the ratio of total losses to building losses in the appropriate year. For 1970 this ratio was 1.45:1, while for 2000, it was 1.99:1.

Area Protection against Fifty Year Floods

Without considering the annual costs of maintenance and operation, the cost feasibility of pre-1970 investments in flood control projects was estimated as follows:

Initial mitigation costs (IMC)	$13,523 million
Annual estimated building loss reductions (BLR), 1970	$999 million
Annual estimated total loss reduction (TLR), 1970	$1,449 million
Payout period (years) (IMC/TLR)	9.33
Fifty year annual repayment cost (0.0%)	$270 million
Ratio of annual loss reduction (ALR) to annual repayment cost (ARC)	5.36:1

Table 6-25. Selected Data for Counties in Which 50 Year Loss Reduction from Use of Natural Hazard Structural Mitigations Is Equal to or Less than the Cost of Applying the Mitigation[a]

Hazard	Mitigation	Number of counties with savings-cost ratio of 1:1 or greater	1970 value of structure wealth at risk (millions $)	Damage rate at midpoint of range for counties in class
Earthquake[b]	UBC Zone 3 standards	2	307.6	0.03340
Storm surge	4 ft. elevation of structures	17	4,957.9	0.01115
Storm surge	2 ft. elevation of structures	18	5,519.4	0.01445
Storm surge	2 ft. floodproofing	25	23,351.3	0.00588
Wind	1.5 UBC	229	200,100.0	0.00213
Wind	3.0 UBC	72	54,650.0	0.00339

[a]Defined as savings–cost ratio of 1:1, or greater.
[b]California counties are not included.

Fifty year annual repayment cost (6.0%)	$854 million
Ratio of ALR to ARC	1.7:1
Fifty year annual repayment cost (9.0%)	$1,231 million
Ratio of ALR to ARC	1.18:1

Using loss projections from the riverine flood model and estimated flood protection costs of $6.6 million per urban community, the cost feasibility of 1970–2000 investments in area facilities to protect against 50 year floods was estimated as follows:

Number of cities to be protected	4,418
Initial mitigation cost	$29,327 million
Annual estimated BLR, 2000	$1,628 million
Annual estimated TLR, 2000	$3,240 million
Payout period (years)	9.05
Fifty year annual repayment cost (0.0%)	$587 million
Ratio of ALR to ARC	5.5:1
Fifty year annual repayment cost (6.0%)	$1,853 million
Ratio of ALR to ARC	1.75:1
Fifty year annual repayment cost (9.0%)	$2,700 million
Ratio of ALR to ARC	1.21:1

Estimated annual building losses, 1970 (millions $)	Estimated value of new construction 1980–2000 (millions $)	Cost of mitigation, 1980–2000 (millions $)	Total 50 year loss reductions expected from use of migitation (millions $)	Net 50 year savings expected (millions $)	50 year savings as percent of cumulative building value (millions $)
10.27	749.00	22.5	85.2	62.7	8.37%
55.28	13,096.36	2,173.0	3,169.0	996.7	7.61%
79.76	14,689.00	1,939.0	3,287.0	1,347.0	9.17%
137.19	69,871.79	8,175.0	10,350.0	2,170.0	3.1%
425.21	475,428.57	9,984.0	16,710.0	6,729.0	1.13%
185.26	156,549.00	7,984.0	13,850.0	5,598.0	3.58%

Based on these crude calculations, both pre-1970 and 1970–2000 investments in area systems to protect against 50 year floods may be considered to be cost feasible at all three interest rates employed (0.0, 6.0 and 9.0%). At the 6.0% interest level, pre-1970 investments in flood control works produced net savings (annual total loss reduction minus 50 year ARC) totaling $595 million per year, or a net annual reduction of 20.5% for projected loss ($2900 million) that would have occurred without the investment. Of course, to the extent that these investments stimulated construction in 50 year floodplains, which would not otherwise have taken place, and stimulated additions to the stock of flood-damageable urban infrastructure (streets, utility lines, sidewalks, etc.), these *net savings* may, in fact, be reduced to values approaching zero. Since losses to urban infrastructure were *not* estimated in this study, no offsetting reduction has been applied to this "net savings" value.

To determine the cost feasibility of avoidance and area flood protection strategies between 1980 and 2000, the data presented in Table 6-30 were used.

Assuming that avoidance strategies result in no direct costs, a shift of all 1980-2000 building growth to Flood Zone F will produce a net total loss reduction per year of $955 million ($480 million × 1.99). In contrast, an area flood protection strategy will produce a *gross* annual loss reduction of $1065

Table 6-26. Expected Building Exposures, Losses, and Building Damage Rates, 1980, by Flood Zone and Flood Protection Level of Flood-Prone Communities

Variable	Zone A–D 2–50 yr. floodplain	Zone E 50–100 yr. floodplain	Zone F area outside 100 yr. floodplain	All Zones
		All flood-prone cities		
1. Structures exposed (1000s)[a]	3,873.4	1,456.7	5,459.0	10,790
2. Average value of structures[a]	79,855.32	80,660.20	80,660.20	–
3. Total value of buildings ($ million)[a]	309,311.60	117,497.71	440,324.03	867,133.34
Percent:	35.67	13.55	50.78	100.0
4. Annual expected building damage losses ($ millions)	1,134.86	131.19	224.76	1,490.81
5. Annual expected building damage rate (%)	0.3669	0.11165	0.05105	0.1719
	Cities protected by area facilities against 50 year floods			
1. Structures exposed (1000s)[a]	3,245.13	1,220.42	4,573.55	9,030.86
2. Average value of structures[a]	79,855.32	80,660.20	80,660.20	–
3. Total value of buildings ($ million)[a]	259,140.89	98,439.32	368,903.46	726,483.67
Percent:	35.67	13.55	50.78	100.0
4. Annual expected building damage losses ($ million)	643.188	109.908	188.325	941.421
5. Annual expected building damage rate (%)	0.2482	0.11165	0.05105	0.1296
	Cities not protected by area facilities against 50 year floods			
1. Structures exposed (1000s)[a]	628.27	236.28	885.45	1,750.14
2. Average value of structures[a]	79,855.32	80,660.20	80,660.20	–
3. Total value of buildings ($ million)[a]	50,170.71	19,058.39	71,420.57	140,649.67
Percent:	35.67	13.55	50.78	100.0
4. Annual expected building damage losses ($ million)	491.723	21.279	36.46	549.462
5. Annual expected building damage rate (%)	0.9801	0.11165	0.05105	0.3907

[a]1970 dollars.

Table 6-27. Expected Building Wealth Exposures, Damage Rates, and Losses in Individual Flood Zones, 1980

Flood Zone	Exposed structures (1000s)[a]	Building wealth ($ million)[a]	Damage rate (%)	Loss ($ million)[a]
	Cities without area protection against 50 year floods			
Zone A	116.16	9,068.223	2.7764	251.77
Zone B	129.10	10,210.140	1.1980	122.31
Zone C	172.14	13,885.331	0.5819	80.79
Zone D	210.84	17,006.961	0.2168	36.87
Zone E	236.28	10,005.390	0.1116	21.22
Zone F	885.45	71,420.570	0.0510	36.46
ΣZone A–C	417.42	33,163.690	1.3716	454.80
ΣZone A–D	628.27	50,170.710	0.9801	491.75
ΣZone A–E	864.69	69,176.050	0.7416	512.97
ΣZone A–F	1,750.14	140,595.615	0.3908	549.43
	Cities with area protection against 50 year floods			
Zone A	600.02	46,838.73	0.4597	215.31
Zone B	666.87	52,881.83	0.2855	150.97
Zone C	889.17	71,720.63	0.2139	153.41
Zone D	1,089.07	87,844.60	0.14088	123.75
Zone E	1,220.42	98,439.32	0.11165	109.90
Zone F	4,573.55	368,903.46	0.0515	188.32
ΣZone A–C	2,156.06	171,296.29	0.3052	519.70
ΣZone A–D	3,245.13	259,140.89	0.2483	643.46
ΣZone A–E	4,466.31	357,580.21	0.2107	753.37
ΣZone A–F	9,039.86	726,483.67	0.1296	941.69
		All flood prone cities		
Zone A	–	–	–	467.09
Zone B	–	–	–	273.30
Zone C	–	–	–	234.21
Zone D	–	–	–	160.63
Zone E	1,456.7	117,497.71	0.1116	131.19
Zone F	5,459.0	440,324.03	0.0510	224.76
ΣZone A–C	–	–	–	974.59
ΣZone A–D	3,873.4	309,311.60	0.3669	1,134.86
ΣZone A–E	5,331.0	–	–	1,266.35
ΣZone A–F	10,790.0	–	–	1,491.13

[a]1970 dollars.

million ($535 million × 1.99). Use of such a strategy would require the flood protection of 2418 communities at a cost of $6.6 million, each. Thus, the net total annual savings from area flood protection between 1980–2000 is as follows:

Initial mitigation cost (2418 × 6.6 million)	$16,051 million
Annual 50 year repayment cost (6.0%)	$1,014 million

Annual loss reduction	$1,065 million
Net loss reduction	$51 million

Although such strategy clearly is cost feasible, much larger net savings are achievable by a strategy that prohibits net new growth in the 100 year flood-plains. Also if an avoidance strategy is adopted, then the provision of flood protection facilities to 1980 structures in Zones A-D will not be cost feasible. Such protection would reduce Flood Zone A-D losses in unprotected communities from $492 million per year for an annual loss reduction of only $367 million.

Building Removal Strategies

To determine the cost feasibility of building removal strategies between 1980 and 2000, the data presented in Table 6-31 were used. As shown there, this strategy is effective only at the 0.0% interest level in Flood Zone A. At 6% it is cost feasible in no flood zone. Of course, if the potential recreational and public use benefits were calculated for the public open space liberated as a result, then the policy might be considered cost feasible in at least Flood Zone A, of "unprotected" communities. In any event, no net savings are realized from this mitigation.

Floodproofing of Structures

The final riverine flood cost feasibility test was applied to a joint strategy involving both structure floodproofing and provision of area protection against 50 year floods. Data utilized in the analysis are shown in Table 6-30. The mitigation involves the 4-foot floodproofing of all structures in Flood Zones A-D under a condition in which all communities have been provided with area pro-

Table 6-28. Selected Attributes of Cities in Flood Model, 1970–2000

Variable	1970	1980	2000
1. Number of cities	5,539	–	6,455
2. Number of buildings (1000s)[a]	9,108	10,790	15,729
3. Building wealth ($ million)[a]	731,977	867,153	1,264,083
4. Structures per city	1,644.0	1,846.0	2,116.0
5. Building wealth per city[a]	132.15	–	195.83
6. Number of cities with area protection against 50 year floods	2,037	4,037	–
7. Percent of all structures provided with area protection against 50 year floods	51.64	83.78	–
8. Population: total (1000s)[a] per city	24,712	29,292	42,769

Source: Lee et al., 1978
[a]1970 dollars

Table 6-29. Building Exposures, Losses, and Damage Rates under Selected Conditions, by Flood Zone, 1970–2000

Variable	1970	1980	2000	1970 No flood protection
1. Number of exposed buildings (1000s)	9,108	10,790	15,729	9,108
2. Building wealth (millions 1970 $)				
Flood Zone A	47,992.50	56,852.52	82,878.4	47,992.50
Flood Zone B	54,020.61	64,000.70	93,292.5	54,020.61
Flood Zone C	73,465.31	87,032.36	126,870.4	73,465.31
Flood Zone D	89 992.59	106,616.65	155,416.1	89,992.50
Flood Zone E	99,179.78	117,489.65	171,273.9	99,179.78
Flood Zone F	367,326.55	435,161.78	634,352.0	367,326.55
Flood Zone A–D	265,471.01	315,502.23	458,457.4	265,471.01
3. Building damage losses (millions 1970 $)				
Flood Zone A	758.3	475.06	380.98	1,332.46
Flood Zone B	392.6	277.49	266.39	647.16
Flood Zone C	287.9	238.07	266.39	427.49
Flood Zone D	159.8	163.31	218.95	195.10
Flood Zone E	111.8	129.88	191.23	110.73
Flood Zone F	190.6	218.61	328.84	187.52
Flood Zones A–D	1,598.6	1,153.93	1,137.65	2,602.23
Flood Zones A–F	1,901.1	1,502.43	1,657.72	2,900.48
4. Building damage rates (%)				
Flood Zone A	1.5800	0.8356	0.4597	2.7764
Flood Zone B	0.7268	0.4336	0.2855	1.1980
Flood Zone C	0.3919	0.2736	0.2139	0.5819
Flood Zone D	0.1776	0.1532	0.1408	0.2168
Flood Zone E	0.1127	0.1106	0.1116	0.1116
Flood Zone F	0.0519	0.0502	0.0510	0.0510
Flood Zones A–D	0.6022	0.3669	0.2481	0.9802

tection against 50 year floods. Also, it was assumed that only 40% of all buildings were floodproofed by 2000 (Lee et al., 1978). Costs of floodproofing were fixed at 16.5% of building value and this cost was applied to 40% of the building value (Zones A-D) shown in Table 6-32.

The cost feasibility of this mitigation was found to be as follows:

Initial mitigation cost	$29,892 million
Annual estimated building loss reduction	$247 million
Payout period required (years), 0.0%	60.8
Fifty year annual repayment cost (0.0%)	598 million
Ratio of ALR to ARC	0.83:1

Table 6-30. Riverine Flood Building Damage Losses, 2000, under Two Mitigations

	Flood Zone			
	Zones A–D	Zone E	Zone F	All Zones
1. Building wealth (1980 distribution)	450,989.5	171,283.3	641,901.5	1,264,083.3
2. 1980 building damage rate (%)	.3669	.11165	.05105	–
3. Losses under 1980 conditions ($ millions)	1,654.35	191.24	327.69	2,173.3
4. Building damage rate with 50 year flood protection extended to all areas	.2482	.11165	.05105	–
5. Losses under mitigation condition	1,119.13	191.24	327.69	1,638.08
Reduction:	–	–	–	(535.22)
6. Building wealth under avoidance strategy	301,311.6	117,497.7	837.274	1,264,083.3
7. 1980 building damage rate (%)	.3669	.11165	.05105	–
8. Losses under avoidance mitigation	1,134.86	131.19	427.23	1,693.48
Reduction:	–	–	–	(479.82)

Table 6-31. Cost Feasibility of Building Removal Strategies in Flood Zone A–C of Flood-Prone Cities Not Provided with Area Protection against 50 Year Floods

	Zone A	Zone B	Zone C
1. 1980 building value[a] ($ million)	9,068.223	10,210.140	13,885.331
2. Removal cost[b] ($ million)	11,657.20	13,125.14	17,849.59
3. Annual building damage loss[a] ($ million)	251.770	122.317	80.799
4. Annual loss reduction (ALR)[a] ($ million)	501.02	243.41	160.79
5. Payout period, in years	23.27	53.92	111.01
6. 50 year annual repayment cost (ARC) at 0.0% ($ million)	233.14	262.50	356.99
7. Ratio of ALR to annual repayment cost	2.15:1	0.93:1	0.45:1
8. 50 year ARC, 6.0%	736.37	NC	NC
9. Ratio of ALR to ARC	0.68:1	NC	NC

[a]1970 dollars
[b]Removal Costs calculated at 1.2855 × Building Value. See Johnson (1978), and Lee et al. (1978).

Table 6-32. Baseline Data for Building Floodproofing Cost Feasibility Analysis

	Computer model condition		Building damage rates and losses: Area protection, only		Building damage rates and losses: 4 ft. flooding at Zones A–D building plus area protection	
Flood Zones	Structures exposed	Building wealth[a]	Damage rate (%)	Loss[a] ($ million)	Damage rate (%)	Loss[a] ($ million)
Zone F	7,722.5	662,898.4	0.0510	317.99	0.0510	317.99
Zone E	2,085.1	168,184.6	0.1116	187.78	0.1116	187.78
Zones A–D	5,637.4	450,176.4	0.2482	1,117.34	0.2000	900.23
All Zones	15,445	NC	–	1,623.11	–	1,406.00

[a]1970 dollars.

Storm Surge Mitigations

As noted above (*Building Strengthening Strategies*), a break-even analysis was conducted to identify those counties in which the application of various building strengthening strategies to new construction between 1980 and 2000 would produce loss reduction to cost ratios of 0.75 to 1.00, or greater. Less than half of the U.S. counties exposed to storm surge events met the test, and only 25 exhibited loss reduction to cost ratios of 1:1 or higher for 2-foot floodproofing of all structures. The latter counties represented only 31.07% of 1970 building damage losses from storm surge ($137.19 million/$441.59 million).

However, the test employed assumed that *all* new structures added to the storm surge prone counties would be subjected to the mitigation, and not simply those located within the surge-vulnerable portions of the county (see Chapter 4, Cost Feasibility Analysis). Accordingly, Table 6-33 was developed. As shown, only 25.28% of the 1970 building wealth of surge-exposed counties was located at an "adjusted" elevation of 20 feet or less above mean sea level. It was that level that was defined in the model as being the surge-damage area (see Chapter 4). It seems clear that the gross-grained break-even analysis probably understates the cost-feasible building mitigations that may be applied to structures within storm surge hazard areas.

The data presented below were then developed, based on the annual expected building damage losses generated by storm surge Mitigation 1, which assumed a near complete 4-foot floodproofing of all hazard zone structures between 1980 and 2000. For the purpose of the analysis, it was assumed that 4-foot floodproofing costs (16.6% of building value) should be applied to the entire building wealth of the storm surge zone. The results of the analysis are as follows:

Initial mitigation cost ($278,999.76 million × 16.6)	$46,313.96 million
Annual 50 year repayment cost, 0.0% (ARC)	$926.28 million
Annual building damage loss reduction	$699.60 million

Table 6-33. Distribution of Population and Building Wealth in Storm Surge Zones, 1970, by Adjusted Elevation above Mean Sea Level

Adjusted elevation above MSL (feet)	Population		Building wealth[a]		Annual expected building damage losses[a] ($ million)	Building damage rate (%)
	N	%	($ million)	%		
1. 0–5 MSL	4,502,369	45.3	47,410.39	45.3	–	–
2. 5–10 MSL	1,363.285	13.7	14,335.52	13.7	–	–
3. 10–15 MSL	2,224,694	22.4	23,426.25	22.4	–	–
4. 15–20 MSL	1,849.753	18.6	19,478.09	18.6	–	–
5. 0–20 MSL	9,940.101	100.0	104,670.25	100.0	441.591	0.4219
6. Other areas of parent counties	29,447.146	–	309,328.48	–	None	None
7. All areas of exposed counties	39,387.247	–	413,998.75	–	441.591	0.1067

[a] 1970 dollars.

Annual total loss reduction (TLR) ($699.60 × 1.99) $1,392.20 million
Ratio of annual total loss reduction to ARC 1.50:1

Although this level of payout performance is not acceptable at interest levels of 6.0 and 9.0%, it does suggest that such mitigations would be cost feasible at these higher interest rates if the mitigations were applied to selected flood frequency and flood height classified sections of the hazard zone. Although such an analysis was not performed, it seems not unreasonable to assume that approximately one-half of the mitigated loss reduction ($349.8) would be cost feasible at the 6.0% interest level and this value therefore was entered on Table 6-34.

Although seawall construction was the "most effective" mitigation shown for storm surge, no cost analysis was performed for this mitigation. It is known that costs of such construction have ranged from $100 to $500 per foot of shoreline protected (U.S. Army Corps of Engineers, 1971). However, determining the cost range appropriate for each of the counties in the exposed area, i.e., the approximate miles of seawall required, was beyond the scope of this analysis.

As in the case of riverine flooding, however, it seems likely that the largest net savings are to be derived from use of avoidance strategies. Lee et al. (1978) show that prohibitions on net new exposures in the 50 year storm surge floodplain would reduce baseline year 2000 losses to $913 million, for an annual loss reduction of $264 million, all of which may be considered *net savings*. Even if seawall construction costs were only $300 per linear foot, and even if only 3000 miles of coastline require such walls in order to produce the loss reductions, the ***net*** annual loss reduction from use of seawalls would be as follows:

Initial mitigation cost (3000 miles × $1.6 million/mile) $4,752 million
Fifty year annual repayment cost (6.0%) $300 million
Annual loss reduction ($1,777 million–$335 million) $842 million
Net savings $542 million

From these net savings, appropriate analysis would require offsetting costs to account for damage to the coastal ecosystem (White, 1976), including the steepening and reduction of beach width produced by modified wave action. Moreover, the above simplified calculation unquestionably understates the costs involved in seawall construction per mile of real coastline protected. Thus, in nearly half of the hurricane and shoreline protection studies completed by the Corps of Engineers since 1955, the Corps found the proposed projects to be economically unfeasible.

Economic Payoffs from Major Mitigation Groups

Annual loss reductions from year 2000 exposures were estimated for each hazard for each of 3 mitigations: (1) the mitigation that produced the *largest gross annual reduction* in building damage losses; (2) the *cost feasible* mitigation that produced the largest gross annual reduction in building damage losses; and

(3) the mitigation that produced the *highest net annual loss reduction* (savings). Table 6-34 provides a summary of the expected building damage for the several mitigation assumptions.

A *cost feasible* mitigation is one whose annual 50 year amortized cost at 6.0% interest is at least equal to the annual loss reduction secured through use of the mitigation. *Net savings* are defined as the difference between the annual loss reductions achievable through use of the mitigation and the annual amortized 50 year costs of the mitigation at 6.0%. The specific mitigations judged to offer the *highest annual net savings* are summarized in Table 6-35. As noted earlier, all avoidance mitigations are assumed to involve zero net *national-level* costs.

Due to the gross quality of the hazard and vulnerability data it was not possible to perform a hazard by hazard, building mix by building mix, and mitigation by mitigation marginal analysis of costs in relation to loss reduction. Thus, the terms *net savings* and *highest net savings* mitigation should be viewed accordingly. Further, constraints on the study prevented the determination of confidence limits on the loss estimates and the performance of formal sensitivity analysis on the computer model outputs. However, in order to assess the stability of the conclusions we parametrically varied the cost to loss reduction ratios (1:0.75 to 1:2.00) and the "true" interest rates (0.0% to 9.0%). The results were determined to be insensitive to the parametric variation. Thus, it can be asserted that the lack of incorporation of uncertainty did not measurably affect the viability of the analysis, and we were able to utilize the model outputs in developing policy statements, and to fix the gross boundaries within which the probable costs and payoffs of technologically feasible mitigations may be judged at the national level. The analysis is not intended to provide a fine-grained perspective of what the economic gains and losses are, through any particular mix of mitigations. With this understanding, and when annual expected hazard costs are defined as including both annual expected losses associated with hazard exposures and the annual amortized costs of mitigating those hazards, then a range of policy useful natural hazard cost data for 2000 can be presented (Table 6-36).

These data are based on the following assumptions: (1) *all* building wealth exposed to hurricane winds also is exposed to severe wind; (2) all building wealth exposed to tornado also is exposed to severe wind; (3) severe wind exposures, for purposes of cost analysis, are limited to areas not also prone to either hurricane winds or tornado; and (4) annual expected losses, loss reductions, and mitigation costs may be apportioned to the aforementioned subsets of counties in proportion to their projected building wealth increase between 1980 and 2000. These assumptions therefore alter the cost analysis presented in Table 6-22, as shown in Table 6-36. Also, the inability to fix a cost for seawall construction resulted in specifying a 4-foot floodproofing of all structures in surge-prone zones (0–20 feet MSL, adjusted) as the "most effective" mitigation for this hazard for purposes of the data in Table 6-36.

Further, in the case of riverine flooding, 2 optional *most effective* mitigations were considered: Option A involves the removal of all structures in Flood Zone

Table 6-34. Annual Expected Natural Hazard Building Damage Losses under Several Alternative Assumptions, 1970 and 2000

Hazard	1970 annual expected loss (millions $)	Annual expected losses in 2000 (millions of 1970 $)				
		Level 1 (no mitigations)	Level 2 (expected losses)	Level 3 (losses under "most effective" mitigation)	Level 4 (losses under "most effective" cost feasible mitigation)	Level 5 (losses under highest "net savings" mitigation)
1. Earthquake	655.2	1378.5	1177.0	1084.0	1174.3	1174.3
2. Expansive soils	798.1	997.1	997.1	760.1	969.2	969.2
3. Landslide	370.3	871.2	871.2	349.4	349.4	349.4
4. Hurricane	685.5	1742.0	1742.0	1280.0		
5. Severe wind	11.4	24.8	24.8	19.1	3217.4	3217.4
6. Tornado	879.8	2058.0	2058.0	1564.0		
7. Riverine flooding	1901.0	2634.0	1594.0	953.4	1653.0	1694.2
8. Storm surge	442.0	1176.0	1176.0	334.8	826.2	912.8
9. Tsunami	8.7	19.8	19.8	19.5	19.5	19.5
Total	5752.0	10901.4	9659.9	6364.3	8209.0	8336.8

Table 6-35. Mitigations Exhibiting Highest Net Savings by Hazard

Hazard	Mitigation
1. Hurricane winds 2. Tornado 3. Severe winds	For new buildings constructed after 1979 in 229 cost-feasible counties (Table 6-24), increase designed wind resistance capability to level equaling 1.5 X the level specified in the UBC. In the cost-feasible subset numbering 72 of these same counties (Table 6-24), increase designated wind resistance capability to level equaling 3.0 X the level specified in the UBC.
4. Earthquake	Continue application of UBC Earthquake Zone 3 lateral force requirements in all California counties and extend requirements to the two non-California counties in which this mitigation is cost feasible (Table 6-24).
5. Expansive soils	Apply construction site moisture control techniques in areas where the building season is marked by wide variability in rainfall levels.
6. Landslide	Require soils testing and improved site-grading techniques in all landslide-prone areas. Apply mitigation to new construction after 1979.
7. Tsunami	Permit zero net residential growth in tsunami-prone areas after 1979.
8. Storm surge	After 1979, prohibit net new building construction on the 50 year floodplain.
9. Riverine flooding	After 1979, prohibit net new construction in 100 year riverine floodplains.

A–C of *all* flood-prone communities. Option B involves the removal of such structures only from these same zones in flood-prone communities that have *not* been provided with area flood control facilities sufficient to protect them against 50 year floods.

The *highest net savings* entry in Table 6-36 for wind hazards was derived by calculating the annual amortized costs of the mitigation at 6.0% from the data in Tables 6-22 and 6-25. This value, plus 1/50 of the net 50 year savings shown for the mitigation in Table 6-25, was subtracted from the baseline loss estimate to secure the loss level associated with the mitigation. An identical procedure was used to estimate earthquake loss for the *highest net savings* mitigation.

As shown in Table 6-36, the zealous application of the *most effective* mitigations to all hazards would escalate net annual expected hazard costs in 2000 to a level ranging from 38 to 90% beyond the level that would be experienced if the nation adopts a complete "do nothing" policy in 1980 and adheres to such a policy through the year 2000. In contrast to the escalation in net hazard costs that result from use of *most effective* mitigations, the use of the mitigations that yield the *highest net savings* will reduce net hazard costs in 2000 by 41.9%

Social Payoffs from Major Mitigation Groups

In the prior discussion, alternative mitigation strategies have been examined primarily from the perspective of *direct* economic gains and losses. Thus, each of the 5 categories of annual expected building loss estimates for the year 2000 (Table 6-34), as well as the annual expected *net* hazard costs shown in Table 6-36, have considered only the economic costs associated with expected damage

Table 6-36. Net Annual Expected Natural Hazard Costs for Year 2000[a]

	Alternative 1 (do nothing after 1980)	
Hazard	Building damage loss ($ million)	Total damage loss ($ million)
1. Hurricane	2,554.4	5,875.3
2. Tornado	1,267.2	2,914.6
3. Severe wind	3.1	7.2
4. Earthquake	1,177.	1,553.7
5. Expansive soils	997.1	997.1
6. Landslide	871.2	871.2
7. Tsunami	19.8	40.4
8. Storm surge	1,176.0	2,342.9
9. Riverine flood		
Option A	2,634.	5,241.7
Option B	2,634.	5,241.7
Totals	10,699.9	19,844.1

		Alternative 2 (apply most effective mitigation)		
Hazard		Total damage loss ($ million)	Annual mitigation cost at 6% ($ million)	Total hazard cost ($ million)
1. Hurricane		4,364.3	2,428.3	6,792.6
2. Tornado		2,215.2	3,748.8	5,964.0
3. Severe wind		5.6	835.7	841.3
4. Earthquake		1,430.9	2,549.5	3,980.4
5. Expansive soils		760.1	292.6	1,052.7
6. Landslide		349.4	0	349.4
7. Tsunami		39.6	0	39.6
8. Storm surge		950.2	2,925.6	3,875.8
9. Riverine flood				
Option A		1,897.3	12,915.5	14,812.8
Option B		2,465.4	2,094.9	4,560.3
Totals	Option A	12,012.61	25,696.	37,708.6
	Option B	12,580.71	14,875.4	27,456.1

Table 6-36. (continued)

	Alternative 3 (apply highest net savings mitigation)		
Hazard	Total damage loss ($ million)	Annual mitigation cost at 6% ($ million)	Total hazard cost ($ million)
1. Hurricane 2. Tornado 3. Severe wind	7,869.7	927.34	8,797.1
4. Earthquake	1,550.1	1.42	1,551.5
5. Expansive soils	969.2	0	969.2
6. Landslide	349.4	0	349.4
7. Tsunami	39.6	0	39.6
8. Storm surge	1,816.5	0	1,816.5
9. Riverine flood			
Option A	3,371.5	0	3,371.5
Option B	–	–	–
Totals	15,966.0	928.76	16,894.8

[a]1970 dollars.

to buildings and their contents, supplier losses, income losses, and the capital costs associated with possible risk-reducing mitigations. No attempt was made to translate into economic terms either the costs or the benefits associated with changes in hazard-induced life loss, injury, illness, homelessness, and unemployment. However, it was noted in the loss analysis that reduction in building damage for the catastrophic-type events, which result in a large number of severely damaged or collapsed buildings, will also reduce social impacts. The real policy question not answered here is what is the appropriate level of expenditures necessary to reduce building loss to achieve an acceptable reduction in social costs.

Similarly, no attempt was made to translate in economic terms the possible benefit associated with the "open space," "aesthetic," and "recreational" gains that might be achieved through avoidance of development in riverine and coastal floodplains and other similar hazard zones. Neither, or course, were the possible disadvantages of such policies calculated. However, the observations, which follow, may be made in respect to these matters.

Life Loss

The procedures used to estimate life loss were based on assumed, but empirically supported, relationships between the magnitude of dollar loss associated with hazardous occurrences and the loss of life also associated with such occurrences (see Chapter 4). These procedures resulted in annual expected life losses substantially greater for 1970 and 2000 than the annual average life loss for natural hazards actually reported for *any* of the decades in the current century (Dacy and Kunreuther, 1969). Moreover, both hazard-induced death rates and the absolute annual average number of deaths from natural hazards have been declining steadily throughout the century. Thus, even though the estimates of life loss were probabilistically derived and therefore reflect the intermittent and large losses of life expected from major catastrophes, the annual expected estimates of life loss nonetheless seem to err on the side of overstating the consequences of natural hazard exposures. We believe any such overstatement to be compatible with the objectives of public policy making. For example, the Working Group on Earthquake Hazards Reduction (U.S. Office of Science and Technology Policy, 1978) has suggested that "the primary objective of an earthquake hazards reduction program is to save lives," and the United Nations Office of the U.N. Disaster Relief Coordinator (1976) similarly has urged that the highest priority target of natural hazards management policies be the protection of human life.

The annual estimates of life loss determined in this analysis, however, are not impressive when compared with deaths from other causes in U.S. society (see Chapter 5). Moreover, it seems clear that the most effective means for avoiding hazard-induced life loss are those involving (a) avoidance of high hazard areas; (b) hazard warning and population evacuation systems; and (c) rapid extension of high quality emergency medical and survival-promoting services to disaster victims. Although the provision of area-protection facilities (dams, levees, seawalls, etc.) to hazard-prone communities may reduce the death rates that otherwise would be experienced in such areas, the very existence of these facilities may lure additional numbers of people into the hazard zone, unnecessarily increasing the sense of security of existing residents, and therefore increase the absolute number of deaths over the level that otherwise would occur in these areas. Also, it is clear that dams sometimes fail, seawalls are overtopped, and the engineered limits of area-protection facilities sometimes are exceeded. On a comparative basis, therefore, logic requires that such mitigations be judged as less effective than avoidance policies. In the case of such hazards as earthquake, of course, avoidance policies offer limited opportunities for life loss reduction and emphasis clearly must be placed on building strengthening and other structure-related strategies if large life losses from major seismic occurrences are to be avoided.

How to value reductions in life loss achievable through any type of mitigation is a perennial problem of analysts. We have approached this task not as one of measuring the value of life, per se, but rather as an "opportunity cost" problem. Thus, it seems most appropriate to ask: What reductions in life loss from other causes must be foregone as a result of investments in natural hazard risk reduction programs? The questions, of course, assumes at least 3 facts: (1) insufficient resources are available to reduce all premature deaths to an effective level of zero; (2) resources available for reducing life loss from natural hazards are, in theory, available for reducing life loss resulting from other causes; (3) the normative requirement governing the total national mix of life loss reduction programs is that of achieving the maximum attainable reduction in life loss per one million dollars invested in such programs. In these terms, it is interesting to note the following: If one were to make the implausable assumption that the $7612 to $17865 million annual escalation in net annual hazard costs resulting from use of the *most effective* mitigations shown in Table 6-36 would purchase the complete elimination of all natural hazard deaths projected for 2000, or 1790 deaths, then the net cost per life saved under this strategy would range from $4.09 to $9.6 million. In contrast, Grosse (1972) has shown that the cost per death averted in 4 different cancer control programs in the period 1966–1972 ranged from a low of $2217 for uterine cervical cancers to a high of $46,181 for colon-rectum cancers.

Examination of this evidence suggests that the cost per life saved in natural hazard risk reduction programs can well be escalated to levels substantially in excess of those associated with other death and injury reducing programs, which currently may be underfunded. Although this inference is not intended to suggest that life loss reduction should not be an objective of natural hazard management programs (indeed, we believe the reverse to be true), neither does it seem appropriate to overstate the benefits or understate the costs associated with such programs.

Other Social Costs

Similar comments can be made in respect to the reductions in homelessness, income, and other psychosocial disbenefits associated with natural hazard exposures. If the minimum "pain" level of social disbenefits is to be the objective of natural hazard management programs, then it seems clear that hazard zone avoidance policies have the edge over all other strategies. As in the case of life loss, it seems clear that the magnitude of these undesirable consequences of natural hazard exposures may easily be overstated. Thus, annual dwelling unit losses associated with highway construction and other public facility construction programs is probably substantially greater than the drawdowns on housing stock that may be charged to natural hazards exposures. Unemployment induced by changing Federal Reserve Board policies unquestionably is substantially in excess of that induced by natural hazards, and federal regulatory policies undoubtedly exert substantially more impacts on supplier costs than do natural hazard occurrences.

Conclusion

Annual expected losses from natural hazard exposures in 2000 may be reduced by more than 40% through application of currently available risk-reducing technologies and policy mitigations. Between 1980 and 2000, the application of selected combinations of building strengthening, area structural protection, hazard zone planning, hazard zone avoidance, and building site preparation technologies can reduce annual expected building damage losses, alone, from \$10,901 to \$6364 million. *Total* annual expected losses may be reduced from \$19,844 to \$12,013 million. This level of mitigation would hold annual building damage losses to a value exceeding that of 1970 by only 11% in spite of the substantial projected 1980–2000 increases in the population exposed to significant natural hazards.

If mitigation costs are ignored, the largest achievable building damage loss reductions in 2000 are, in order of their magnitude (millions of dollars), as follows: (1) riverine flooding (\$641 to \$1681); (2) storm surge (\$841); (3) landslide (\$522); (4) tornado (\$462); (6) earthquake (\$295); (7) expansive soils (\$237); (8) severe wind (\$6); (9) tsunami (\$0.4).

Overzealous application of strengthened building codes and standards can substantially increase net annual natural hazard costs in the year 2000. The data suggest that overzealous application of risk-reducing mitigations can increase net annual expected natural hazard costs in 2000 by 38 to 90% above baseline, or "unmitigated" levels. Net annual expected hazard costs are defined as annualized losses produced by hazard exposures plus the annual amortized costs associated with efforts to reduce those losses.

In terms of building strengthening strategies, in only a comparatively small number of U.S. counties are the annual expected loss reductions associated with the strategy equal to the annual amortized cost of the building strengthening mitigation at 6.1% interest. Only 2 non-California U.S. counties meet this test in respect to earthquake-related strategies, and only 229 for all wind hazards. In those counties in which loss reduction to cost ratios were 1:1 or greater for the "most effective" building mitigation, estimated annual building damage losses in 1970 totaled only 11% of all national losses in the same year for all wind hazards, 5% for all non-California earthquake hazards, and only 13% for all storm surge losses.

With respect to riverine flooding, risk-reducing mitigation strategies that involve both the provision of area protection against 50 year flood heights and requirements for 4-foot floodproofing of all structures in 50 year floodplains do not appear to be cost feasible. Similarly, the elevation or floodproofing of all new structures in counties influenced by storm surge flooding does not appear to be economically feasible, but such strategies are feasible in sections of the coast lying below 20 feet, adjusted mean sea level.

Although building-strengthening strategies are important loss-reducing tools, the study suggests that these strategies should not dominate the nation's natural hazards management policy. Indeed, these findings contain within them the "hint" that current model building code requirements may well exceed economically justifiable levels if applied in numerous U.S. counties. Examination of this possibility was beyond the scope of this analysis but seems worthy of future investigation.

Riverine and coastal area flood control facilities constructed in compliance with post-1936 economic criteria appear to be cost-effective methods for controlling losses at specified levels of hazard zone occupancy, but contribute to temporal increases in annual expected losses within the hazard zones. For most decades in the current century, the construction of area protection facilities (dams, levees, seawalls, etc.) has, together with warning and population evacuation systems, comprised the primary U.S. public policy approach to the mitigation of risks associated with riverine and coastal flooding. Based on the crude cost feasibility tests applied to this policy, it appears that, at the national level, the following conclusion is warranted. Under a condition where community values decree that flood-prone lands are to be developed to the level currently exhibited by the sum of U.S. flood-prone urban communities, then the provision of these facilities seems to be cost feasible, at 6.0% interest, i.e., the annual loss reductions under conditions of development appear to be less than the annual amortized cost of the facilities. *Optimistically,* net annual savings from seawall protection of surge-prone coastal areas may be as high as $480 million per year in 2000. In 1970, annual net savings from construction of all prior riverine flood control works may have been as high as $595 million; and the 1970–2000 investments in flood control works for 4418 flood-prone communities could produce annual net savings in 2000 of approximately $570 million (at 9.0%) to $1400 million (at 6.0%).

These estimates should be viewed as optimistic. They do not consider such potential disadvantages as ecological damage and the increase that might result in the absolute number of persons exposed to the hazards of dam failure and flood heights above the 50 year level. Neither do they consider the possible downstream increases in water volumes resulting from higher runoff and lower water retention in flood-protected areas. Also, although tests of feasibility were based on both 6.0 and 9.0% interest levels, future annual net savings were not discounted to their current value. Also, the application of this strategy does not appear to be *generally* cost feasible when combined with a floodplain avoidance strategy for new construction and, if applied above, would result in temporal increases in absolute annual losses from flooding.

The large "opportunity costs" associated with appliction of "most effective" mitigations suggests that the use of such mitigations is not an economically justifiable approach for curbing the life loss associated with natural hazard exposures. Even though the life loss estimates generated in this study are on the side of overstatement, annual expected life loss resulting from natural hazard exposures is not high as compared with other causes of controllable premature death in U.S. society. Moreover, the use of "most effective" natural hazard mitigations, even under implausibly optimistic assumptions concerning their death reduction effectiveness, results in extraordinarily high costs per death averted.

If a paramount objective of the total mix of national-level policies concerning environmental quality, workplace safety, natural hazards, and public and environmental health is to achieve maximal reduction in deaths from these causes at any constrained annual level of public and private investment, then other causes of premature death would warrant a higher policy priority. This conclusion seems justified even if the deaths-avoided findings are doubled and costs of death aversion are halved for the "most effective" natural hazard mitigations.

Also, logic suggests that hazard zone avoidance, hazard warning, and population evacuation systems and policies would prevent more hazard-related deaths than the across the board application of the "most effective" mitigations identified in this study.

In any event, the interests of rational, comprehensive, and compassionate public policy making suggest that further intensive and quantitatively oriented analysis be directed at the costs and payoffs of alternative death and injury avoiding environmental hazard programs, when such hazards are defined as including both those of "natural" and "man-induced" origin.

References

Dacy D. C., and H. Kunreuther. 1969. The Economics of Natural Disasters, Implications for Federal Policy. Free Press, New York.

Grosse, R. N. 1972. Cost benefit analysis of health service. Ann. Am. Acad. Political Soc. Sci. 399:89–99.

Hart, G. C. 1976. Natural Hazards: Tornado, Hurricane, Severe Wind Loss Models. Prepared for the National Science Foundation (NTIS No. PB 294594/AS). J. H. Wiggins Company, Redondo Beach, California.

Johnson, W. K. 1978. Physical and Economic Feasibility of Nonstructural Flood Plain Management Measures. Institute for Water Resources, U.S. Army Corps of Engineers, Ft. Belvoir, Virginia.

Lee, L. T., J. D. Chrostowski, and R. T. Eguchi. 1978. Natural Hazards Riverine Flooding, Storm Surge, Tsunami Loss Models. Prepared for the National Science Foundation (NTIS No. PB 294682/AS PCA10). J. H. Wiggins Company, Redondo Beach, California.

Miller, C. H. 1978. Flood Plain Management; Issues and Alternatives for the Congress. Office of Technology Assessment, U.S. Congress, Washington, D.C.

United Nations Office of the U.N. Disaster Relief Coordinator. 1976. Guidelines for Disaster Prevention. 3 vols. U.N., Geneva.

U.S. Army Corps of Engineers. 1971. National Shoreline Study; Shore Protection Guidelines. U.S. Govt. Printing Office, Washington, D.C., p. 33.

U.S. Bureau of the Census. 1972. Statistical Abstract of the United States: 1972. 93rd edition. U.S. Govt. Printing Office, Washington, D.C.

U.S. Office of Science and Technology Policy, Executive Office of the President. 1978. Working group on earthquake hazards reduction. Earthquake Hazards Reduction: Issues for an Implementation Plan. U.S. Govt. Printing Office, Washington, D.C.

White, G. F. 1976. Natural Hazard Management in Coastal Areas. Prepared for the Office of Coastal Zone Management, U.S. Department of Commerce, Washington, D.C.

Wiggins, J. H., J. E. Slosson, and J. Krohn. 1978. Natural Hazards: Earthquake, Landslide, Expansive Soils Loss Models. Prepared for National Science Foundation (NTIS No. PB 294686/AS). J. H. Wiggins Company, Redondo Beach, California.

7
Catastrophic Natural Hazard Occurrences

Introduction

The cumulative natural hazard losses that may be charged to a particular place over any extended time frame are the product of 2 major types of events: (1) those that recur at a comparatively high level of frequency; and (2) those that occur rarely. Over long time lines (100-200 years), the cumulative absolute and per capita losses charged to 2 different places may be the same. However, in one place the losses may be concentrated in a few years, while in another place the losses may vary little from one year to another.

In places where large fractions of cumulated losses may be incurred over a long time frame as a result of infrequently occurring catastrophic events, policy makers or exposed citizens may wish to consider not the probabilistically derived annual expected loss level associated with a particular type of hazard exposure, but rather the episodic level of loss associated with the low probability event. In effect, the exposed population may wish to place a probability of 1.0 on the occurrence of a catastrophe and then seek an optimal approach to the minimization of loss.

Accordingly, this analysis focused on development of such loss estimates for 2 low-probability events: a recurrence of the "great" San Francisco earthquake, and a recurrence of Hurricane Camille.

Approach to Scenario Development

Any useful scenario must be believable. Thus, in examining the consequences in the year 2000 of a recurrence of 2 of the most severe catastrophes in the nation's history, no alterations in the significant geophysical facts associated with the original events were made. The scenarios alter only the demographic and economic attributes of the disaster area.

Thus, neither what the consequences of Camille would be if her landfall were shifted slightly westward so that she would strike with full force on New Orleans instead of Gulfport, Mississippi, nor what would happen if the San Francisco earthquake were to recur at noon instead of 5:13 A.M. was addressed. Yet, either of these alterations in the essential facts of the original occurrence is both believable and possible, and would *substantially* alter the estimates of life and property loss presented in the following scenarios.

Camille II

Approximately 3000 miles of the U.S. coastline are vulnerable to either "land falling" or "land existing" storms of hurricane forces (Hart, 1976). Reaching from Port Isabel, Texas, to Eastport, Maine, this hurricane-vulnerable stretch of coastline contained a 1970 population of 62.74 million persons and $682 billion in building wealth which were at risk of exposure to winds of hurricane velocity. Also, a population of 38.39 million and a building wealth of $438 billion were at risk of exposure to hurricane-related storm surges (derived from Hart, 1976; Lee et al., 1978).

However, between 1931 and 1972, only 77 hurricanes touched any part of the coastline, and only a few of these could be classed as "great" hurricanes (wind velocity in excess of 125 mph). Given that fewer than 2 hurricanes per year of any type have visited this coastline over the 41 year period and that the typical hurricane spins winds of hurricane force over an area of only 100 square miles in diameter and gale-force winds over an area only 400 miles in diameter (Hart, 1976), the annual probability that any given coastal segment will experience a hurricane is low indeed.

As shown in Figure 7-1, the probability in any given year that any given 50 mile segment of the coastline will experience a hurricane-force storm varies from 0.0 to 1.0% for seven segments to 14.0% or more for four segments. The annual probability of a great hurricane varies from 0.0 to 1.0% for 37 segments to 4.0% or more for 13 segments.

As depicted in Figure 7-2, similar low probabilities are exhibited for hurricane-induced storm surges. Of course, no matter what the probability of a hurricane strike may be for any specific place, when such an event does occur, the results can be devastating, as was the case with Hurricane Camille in August 1969. One of the most intense hurricanes ever known to enter the United States, Camille resulted in 262 deaths and the destruction of thousands of buildings; estimated losses exceeded one billion dollars.

From the point of landfall, Hurricane Camille moved north across Mississippi and continued her destructive path inland. In addition to coastal zone damage resulting from storm surge and high winds, Camille contributed to heavy rains in Tennessee, Kentucky, West Virginia, and Virginia, which, in turn, caused flash floods and mudslides. The James River rose to a record high, wreaking havoc on river towns from Lynchburg to Richmond, Virginia.

In this analysis, an exact repeat of Camille has been examined, but under the

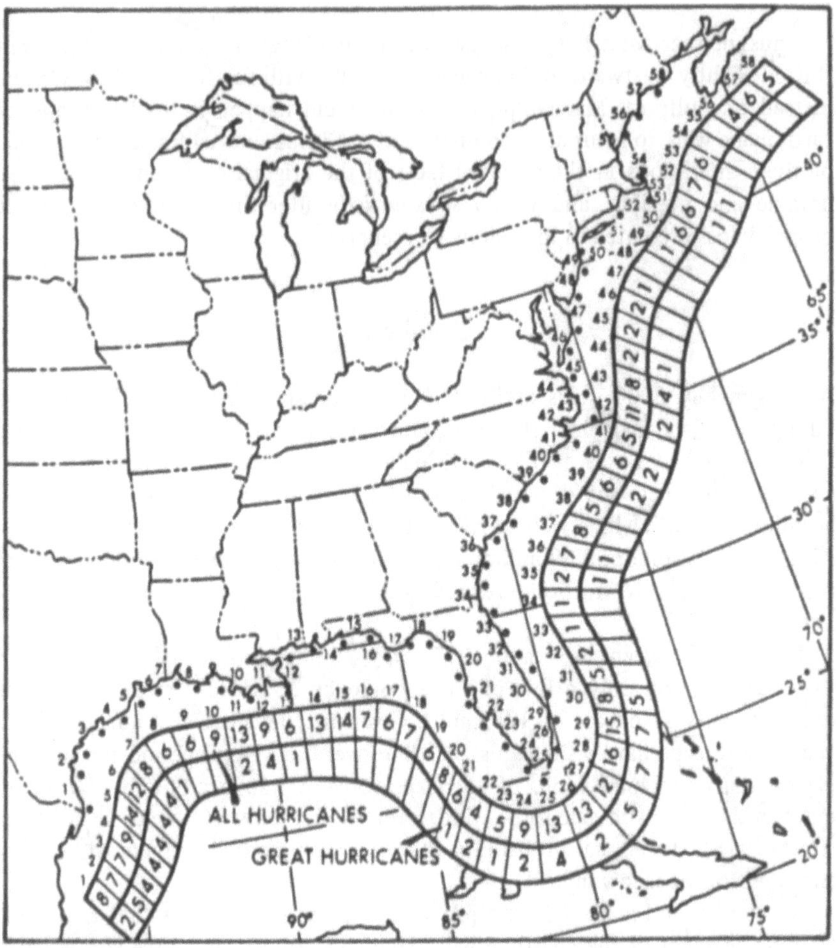

Figure 7-1. Hurricane probability map. Probability (percentage) that a hurricane (winds exceeding 73 mph) or great hurricane (winds exceeding 125 mph) will occur in any 1 year in a 50 mile segment of the U.S. coastline (after Simpson and Lawrence, 1971). From U.S. Department of Commerce, 1976.

demographic and economic conditions predicted for the target area in the year 2000. Also, for purposes of this scenario, only the results of Camille-related storm surge and hurricane winds will be examined; the aforementioned riverine flooding and mudslides are not included. Figures 7-3, 7-4 and 7-5 indicate the nature and severity of the hurricane. These exhibits portray the landfall area and intensities of storm surge on a theoretical Hurricane Camille, circa 2000. They were based on data in a U.S. Army Corps of Engineers' report on Hurricane Camille (1969).

Figure 7-3 shows the path of Hurricane Camille. The distribution of wind velocities at point of landfall is shown in Figure 7-4, e.g., winds will reach well above 175 miles over an approximate 20 mile coastal section from Biloxi to

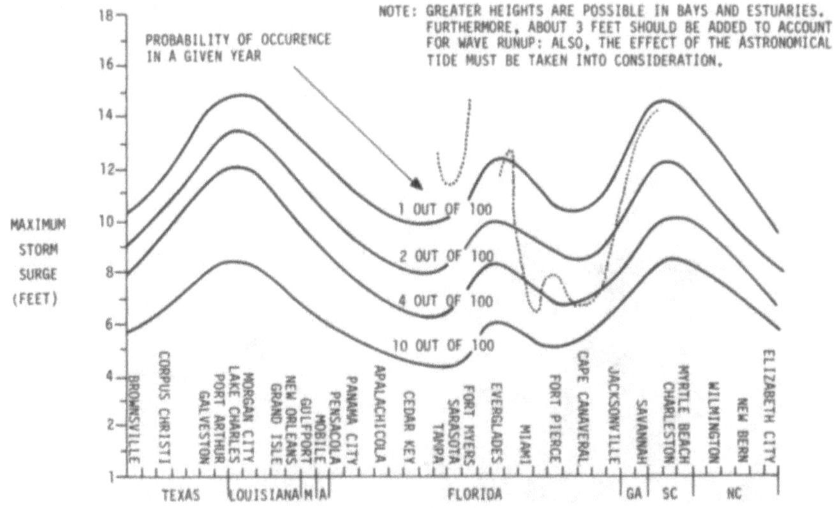

Figure 7-2. Composite estimates of expected surge height. Solid lines based on U.S. Army Corps of Engineers and University of Florida probability estimates of annual occurrence of a storm surge on an open beach area; dotted lines based on National Oceanic and Atmospheric Administration, probability estimates of occurrence of 1 out of 100. From U.S. Department of Commerce, 1976.

Mississippi City, Mississippi. At their most severe, winds will exceed 200 mph. Thereafter, they will diminish to about 75 mph, 50 miles to the west and 60 miles to the east of the storm center. Within the storm center two counties receive major velocities above 200 mph: Hancock and Harrison Counties, Mississippi. Both are at the point of landfall. Figure 7-5 shows the high-water profile associated with the coastal storm surge in the impact area.

Were a repeat of Hurricane Camille to occur in 2000, a 28-county impact area will be affected (Figure 7-6), covering approximately 400 miles of coastline in Louisiana, Mississippi, Alabama, and Florida. Twenty-three of the 28 counties would be subject to wind damage, 10 of the counties subject to flood damage, and 5 of the counties subject to both. In reality, only the coastal portion of the 10 "flood-damage counties" would be subject to flooding. However, for purposes of this analysis, the total county is considered as an economic impact area for coastal flooding.

For each of the 28 counties in the total impact area, 1970 values of economic parameters were projected to the year 2000 utilizing the economic model discussed earlier. Base measures projected include population, per capita income, total economic value at risk, and building valuation at risk. The 28 county area population in the year 2000 is projected to be approximately 2,335,200, an increase of 15% over 1970 population of 2,026,600. Similarly, average per capita income in the year 2000 of $6332 (in 1970 constant dollars), versus $2547 for 1970, reflects an increase of real income in the 30 year period. (Per capita income was computed from *County and City Data Book–1972* (U.S. Bureau of

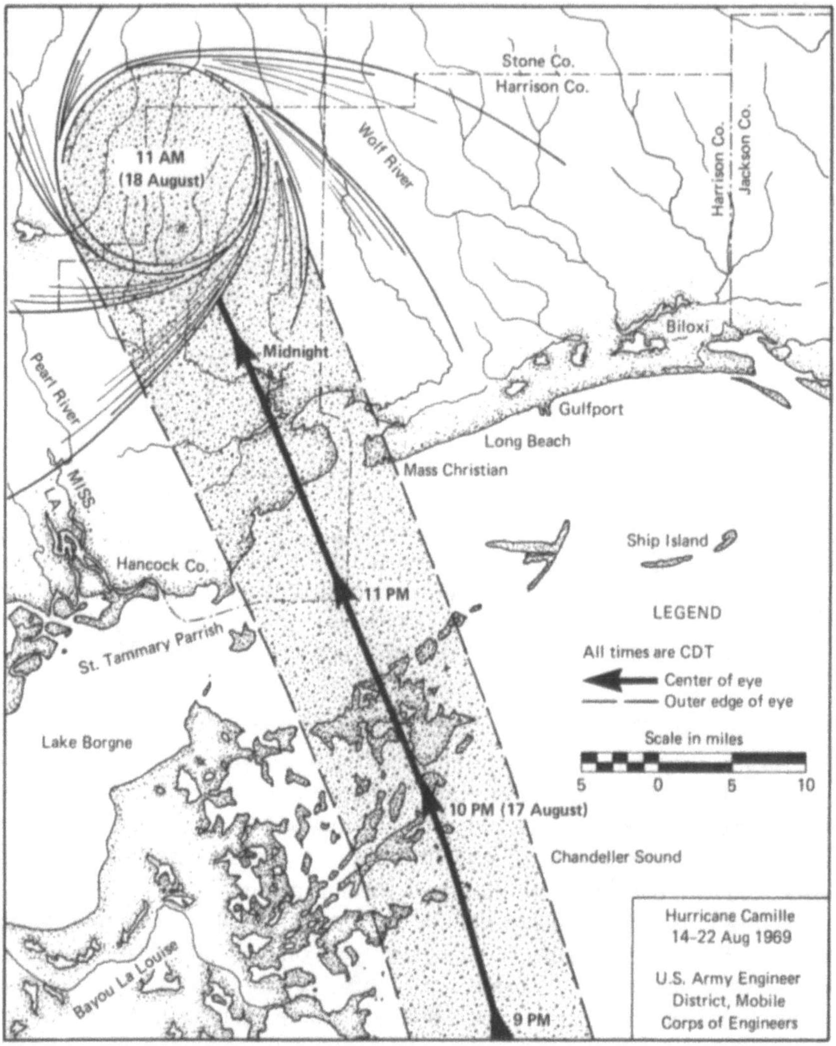

Figure 7-3. Path of Hurricane Camille at landfall.

the Census, 1973); U.S. Department of Commerce (1976); and adjusted according to *OBERS Projections* (U.S. Water Resources Council, 1974, Vol. 2), for 1970 and 2000 expressed in 1970 dollars.

Figure 7-7 shows total economic value of buildings at risk in the 28 county area as approximately $44 billion in the year 2000 (in 1970 constant dollars). Total value at risk for the full 28 county impact area is approximately $18,900 per capita (expressed in 1970 constant dollars for the year 2000). This figure varies only slightly from county to county within the full 28-county impact area (Figure 7-8).

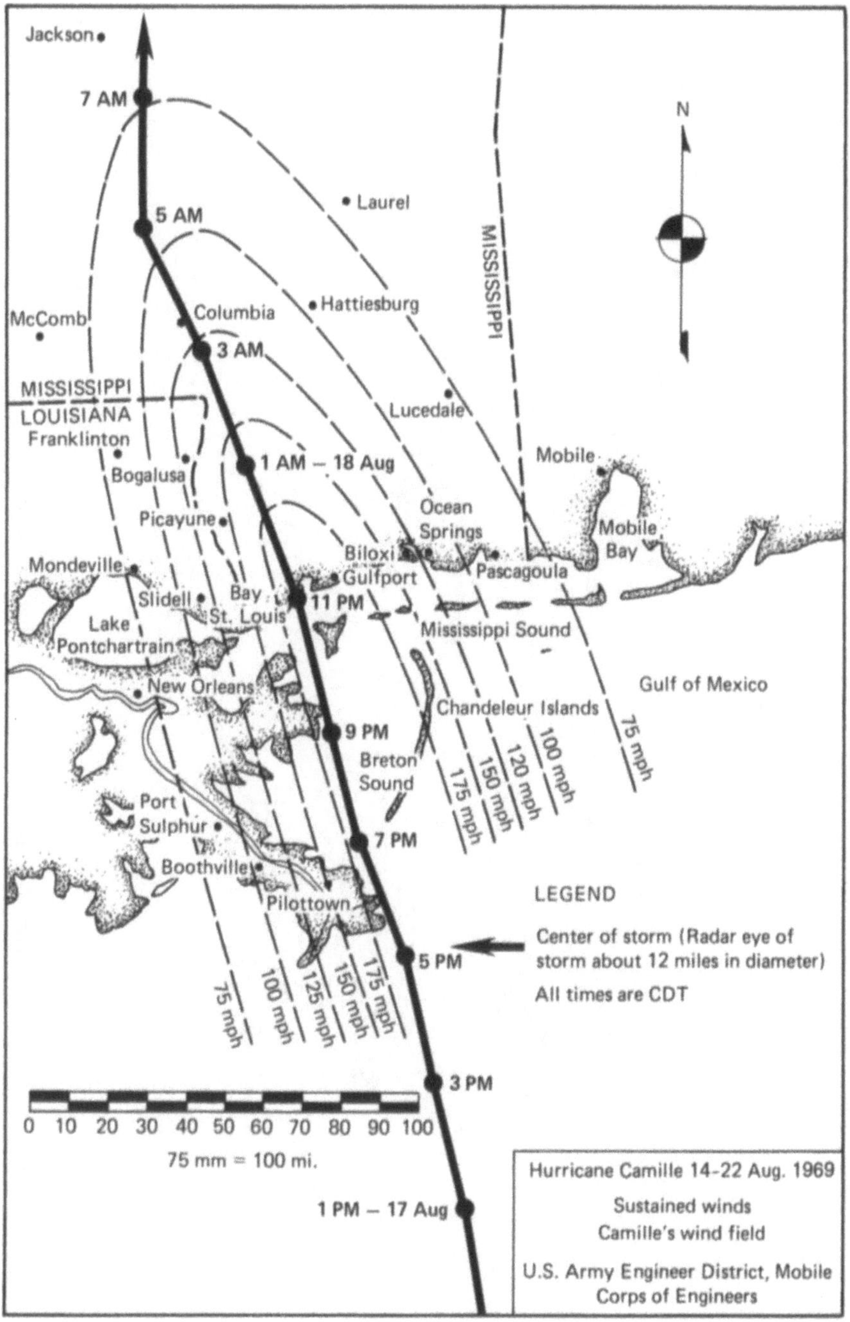

Figure 7-4. Hurricane Camille's wind field.

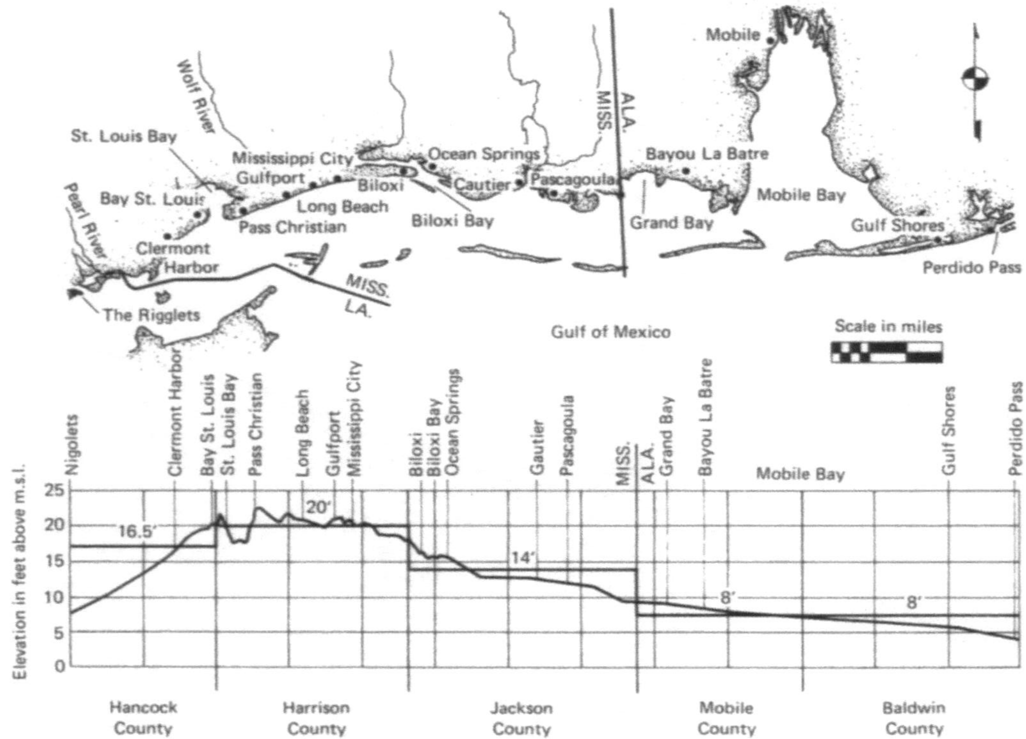

Figure 7-5. Hurricane Camille's high water profile.

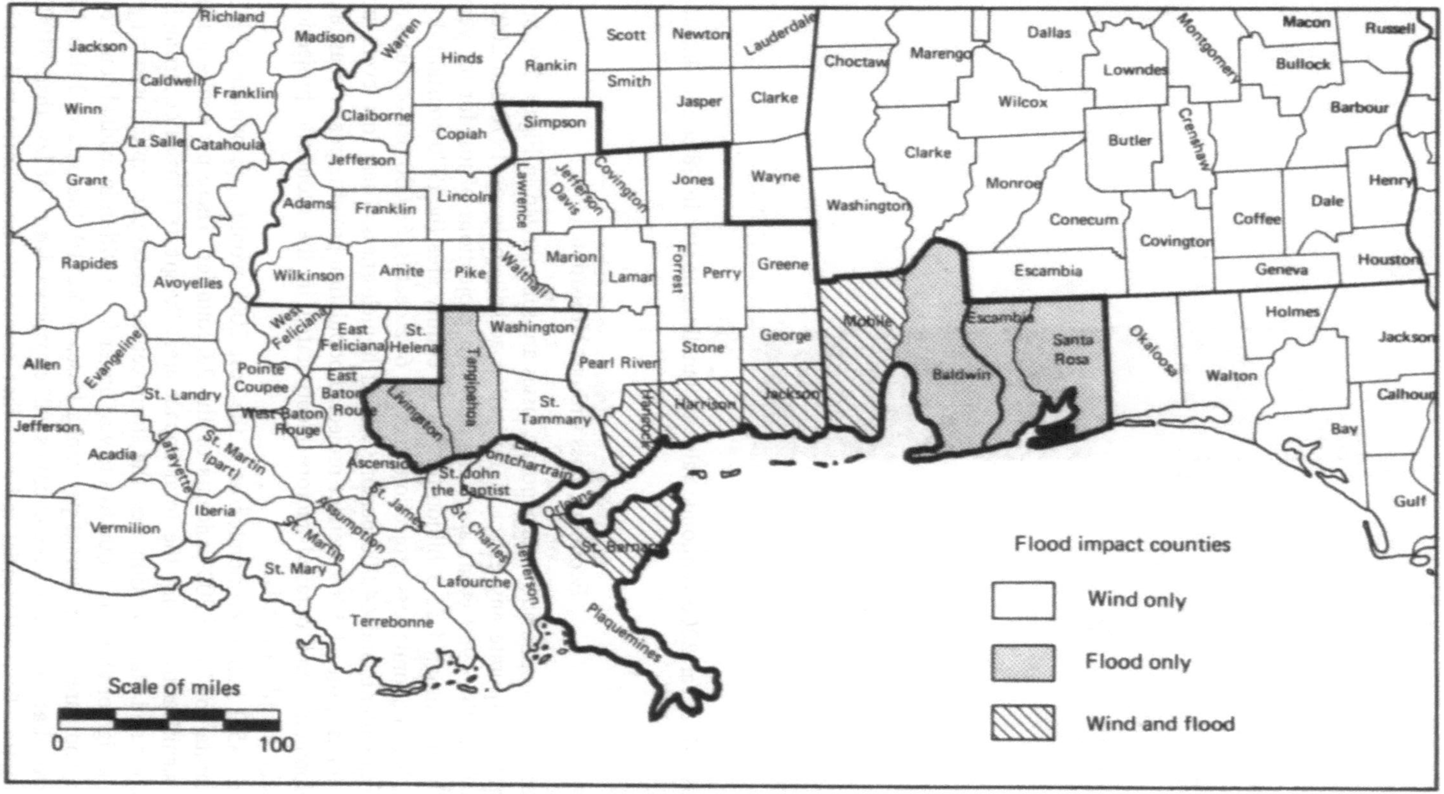

Figure 7-6. Twenty-eight county impact area.

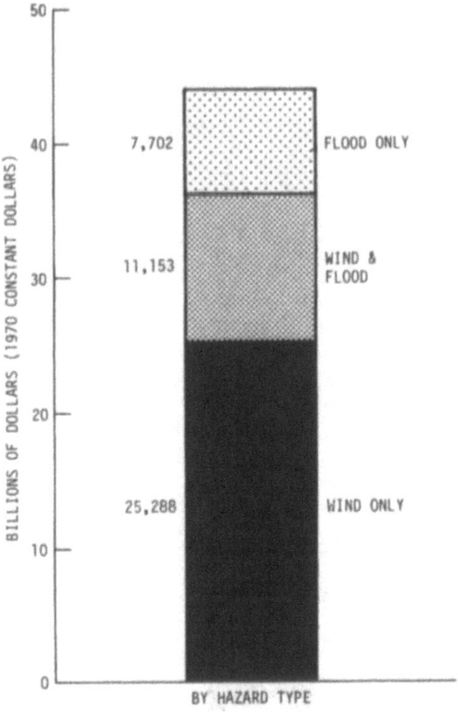

Figure 7-7. Year 2000 economic value of buildings at risk in 28-county impact area.

Economic Loss without Mitigation

Projected economic and noneconomic losses are summarized in Figure 7-9, but do not incorporate any effect of mitigation activity between 1970 and 2000. Losses for the 1969 event are compared with projections of loss for the year 2000—an estimated \$5.9 billion for the 28 county area (in 1970 constant dollars). Noneconomic loss projections include 642 deaths, 68,000 person-years of unemployment, and 64,000 person-years of homelessness. These loss levels—up to 4 times those of 1969—reflect projected population growth, greater coastline development, and detailed factors incorporated in the economic projection model on a subarea basis.

The \$5.9 billion economic loss in the year 2000 would approximate \$2510 per capita for 2,335,200 residents of the 28 counties. These per capita losses vary significantly within the total impact area; the greatest losses occurring in Hancock and Harrison Counties, the storm center. Total per capita losses range from a high of about \$16,000 to a low of under \$100 within individual counties.

As compared with average per capita losses of \$2510, the data in Table 7-1 indicate that one-half of the counties will experience losses of less than \$1000 per capita. This differential indicates that several of the counties will suffer

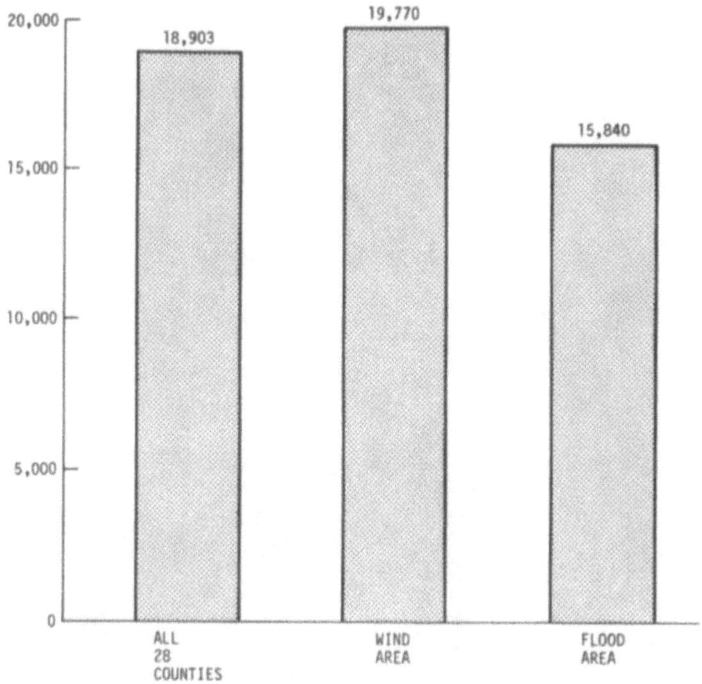

Figure 7-8. Year 2000 per captia economic value at risk in 28-county impact area.

extremely high losses. The heaviest per capita losses by far would be in the three counties in the direct path of the hurricane: Hancock, Harrison, and Pearl River.

Projected economic loss of $5.9 billion for the 28 county impact area equals 39.6% of per capita income (projected to be $6332 in 1970 constant dollars). Table 7-2 indicate these percentages vary widely, from a high of 294% in Hancock County to below 10% in 11 counties.

Composition of economic loss for the 28 county area is indicated in Figure 7-10. Building losses will be $2.7 billion, about 46% of total economic losses. The principal contributor to building loss would be hurricane winds (roughly two-thirds of the total) with the remainder caused by floodings. "Other" includes supplier loss, which covers the extra costs of acquiring goods and services which would ordinarily be from a closer source. Building contents includes all contents except food and other nondurable goods; the exception is inventories for food stores and eating establishments. Income loss is loss in take-home pay caused by a company or one of its buyers closing.

Table 7-3 compares population distribution to distribution of economic loss by family income group. The projected economic loss would be roughly in proportion to population; no single income group would bear an inordinately high share of the loss. In this analysis, however, no attempt has been made to determine the amount of loss that various income groups could "afford." While a high

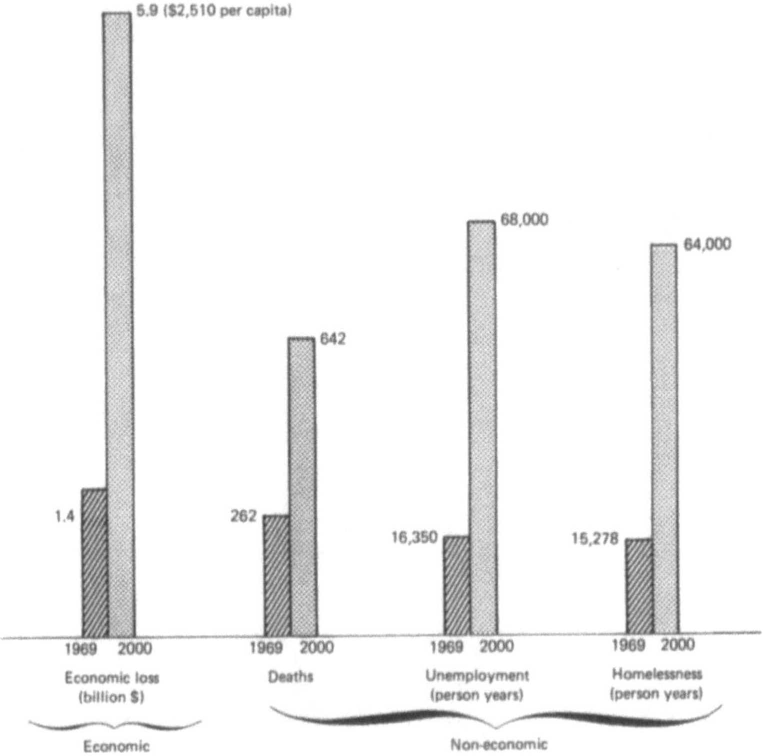

Figure 7-9. Loss comparisons for 1969 event versus year 2000 scenario event without mitigation.

Table 7-1. Year 2000, Per Capita Economic Loss without Mitigation (1970 Dollars)

Dollar loss	Number counties	Percent total
under 100	4	14.3
100– 499	7	25.0
500– 999	3	10.7
1,000–1,999	4	14.3
2,000–2,999	2	7.1
3,000–3,999	1	3.6
4,000–4,999	3	10.7
5,000–9,999	–	–
10,000 and over	4	14.3
Total	28	100.0
Median 1,000		
Average 2,510		

Table 7-2. Year 2000, Per Capita Economic Loss as Percent of Per Capita Income (1970 Dollars)

Percent of income	Number counties	Percent total
Under 1.0	3	10.7
1.0– 4.9	5	17.9
5.0– 9.9	3	10.7
10.0– 24.9	4	14.3
25.0– 49.9	5	17.9
50.0– 99.9	3	10.7
100.0–199.9	2	7.1
200.0–300.0	3	10.7
Total	28	100.0
Median	17.0	
Average	39.6	

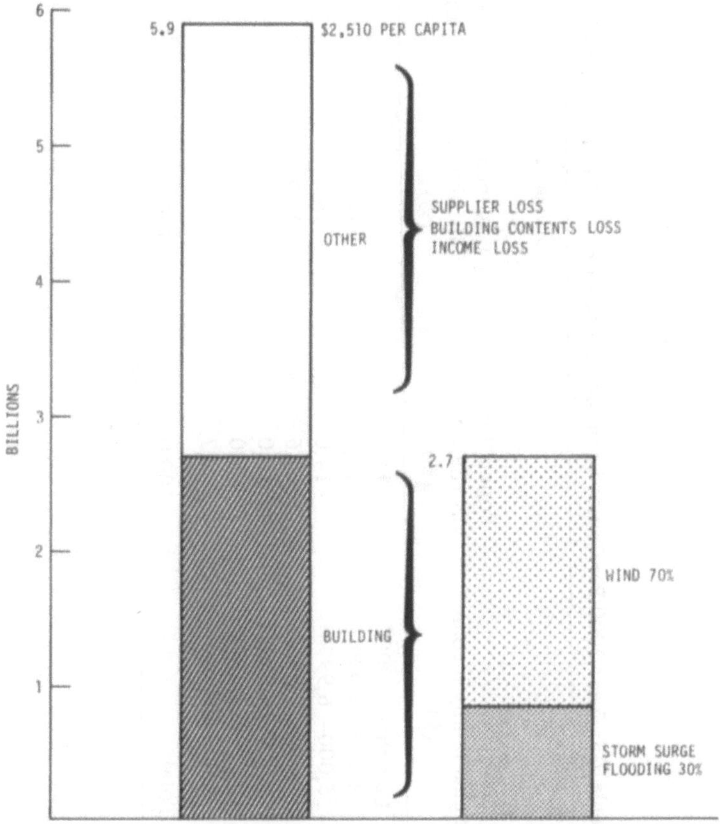

Figure 7-10. Year 2000 economic loss without mitigation for the 28-county impact area.

Table 7-3. Year 2000 Per Capita Economic Loss as a Percent of Per Capita Income, without Mitigation

Annual family income ($)	Population distribution (%)	Storm surge and hurricane total economic loss distribution (%)	Difference (%)	Storm surge loss distribution (%)	Hurricane loss distribution (%)
1. Less than 3,000	16.6	15.9	(0.7)	12.6	17.4
2. 3,000– 4,999	13.2	13.7	0.5	12.0	14.5
3. 5,000– 6,999	14.0	16.4	2.4	15.9	16.6
4. 7,000– 9,999	25.6	21.9	(3.7)	23.9	20.9
5. 10,000–14,999	19.0	20.9	1.9	23.7	19.7
6. 15,000–24,999	8.0	9.2	0.3	10.0	8.9
7. 25,000 and over	2.7	2.0	(0.7)	1.9	2.0
Total	100.0	100.0	0	100.0	100.0

income family might have sufficient reserves to cover short-term losses, a low income family might be ruined by a loss situation equal in amount.

As shown earlier in Figure 7-9 3 types of noneconomic loss have been estimated for this scenario: mortality, unemployment, and homelessness. Six hundred forty-two deaths are projected for the year 2000—approximately two and one-fourth times 1969 fatalities. As with economic losses, the high loss figure results from population growth and greater utilization of coastline areas. This means approximately one person out of each 3600 in the 28 county impact area would die from the hurricane. This compares unfavorably to other accidental mortality figures. For example, the 642 expected deaths from Hurricane Camille in the year 2000 is equal to 40% of the average aircraft deaths per year in the United States over the last 10 years (National Safety Council, 1975).

Unemployment for the 28 county area is projected at 68,000 person-years. In the detailed economic model, it is assumed unemployment would not extend beyond one year. However, this understates total unemployment, as some facilities may be shut down permanently as a result of the disaster. The amount of unemployment as a percentage of the total work force would depend on the duration of unemployment; this obviously varies from job to job. Figure 7-11 shows that a full year of unemployment (quite high), would affect approximately 7.3% of the 28 county labor force. This unemployment would be added to "normal" unemployment of approximately 6%. As a consequence, total unemployment for the region, assuming this one year duration, would be above 13%. Assuming a three month duration of unemployment, over 29% of total work force would be affected, forcing total unemployment above 35%. As with mortality, no measure of unemployment can fully represent social impact, but these figures are significant.

An estimated 64,000 person-years of homelessness would result from Hurricane Camille. Immediately after the storm, an estimated 800,000 people would be homeless for at least one day, 500,000 for more than one week, and more than 100,000 for more than one month. Average homelessness duration for the population is estimated at two months. These numbers are significant in that they indicate approximately one-third of the total population would be homeless for at least one night, and more than 10% for over a week. It is impossible to estimate the consequences of this displacement in terms of interruption of work, education, and other activities.

Mitigation Measures and Costs of Implementation

The ability to exercise control over and reduce losses from such a natural event as Hurricane Camille exists. In this scenario, 4 mitigation measures have been considered:

1. *Floodproofing*. After 1980, floodproof all structures within a defined storm surge zone to the four-foot water level.
2. *Increased Wind Design Code Provisions*. Increase building codes for the region. All new structures built within the affected counties must meet a

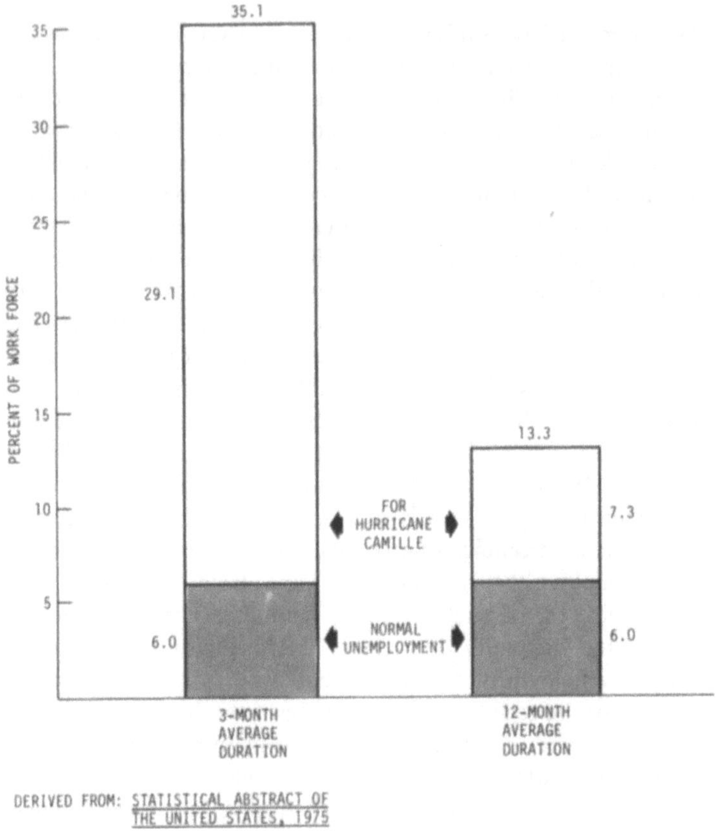

Figure 7-11. Unemployment comparisons for the year 2000. From the U.S. Bureau of the Census, 1975.

lateral wind force standard equivalent to 3 times that of the 1973 Uniform Building Code.

3. *100 Year Storm Seawall.* Protect all counties within a defined storm-surge zone through construction of a seawall capable of withstanding 100 year storm conditions.
4. *No Growth.* Allow no further growth on the 50 year floodplain after 1980.

These 4 measures are by no means the only ones available. However, they reflect enough variety for analytic evaluation.

In this scenario, costs have been projected for only the first 2 mitigations (floodproofing and wind design), which are of a structural nature. The total cost for implementation of these mitigations throughout the 28 county area (23 counties for wind and 10 counties for flood protection) is estimated at $6.0 billion (Figure 7-12). It is assumed this money would be expended over a 20 year period as new construction occurred. To calculate the overall cost of mitigation (structural only) in 1970 dollars, construction cost increases were projected between years 1980 and 2000, and compounded using a 6.1% interest

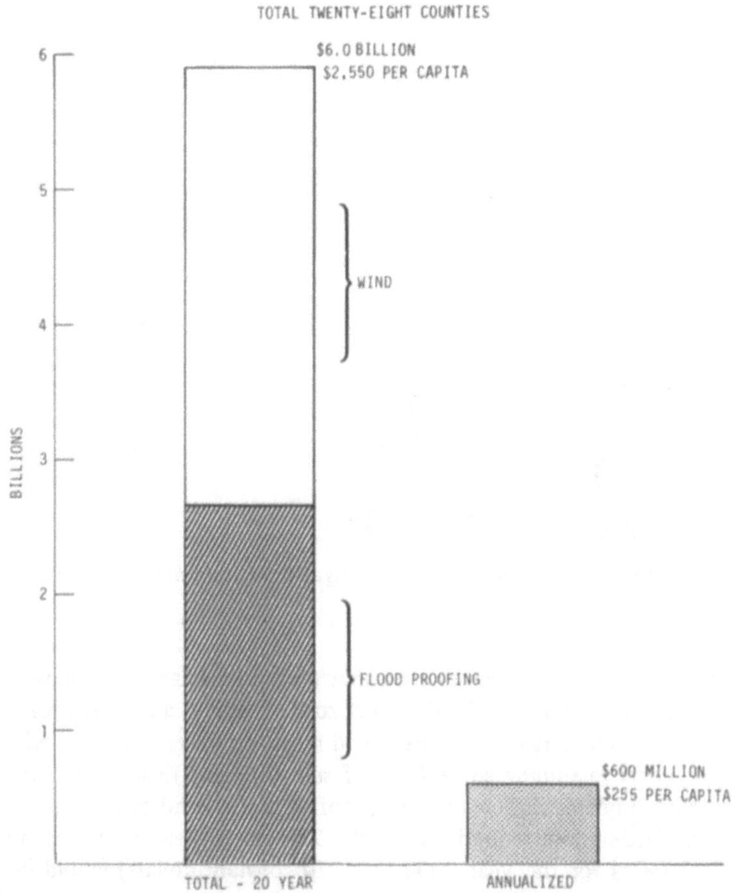

Figure 7-12. Costs of mitigation (buildings only) for Hurricane Camille, year 2000.

rate. This accounts for the corresponding appreciation in value that would occur if the funds spent on extra building costs had been invested at an annual rate of 6.1% interest.

Figure 7-13 shows annual mitigation costs equal to about \$255 per capita, or about 4.0% of per capita income. Figure 7-14 fixes these annual costs at an average of about \$115 per capita for floodproofing (within the 10 flood-vulnerable counties) and \$140 per capita for wind protection (within the 23 wind-vulnerable counties). This might be compared to the cost of municipal service: about \$170 per capita. (Municipal service covers the cost of police, fire, streets, sanitation, trash collection, library, parks and recreation, and city administration). The total estimated cost of \$6.0 million for mitigation equals 13.4% of the total value at risk (building values). The annual mitigation costs of \$255 equals 4.0% of per capita income. When comparing mitigation costs of \$255 per capita to the "normal" costs of providing municipal services (\$170 per capita), the cost of mitigation is substantial.

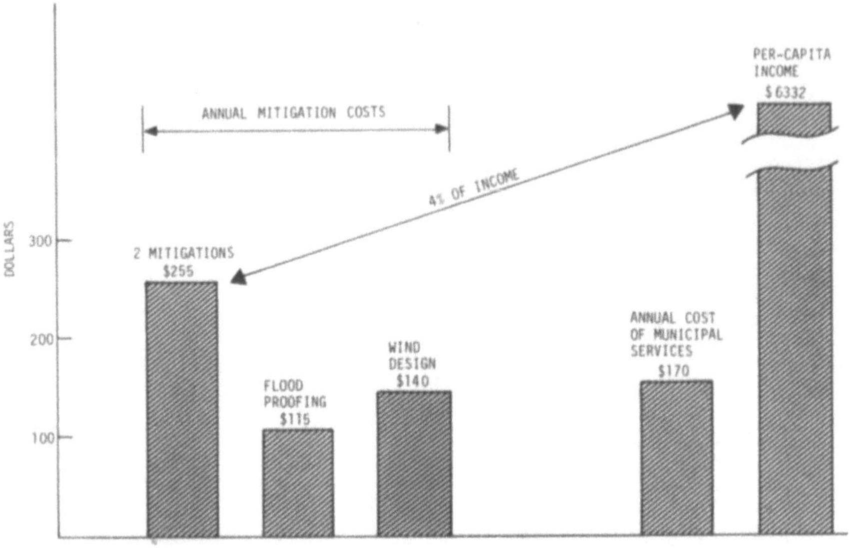

Figure 7-13. Annual average costs of mitigation (buildings only), per capita comparisons, year 2000.

In this analysis, it has been assumed that precautions against flood would be implemented throughout the flood-impact zone. Possibly, a smaller area would actually be protected. However, this would mean a smaller base over which to spread the costs (assuming some form of self-sufficient financing). Table 7-4 indicates that implementation of floodproofing and a wind design code would result in significant loss reductions for the 28 county area. With these mitigations, total losses for the year 2000 (in 1970 constant dollars) would be \$2.5 billion, whereas losses without mitigation would be \$5.9 billion. This reflects a gross savings or loss reduction of \$3.4 billion, or 58%. Floodproofing will reduce loss 47.5%; protection against wind will reduce loss by 31.3%.

Cost and loss reduction ratios were established for each county and for each structural mitigation. Estimates presented in Figure 7-14 indicate summary cost/loss reduction ratios:

Total (two mitigations)	1.74
Wind design	1.31
Floodproofing	3.23

Cost/loss reduction ratios above 1.0 indicate higher costs than loss reduction, and thus, an unfavorable condition for the 28 county area. Within the impact area there is significant variation. However, only a few counties (those hardest hit) show favorable cost feasible ratios (under 1.0), for wind design only. These ratios imply that is may not be practical, on a purely economic basis, to implement the 2 structural mitigations. However, other justifications may be offered, including the alternative costs for insurance, and the value associated with prevention of an impact on a related set of communities.

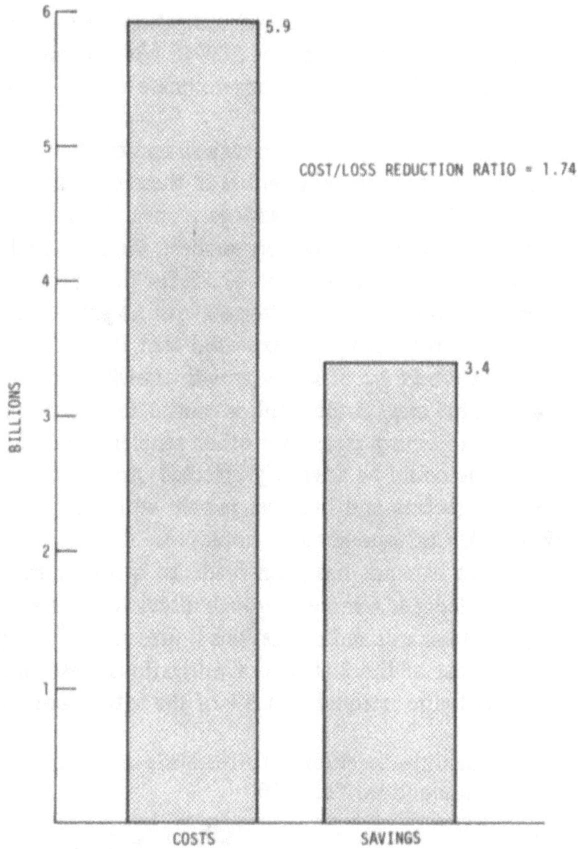

Figure 7-14. Mitigation costs and loss reduction.

Table 7-4. Hurricane Camille, Loss Reductions after Implementation of two Structural Mitigations, Year 2000

	Losses without mitigation ($ millions)	Losses with mitigation ($ millions)	Loss reduction	%
Buildings	2,390	1,515	875	37
Other	3,471	966	2,505	72
Total	5,861	2,481	3,380	58
Buildings only				
Floodproofing	790	415	375	48
Building code	1,600	1,100	500	31
Total	2,390	1,515	875	37

As shown in Table 7-5, the 2 structural mitigations (wind design and floodproofing) will result in loss reduction of 58% to 63% in mortality, unemployment, and homelessness. These figures approximate overall percent dollar loss reductions.

Loss reductions for the 100 year storm seawall and the "no growth" measures are shown in Table 7-6. It is apparent neither of these measures would be as cost feasible as floodproofing (considering savings alone). Loss reduction from the 50 year floodplain mitigation is relatively modest: an estimated 8%, compared to 35% for the 100 year storm seawall and 47.5% for floodproofing.

No cost estimate of the mitigation alternative of no growth has been made in this scenario. It might reasonably be expected that lack of growth in a given flood protection area would be offset by growth elsewhere in the region. Higher costs in small geographic areas could result in considerable economic dislocation, higher population in adjoining areas, and other problems. In addition, property values within the zone could be adversely affected. And, within a microregional setting, building contractors and business people within flood zones could be negatively influenced. Correspondingly, employees of impacted firms could be affected. Although no attempt has been made to quantify these social costs; they must be acknowledged. A more thorough discussion of specific stakeholder impacts resulting from hazards and mitigations is presented in Chapter 8.

As noted earlier, cost of the 2 structural mitigations is estimated at $255 per year per capita. This figure is equal to 1.3% of the total economic value at risk,

Table 7-5. Hurricane Camille Total Impact Area Mitigation Summary—Floodproofing and Wind Design Loss Reduction, Year 2000

Total economic loss	$ 5,861	
Total loss with mitigation	2,480	
Total loss reduction	$ 3,381 million	58%
Overall cost-loss reduction 1.74		
Total unemployment (person-years)	67,732	
Total unemployment with mitigation	24,840	
Total unemployment saved	42,892 person-years	63%
Total homelessness (person-years)	63,543	
Total homelessness with mitigation	24,909	
Total homelessness saved	38,634 person-years	61%
Total number of expected deaths	642	
Total number of expected deaths with mitigation	262	
Total reduction in deaths	380 lives	59%
Overall per capita income loss	$ 2,510	
Overall per capita income loss with mitigation	1,062	
Overall per capita income loss saved	$ 1,448	58%

Table 7-6. Building Loss Reductions from Area Structural Mitigation Measures

	Losses without mitigation ($ millions)	Losses with mitigation ($ millions)	Loss reduction ($ millions)	%
Floodproofing	790	415	375	47.5
100 year storm seawall	790	513	277	35.1
No growth on 50 year floodplain	790	727	63	8.0

and therefore, might be considered a form of insurance, i.e., spending money to save money as opposed to spending money to redistribute losses. In these terms, the cost does not appear to be out of line with current costs for hurricane insurance in Gulf Coast states, or with earthquake insurance in California.

San Francisco Earthquake II

The San Francisco Earthquake of 1906 occurred on April 18, at 5:13 A.M. The principal quake had a duration of one minute, five seconds, and was followed by 30 or more minor tremors before 7:00 A.M. The epicenter was about 10 miles from the center of downtown San Francisco, near Point Reyes. The magnitude was recorded at 8.3, one of the highest ever measured in California. Maximum intensities ranged from IX to XI MMI. Figure 7-15 shows the intensities in various counties. Each county was assigned the intensity recorded at the county seat, using a computer-generated intensity map of the area (see Figure 7-16).

Seismologists assume that this earthquake resulted from an accumulation of stresses that ordinarily would have been relieved by smaller movements (Gilbert et al., 1907; Iacopi, 1971). Fault movement ripped the surface for almost 200 miles, the longest surface rupture caused by a single fault movement during recorded history. Horizontal displacement of 15 to 20 feet occurred; it was later found that this was common to the San Andreas fault. Before this time, earth movements were thought to be predominantly vertical. Figure 7-17 presents a graphic display of the major fault system in California, including the San Andreas fault.

Three days after the earthquake, 2593 acres had burned, destroying 490 city blocks completely and 32 partially. Three hundred fourteen of these acres constituted the congested district of the city. A 50 mile wide zone of destruction was produced, extending 150 to 200 miles from San Francisco. This zone began at Point Arena, went south to Hollister and included San Francisco Bay, Russian River, and the Sonora, Santa Clara, and Salinas valleys.

Over 700 people died in the earthquake and ensuing fires. Four hundred million dollars of property damage was recorded, two hundred fifty million of

Figure 7-15. MMI intensities of counties as recorded at county seats of the counties impacted by the San Francisco Earthquake.

which was in San Francisco. The hardest hit area was from San Jose to Santa Rosa (Iacopi, 1971). Over 200,000 people were left homeless. Loss of life was low, however, due to the time of day that the earthquake occurred. Most people were sleeping, many in frame houses. If the earthquake had occurred four or five hours later when people were actively occupied, many more could have died.

The exact number of deaths is unknown since many were buried quickly and official records were not kept. Many structures were completely consumed by fire. One source claims that 112 patients at Agnew State Hospital perished due to the flimsily constructed building; 61 died in Santa Rosa and 19 in San Jose. Outside of San Francisco, there were 189 known deaths (Richter, 1958).

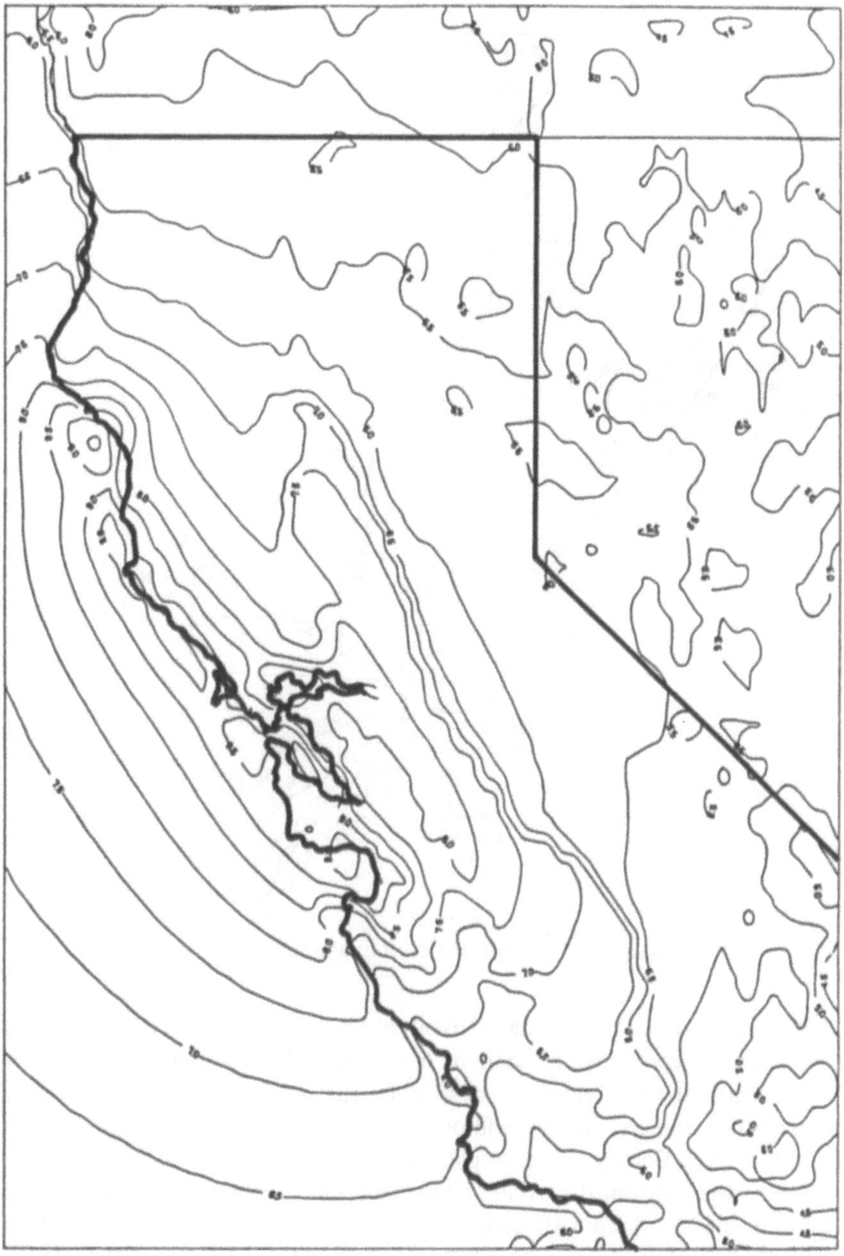

Figure 7-16. San Francisco earthquake scenario intensities due to an M = 8.25 shock.

The buildings hardest hit were those on filled ground. It was difficult to separate the earthquake and fire damage, but estimates attribute 20% to earthquake and the rest to fire. The water supply was crippled by breaks in the distributing mains within the city limits, which led to failure to control the fire, rather than to a lack of water coming into San Francisco (Iacopi, 1971).

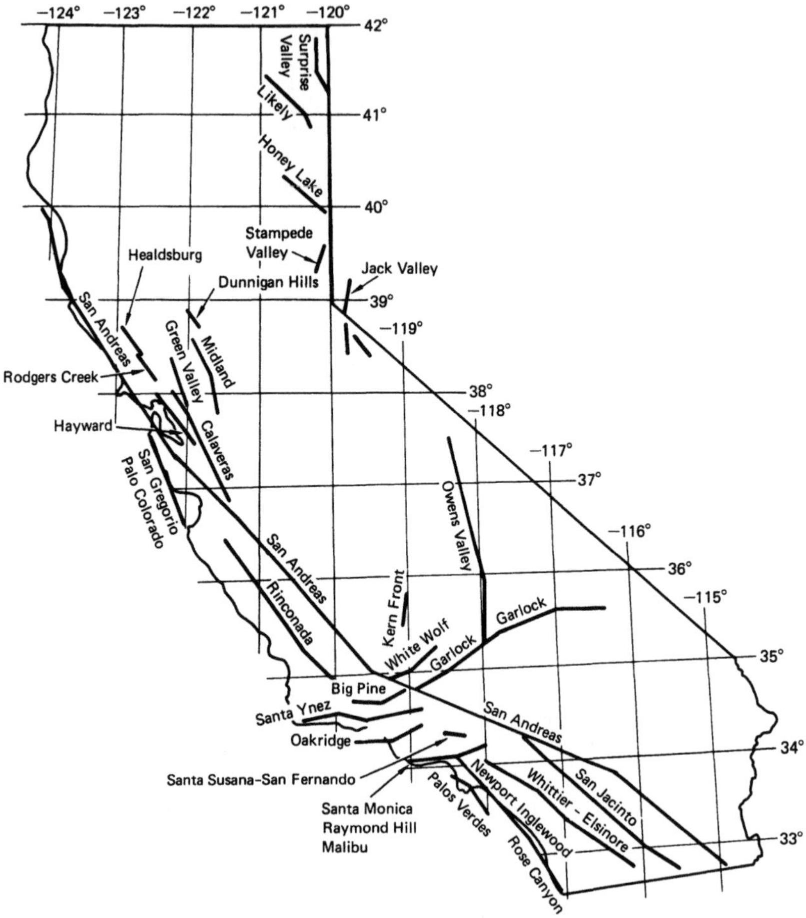

Figure 7-17. California earthquake fault system.

This scenario repeats the original event, but is limited to those counties whose county seats experienced an MMI of 6 or greater in the original event (Table 7-7). Further, the scenario does not consider fire potential or fire losses. Under the scenario, principal damage will occur near the coast along the San Andreas fault. Monterey County, for example, was included because Salinas (the county seat) is at the uppermost part and on the fault line. Projected per capita income figures in the 39 counties affected by the scenario earthquake vary from a high of \$11,060 in Marin County to a low of \$5776 in Kings County. Total population for the 39 counties in the year 2000 is projected to be 12,537,000 persons, an increase of 55% over the 1970 population of 8,079,660. Average per capita income is projected to increase 111% during that period. Total economic value of the buildings in the 39 county area is estimated at \$236 billion in the year 2000, which is equivalent to almost \$18,900 per capita, as an average value for the area (derived from Wiggins, et al., 1978).

Table 7-7. Percentage of Population Distribution by MMI, Year 2000

MMI 9 and over	20.8%	*MMI 8 to 8.9*	46.2%
Monterey		Alameda	
San Benito		Contra Costa	
San Francisco		Marin	
San Mateo		Napa	
Santa Cruz		San Joaquin	
		Santa Clara	
		Solano	
		Sonoma	
MMI 7 to 7.9	21.2%	*MMI 6 to 6.9*	11.8%
Colusa		Amador	
Fresno		Butte	
Glenn		Calaveras	
Kings		El Dorado	
Lake		Humbolt	
Madera		Kern	
Mendocino		Mariposa	
Merced		Placer	
Sacramento		San Luis Obispo	
Stanislaus		Tehama	
Sutter		Tulare	
Yolo		Tuolumne	
Yuba		Douglas, NV	

Economic Loss without Mitigation

Reported losses from 1906 vary and are complicated by the inclusion of losses due to fire. However, the following figures seem to be a reasonable compromise for earthquake only losses: 700 deaths, 200,000 people left homeless, and damage in the range of $400 million to property. The numbers available of those injured are too variable to support an estimate.

Projected losses for the scenario years 1970 and 2000, economic and noneconomic, are shown in Figure 7-18. Noneconomic losses will be noticed more in those counties with the highest intensities. The greatest unemployment is found in San Francisco, San Mateo, and Santa Clara: over 3000 per county in the year 2000. Also, over 500 deaths might be expected in San Francisco County alone. Many counties will not suffer noneconomic losses. These counties are generally the ones with MMI's in the range of 6 to 7, farther away from the epicenter of the earthquake. Since the greater losses are concentrated in a relatively small area, care must be taken in assessing the impact of total losses from such a scenario.

Economic loss for the year 1970 is estimated to be over 21 billion and in the year 2000 to be almost twice that much. These estimates yield per capita losses

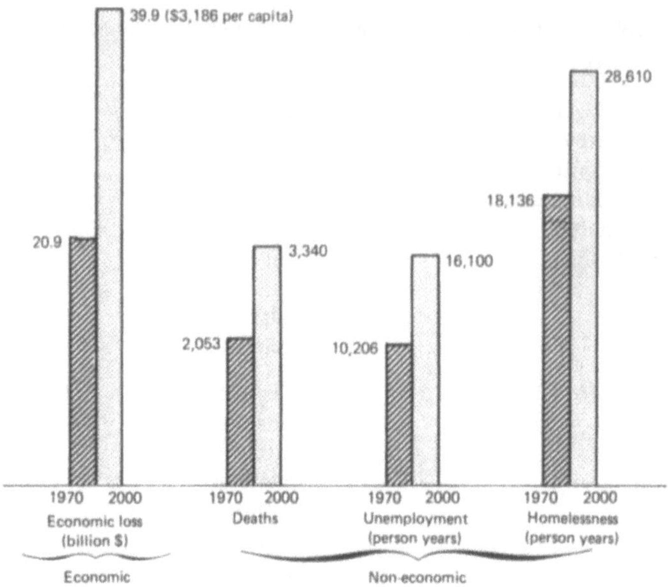

Figure 7-18. Loss comparisons, 1970 and 2000 scenario without mitigations.

of $2584 in 1970 and $3186 in the year 2000. However, losses vary greatly from county to county. In dollar amounts, the greatest losses will occur in San Mateo ($7555 per capita) and Santa Cruz ($7044 per capita). Table 7-8 shows that half the counties will experience a loss of under $1000, less than one-third the average figure. Because of high exposure values and concentrated population, only nine counties will experience losses in excess of the average, four of these over $5500 per capita.

Table 7-8. Year 2000 Per Capita Economic Loss without Mitigation (1970 Dollars)

Economic loss	% of total	Number of counties
Under 100	7	17.9
100– 499	8	20.5
500– 999	5	12.8
1,000– 1,999	9	23.1
2,000– 2,999	0	0.0
3,000– 3,999	4	10.3
4,000– 4,999	2	5.1
5,000–10,000	4	10.3
Total	39	100.0
Median	913	
Average	3,186	

Table 7-9. Year 2000 Per Capita Economic Loss as a Percent of Per Capita Income, without Mitigation

Annual family income ($)	Population distribution (%)	Loss distribution (%)	Difference (%)
1. Less than 3,000	8.4	7.4	(1.0)
2. 3,000– 4,999	9.0	7.5	(1.5)
3. 5,000– 6,999	10.2	9.1	(1.1)
4. 7,000– 9,999	18.1	17.1	(1.0)
5. 10,000–14,999	28.0	28.9	.9
6. 15,000–24,999	20.5	23.1	2.6
7. 25,000 and over	5.8	6.9	1.1
Total	100.0	100.0	0

Projected economic loss due to a San Francisco earthquake in the year 2000, without new mitigation measures, would be close to 40 billion dollars in 1970 dollars in the 39 county area. This is equivalent to about 38.3% of the per capita income in the area (projected to be $8320 in 1970 dollars). Table 7-9 show the variance in losses among the counties as a percentage of per capita income.

In the 39 county area, approximately 45.5% of the $39.9 billion loss represents the value of building damage, the balance representing losses of building contents, except for food. Food would be considered part of building contents only if the building were a food establishment. Supplier loss and income loss are also included in the 54.5%.

Table 7-10 shows the percentage of the population in different income categories for the 39 counties. The loss is also distributed using the same percentage

Table 7-10. Year 2000 Per Capita Economic Loss as a Percent of Per Capita Income without Mitigation (1970 Dollars)

% of income	Number of counties	% of total
Under 1.0	5	12.8
1.0– 4.9	8	20.5
5.0– 9.9	2	5.1
10.0– 24.9	13	33.4
25.0– 49.9	6	15.4
50.0–100.0	5	12.8
Total	39	100.0
Median	14.0	
Average	38.3	

distribution by county as that used for the population. The higher income groups are experiencing a slightly greater loss than the lower income groups; however, the percentage is very close. Actually, no single income group appears to suffer an extraordinary loss.

Only 3 types of noneconomic losses have been considered in this scenario: mortality, unemployment, and homelessness. However, specific stakeholder impacts from general hazard and mitigation impacts are examined in Chapter 8.

Mortality is projected at 2050 for the year 1970, and 3340 for the year 2000. This means one in every 3750 persons may die as a result of the earthquake in the year 2000. It is estimated that between 450 and 700 persons died as a result of the 1906 earthquake. It must be noted that the time of day when the event occurs is extremely important in estimating the number of deaths. For example, in a U.S. Department of Commerce (1972) study of earthquake losses for San Francisco, it is estimated that a change in the time of occurrence from early morning hours to midafternoon hours would result in an increase in deaths by a factor of 4. Thus, if San Francisco Earthquake II were to occur in midafternoon, it is reasonable to expect life loss to be nearly 12,000. Moreover, NOAA also estimates an increase in hospitalized injuries from 10,800 to 40,360 for the same shift in time of occurrence.

Projections of unemployment due to the earthquake are unimpressive: 10,200 person-years in 1970, and 16,100 person-years in 2000. In a population of 12.5 million, normal unemployment at 5% would be about 250,000 persons per year.

As with unemployment, the projected figures for homelessness are low: about 18,000 person-years in 1970, and almost 29,000 person-years in 2000. These figures may be viewed in different ways: (1) that almost 29,000 people will be homeless for a year, which is unlikely, or (2) that over 343,000 people will be homeless for an average of 30 days, a more realistic conclusion. Thus, 2.7% of the poeple in the affected area will be separated from their homes for one month in the year 2000.

Mitigation Measures and Costs of Implementation

The scenario mitigation used to reduce damage involves an upward adjustment in the lateral force specifications contained in the 1973 Uniform Building Code for Zone 3 structures. Present design parameters require earthquake resistance above magnitude 6 (based on a Modified Mercalli scale). Mitigation measures for this scenario require increasing this strength requirement by 2 for all new construction starting in 1980.

Because it is estimated that existing construction will represent about 40% of the total value at risk in the year 2000 ($94.1 of $236.1 billion), and because the scenario counties that will experience growth are not currently highly developed, mitigations were not considered for pre-1980 buildings. Figure 7-19 shows cost of the mitigations when applied to new construction as $33.7 billion, a sum expended over a 20 year period (1980–2000). At 10% interest, annual expendi-

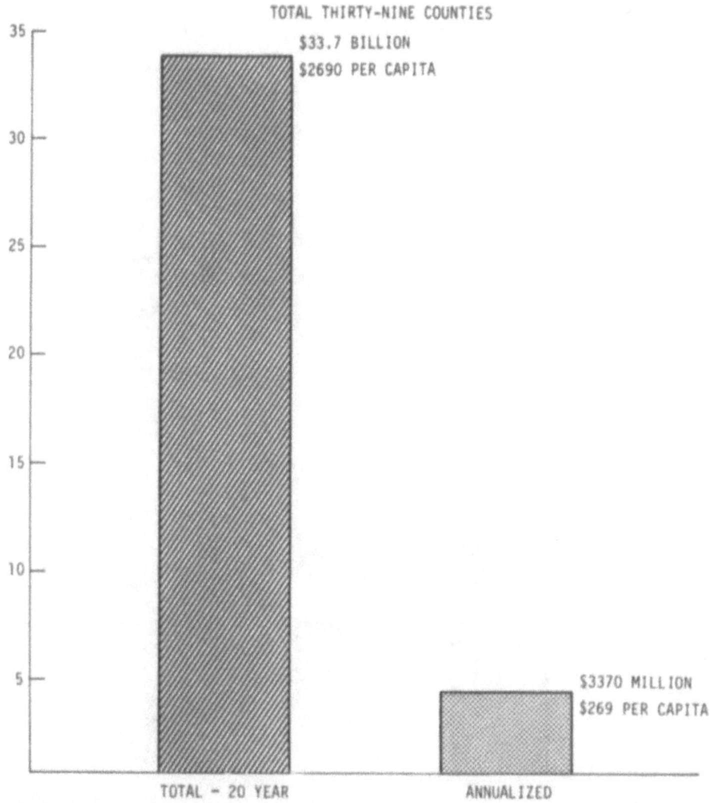

Figure 7-19. Costs of mitigation (structure only) for San Francisco Earthquake scenario, year 2000.

tures would total $3.37 billion. This amount when spread over the projected population comes to $269 per person.

The total estimated cost of mitigation ($33.7 billion) is approximately 14.3% of total building value at risk ($236 billion) in the year 2000. Annual mitigation cost of approximately $269 per person constitutes 3.2% of average per capita income. This mitigation cost compares favorably with the "normal" costs of providing municipal services: approximately $170 per capita (Figure 7-20).

Expected dollar losses without new mitigations from a reoccurrence of the San Francisco Earthquake in the year 2000 would total $39,943 million. By increasing building standards, losses could be reduced to $28,595 million, thus producing a savings of $11,348 million, or 28% (Table 7-11).

The ratio between cost of mitigation and gross loss reduction is 2.97 (presented graphically in Figure 7-21). Because projected costs are significantly higher than expected loss reduction, the mitigation requiring increasing the design lateral force requirements of the UBC by a factor of 2 for Zone 3 would seem to be *economically* disadvantageous to the vulnerable area.

Constructing buildings to meet the more stringent code is estimated to cost

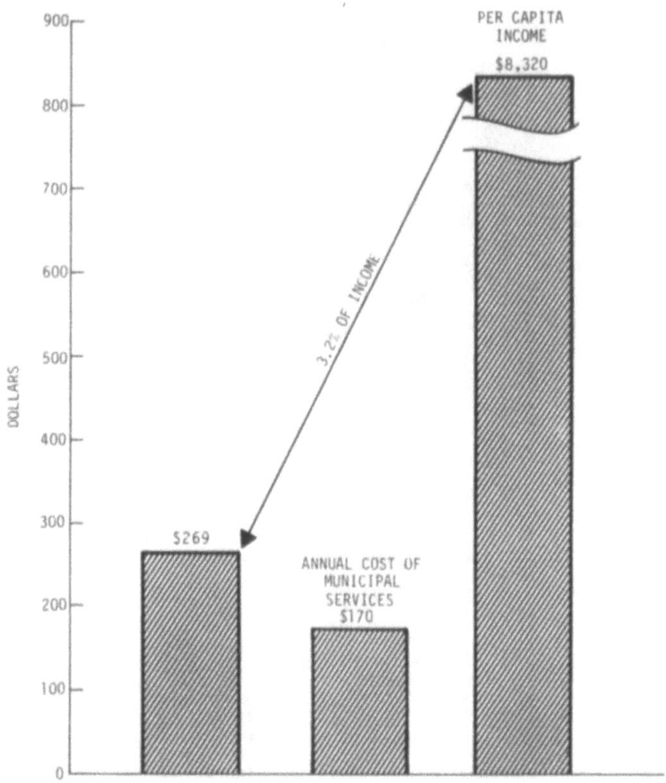

Figure 7-20. Annual average costs of mitigation (buildings only), per capita comparisons, year 2000.

$269 per person per year, 1.4% of the estimated $18,836 per capita at risk. This appears to be approximately the same as that for earthquake insurance available in California at the present time. Thus, the purchase of insurance at the same cost as the annualized cost of the mitigation has an interesting public policy implication. Specifically, an insurance purchase is an individual choice selected by an individual for the purpose of reducing his personal risk of loss, whereas a structural mitigation embodied in a building code is a public choice requiring compliance by all. When considering a cost/loss reduction ration of greater than 1 for all affected counties, it appears that insurance may be an appropriate approach to individual loss mitigation.

Table 7-11. Year 2000 San Francisco Earthquake Building Loss Reductions, after Implementation of Structural Mitigation in the Year 1980

Loss without mitigation[a]	$39,943
Loss with mitigation[a]	$28,595
Loss reduction[a]	$11,348
Percent	28%

[a]Millions 1970 dollars.

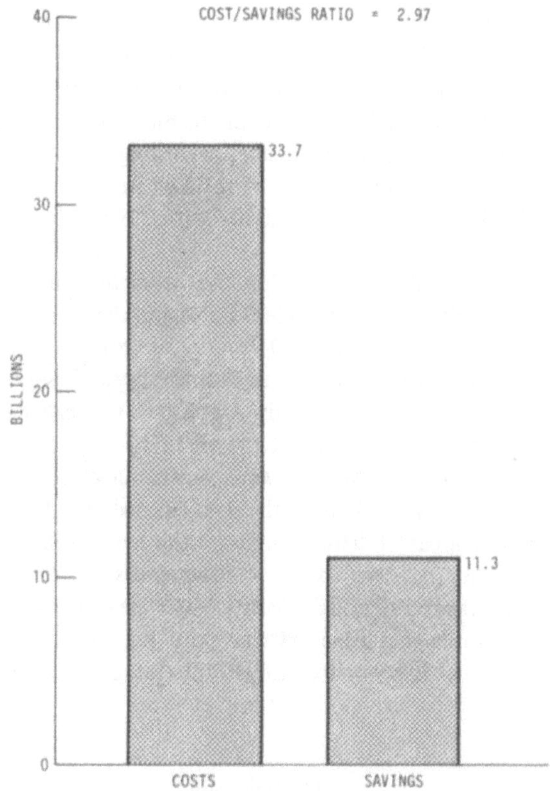

Figure 7-21. Mitigation costs and loss reductions for San Francisco Earthquake scenario, year 2000.

Other Scenarios

During the course of the study other scenarios also were prepared. Thus, the computer models predict that Cook County, Illinois, could experience \$21 billion damage in the year 1980 and \$35 billion damage in the year 2000 from a downtown center strike of a Fujita Level 4 tornado (Hart, 1976); and that a repeat of the riverine flooding associated with Hurricane Agnes (1972) conditions under projected actual conditions of flood protection in 2000 in multistate region 2 would produce damage to more than 151,000 structures and an episodic building damage loss of \$1008 million (Lee et al., 1976).

Other researchers also have produced catastrophic loss scenarios for a variety of hazards and have predicted similar high episodic losses. White and Haas (1975) have prepared qualitative scenarios for the following conditions: (1) a major hurricane strike on Miami; (2) a riverine flood in Boulder, Colorado; and (3) a repetition of the San Francisco Earthquake. The U.S. Department of Commerce, (1972, 1973) has prepared, for the Department of Housing and Urban Development, quantitatively oriented scenarios showing the consequences of major earthquakes in San Francisco and Los Angeles. The Los Angeles scenario predicts 2790 (2:30 A.M.) to 12,385 (4:30 P.M.) deaths from an 8.3 magnitude earthquake on the San Andreas fault and 4090 (2:30 A.M.) to 20,728 (4:30 P.M.) deaths from a 7.5 magnitude earthquake on the Newport-Inglewood fault. Dam ruptures could add 7520 deaths to the San Andreas quake and 2200 deaths to the Newport-Inglewood quake, according to NOAA. Thus, a major quake at the appropriate hour on one of the Los Angeles area fault lines could, according to NOAA, cause up to 25,000 deaths. Roughly comparable results were shown for San Francisco.

Although no modern scenario writer has predicted hurricanes/storm surge deaths at a *rate* recorded during the 1900 Galveston disaster (6000 deaths out of a population of approximately 60,000) it does not stretch one's imagination to assume that 25,000 deaths could result from a "great" hurricane strike on a major Gulf or Atlantic Coast metropolis. A question that remains unanswered is: What is the capacity of our society to respond to and recover from the possibility of two or more disasterous events occurring during a single 12 month period, or one single event such as the 1900 Galveston disaster, should they strike major SMSA areas? Further, what are the nationwide impacts of such a level of devastation, in terms of the consequences for financial institutions, industrial enterprises, and national security? It is clear that the consequences of certain natural hazards that have extreme potential for destruction need to be better understood and appropriate mitigation strategies adopted, implemented, and enforced.

References

Friedman, D. G. 1971. The Storm Surge Along the Gulf and South Atlantic Coastlines. Travelers Insurance Company, Hartford, Connecticut.

Gilbert, G. C. 1907. The San Francisco Earthquake and Fire of April 18, 1906, and their Effects on Structures and Structural Materials. U.S. Govt. Printing Office, Washington, D.C.

Hart, G. C. 1976. Natural Hazards: Tornado, Hurricane, Severe Wind Loss Models. Prepared for the National Science Foundation (NTIS No. PB 294594/AS). J. H. Wiggins Company, Redondo Beach, California.

Iacopi, R. 1971. Earthquake Country; How, Why and Where Earthquakes Strike in California. Lane Books, Menlo Park, California.

Lee, L. T., J. D. Chrostowski, and R. T. Eguchi. 1978. Natural Hazards: Riverine Flooding, Storm Surge, Tsunami Loss Models. Prepared for the National Science Foundation (NTIS No. PB 294682/AS). J. H. Wiggins Company, Redondo Beach, California.

National Safety Council. 1975. Accident Facts 1975. NSC, Chicago, Illinois.

Richter, C. F. 1958. Elementary Seismology. Freeman Press, New York.

Simpson, R. H., and N. B. Lawrence. 1971. Atlantic Hurricane Frequencies Along the U.S. Coastline. NOAA Technical Memorandum No. NWS SR58. U.S. Department of Commerce, Washington, D.C.

U.S. Army Corps of Engineers, 1969. Report on Hurricane Camille, August 14–22, 1969. U.S. Army Engineer District, Mobile, Alabama.

U.S. Bureau of the Census. 1973. County and City Data Book–1972. U.S. Govt. Printing Office, Washington, D.C.

U.S. Bureau of the Census. 1975. Statistical Abstract of the United States: 1975. U.S. Govt. Printing Office, Washington, D.C.

U.S. Department of Commerce. 1972. A Study of Earthquake Losses in the San Francisco Bay Area. USDC/NOAA, Environmental Research Laboratory, Washington, D.C.

U.S. Department of Commerce. 1973. A Study of Earthquake Losses in the Los Angeles, California Area. USDC/NOAA, Environmental Research Laboratory, Washington, D.C.

U.S. Department of Commerce. 1976. Natural Hazard Management in Coastal Areas. USDC/NOAA, Washington, D.C.

U.S. Water Resources Council. 1974. OBERS Projections, Regional Economic Activity in the United States, Series E, Populations. Vols. 1–7. U.S. Govt. Printing Office, Washington, D.C.

White, G.F., and E.J. Haas. 1975. Assessment of Research on Natural Hazards. MIT Press, Cambridge, Massachusetts.

Wiggins, J. H., J. E. Slosson, and J. Krohn. 1978. Natural Hazards: Earthquake, Landslide, Expansive Soils Loss Models. Prepared for the National Science Foundation (NTIS No. PB 294686/AS). J. H. Wiggins Company, Redondo Beach, California.

PART III

NATURAL HAZARDS POLICY PLANNING AND ADMINISTRATION

In a 1980 opinion the United States Supreme Court observed that "the factual finger points, it does not conclude," and went on to state that "factual determinations can at most define the risk in some statistical ways; the judgment whether that risk is tolerable cannot be based solely on a resolution of the facts" (Industrial Union Department, 1980).

This observation is of extraordinary relevance to any effort to devise rational public policies for the mitigation of the range of effects produced by hazardous natural occurrences. To what extent a particular set of occurrences may be viewed as constituting a public problem involves the resolution of as many questions of value as it does issues of fact.

Accordingly, this part of the book is devoted to an examination of the various perspectives that may be used to define those candidate public problems that may result from natural hazard occurrences, and to an explication of the variety of policy approaches that may be taken to resolve such problems. Three separate but related subjects are addressed: the stakeholders whose interests are influenced by hazardous natural occurrences; the constraints that inevitably confront those who seek to resolve the problems faced by the various stakeholder groups; and the various policy options and action recommendations appropriate for consideration by the U.S. government, as well as state and local governments.

8
Policy Makers, Stakeholders, and Candidate Public Problems

The Plight of the Policy Maker

Citizens often casually refer to "public problems." However, these problems can be difficult for the public policy maker to define before taking appropriate action. For example, a situation, impact, or set of effects defined as a *problem* by one group may be regarded as a *solution* by another; one person's cost can be another person's benefit; an action that solves one problem besetting a given group can impose a different problem on that same group. It is the function of the policy maker to resolve these dilemmas and determine whose interests are to be served and whose are to be ignored, who is to receive a benefit and who is to bear the cost, and if and when it is appropriate to act.

Problems affect real people, at some time, in some real location, under some real condition. However, most problem situations are not well defined, but tend to be characterized by clusters of related phenomena. Therefore, it is necessary to specify problems carefully so that they can be dealt with. The problems must be operationalized; that is, they must be defined in terms of what the disvalued attributes of a situation are and how important those elements may be to specific people in specific places under specific conditions.

The first matters to consider in defining problems operationally are what potentially adverse impacts are occurring, who is being affected by those impacts, the geographic areas within which the impacts are being felt, and the values that are being affected or are otherwise involved in these situations.

Who Is Affected?

Who or what is being affected by some set of impacts or effects? Home owners? Renters? Investors? Employers? By specifying as precisely as possible who is affected, one begins to specify the possible character of the problem. If it is residents of the Gulf Coast or farm families in Kansas, then one can begin to add a geographic dimension to the analysis. By assessing how many people are affected directly and how many are affected by second-order effects, and how many can be expected to be affected in the future if current trends continue, ultimate definition of the problem is further operationalized.

What Is the Extent of the Impact?

With natural hazards, as with most other situations with which public officials concern themselves, there is a geographic dimension. Wisconsinites don't worry much about being injured by hurricanes and earthquakes; Southern Californians don't worry much about tornadoes and ice storms. Impacts and potential problem situations may be defined in terms of specific types of locations, such as all inner cities or all peripheral suburban areas; or they may be location-specific, such as the Gulf Coast, the Atlantic seaboard, or the Pacific Coast. They may be confined to coastal counties, and within such counties to specific zones such as tsunami inundation areas.

Which Values Are Being Affected?

It is one thing if a household is inconvenienced. It is quite another if a member of that household dies or is threatened with death. In order to help specify problems, we consider which values are being affected by some impact or effect and with what intensity. Among values—or things that are valued—are economic, physical, and mental well-being. Effects on values may range from minor inconvenience to severe deprivation. We may be able to tolerate a fairly large number of persons being slightly inconvenienced, when we would not tolerate situations where several are likely to be killed.

What Will Happen if Nothing Is Done?

It is important to incorporate a time dimension when defining public problems. In most instances, it is possible to forecast what is likely to happen in a situation if current trends continue and if no policies are changed. Surprisingly, perhaps, no action is sometimes preferable to some action. A principal question to ask is whether the situation will get better or worse all by itself, without any additional public action. How much better or how much worse would the situation become without additional action? Is there a significant benefit to acting now as opposed to acting at some later time? That is, is it possible to exert a small amount of leverage at this time on the impact-causing or problem-causing system that will have a significantly greater influence on mitigating the situation than leverage exerted at some later time, or that will produce fewer unwanted side effects than later action of some other type.

What Are the Elements of the Suspected Problem-Causing System?

Mitigating the effects of natural hazards exposures requires intervention in extremely complex social, geophysical, and technologic systems involving the natural environment, social interaction, and the distribution of costs and benefits. As a rule, it is better not to intervene in a system if one cannot accurately predict the consequences. Unless one understands the cause and effect relationships within a system, intervention can lead to adverse consequences and to unanticipated and undesirable by-products that may far outweigh the intended consequences and that may sometimes be sufficient to lead to total destruction of the system being influenced. Therefore, it is critically important to understand situations in their systemic relationships: how they fit together, what causes what, and the extent to which interdependencies exist. If one does not understand systemic interrelationships, then it is impossible to predict, with any degree of certainty, the consequences of action or inaction. Unless the probable consequences of a recommended action are known, it does not make sense to implement that recommendation.

Where one does not understand systemic relationships or where the probable consequences of action are unknown, it makes sense to conduct research to learn more about those systems. If one is in doubt about what the *real* problems are, and what causes them, the best action may involve the development of further knowledge about the system.

Can the Impacts of Some Obvious Solutions Be Predicted?

Frequently, this preliminary stage in a policy maker's examination of a situation can yield suspicions or assumptions concerning some obvious types of possible solutions. A listing of these can then be used to make some guesses about the impacts that may be produced through the use of such solutions, who will experience them, to what extent, where, and what values may be affected by those impacts.

All these interrelated, first-pass inquiries into situations in which "impacts" from some set of causes have been observed are intended for but one purpose: to gain some insights as to who is being affected, or is likely to be affected as a result of problem-solving efforts, and what values, motivations, attitudes, and interests of these parties are likely to be of governing importance to the specification of selected impacts as "problems." This process leads to the identification of stakeholder groups.

Stakeholders in Natural Hazard Policy

The interests of several distinguishable groups, or stakeholders, are involved in the many kinds of situations in which the exposure losses and mitigation costs occur, or in which these losses and costs are transferred to nonexposed parties.

In general, these stakeholders may be grouped into 3 classes: *loss experiencing parties, mitigation-involved parties, and mitigation-constraining parties.*

Loss Experiencing Parties

Loss experiencing stakeholders include all those groups which bear the losses arising from natural hazard exposures and/or the costs arising from efforts to mitigate the effects of such exposures.

1. Residents of Natural Hazard Zones. For each of the 9 natural hazards identified in this analysis there is a companion set of identifiable hazard zones within the United States, the residents of which may be viewed as being at risk of experiencing the consequences associated with the occurrence of the hazard. The sum of the individuals who reside, own property, or have other financial interests within these zones may be viewed as constituting the current population at the risk of exposure to hazardous natural events. This major stakeholder group consists of at least 2 component sets of parties: (a) individuals who currently are exposed to the hazards as a result of voluntary choice, and (b) parties who are *involuntarily* exposed to the hazards. The former group includes those who, having knowledge of the fact that they were entering into or remaining within a hazard zone, nonetheless chose to do so; while the latter includes parties who (a) either entered into or are remaining in the hazard zone out of ignorance concerning the risk they are running, or (b) are aware of the risk they are running but lack an ability to remove themselves from the hazard zone because of such factors as the demands of their employment or their financial inability to move elsewhere.

It has been stated by others that substantial numbers of individuals who are at risk of exposure to hazardous events are ignorant of the risk they are incurring. Studies by White et al. (1958), and Kates (1971), suggest that a significant fraction of occupants of riverine floodplains are comparatively unaware of the risks and the adjustments they could make to reduce those risks, whereas residents of coastal areas exhibit comparatively good knowledge of their risks and the effects of past storms. Mileti et al (1975) suggest that "most persons simply do not know the character and extent of the hazards(s) for the area in which they reside or work." They further comment that "realtors and civic leaders tend to suppress discussion of flood hazards, refuse frequently to recognize, even privately, the dangers of encroaching development, and sometimes reject flood protection works to avoid admission of the hazard's existence." Although much more needs to be known concerning this subject, it seems safe to state, especially with regard to less studied hazards (i.e., expansive soil), that comparatively few occupants of significant hazard zones are fully aware of the magnitude of the risks they are incurring and of the adjustments that they can make to mitigate those risks. Similarly, it seems clear there is little impulse on the part of sellers and transferrers of property to inform purchasers and users of the risk they are running by locating within a hazard zone.

The threat of a substantial loss of property or life is faced by occupants in many hazard zones, such as high-frequency coastal and riverine floodplains, tsunami inundation zones, and tornado-prone areas. Expansive soils, landslides, earthquakes, and other hazards all exhibit varying annual expected loss burdens to occupants of the hazard zones. However, hazard zone occupants who have experienced the consequences of hazardous occurrences do not, in any great numbers, exhibit a willingness to remove themselves from the hazard zone. Instead, they wish to get on with life as it was before the event occurred, tend to deny that a recurrence of the event will heap any substantial consequences on them, and seek support from the government and other parties to make the area "safer" should the future bring a recurrence of the event.

Personally catastrophic losses may befall both voluntary and involuntary risk takers within natural hazard zones, and there is evidence that such risk takers have a strong propensity to seek means for externalizing those cost burdens. For comparatively few hazards is insurance available to "spread the risk" of loss among the national or regional pool of hazard-specific risk takers. However, even when insurance is available there are comparatively few subscribers, either because of the cost or a persistent tendency to deny that adverse consequences will befall them. Although Federally subsidized flood insurance has substantially increased the number of holders of this type of policy, far fewer occupants of flood zones opt for this protection than economic wisdom might lead a neutral analyst to predict.

2. Potential Future Residents of Natural Hazard Zones. However large the size of the current population at risk of exposure to natural hazards, current patterns of migration and rates of population increase suggest that the size of the hazard-exposed population in 2000 will be substantially greater than now. Indeed, the projected *increase* in annual expected natural hazard losses for the period 1970–2000 is greater than the total estimated losses for 1970.

Three demographic facts about the U.S. population suggest the quantitative importance of this stakeholder group: (a) the population is growing older and the proportion of the population above the "working age" is increasing; (b) the population, in all age groups, is increasingly mobile and given to comparatively frequent changes in their place of residence; (c) much interstate migration has been occurring, particularly from inland to coastal and from cold weather to warm weather climates.

A large fraction of the mobile portion of our national population is being attracted to those high-hazard areas that exhibit a higher than normal mix of biophysical and recreational amenities. Precipitous and slide-prone slopes may offer occupants stunning vistas of cities and natural terrain; coastal and riverine floodplains may offer relief from the congestion and polluted air of cities and easy access to water-based recreational activities; and areas of high seismic activity may offer scenes of mountain vistas and easy access to a variety of recreational areas. Although lured by the amenities, the migrant may lack knowledge concerning the characteristics of the hazards within the area and the magnitude of the risks incurred by residents. Also, the migrant may not know the types of adjustments that can be taken. The conviction that a benevolent and all-knowing government will protect them from harm may dampen the concern of those who have some meager knowledge of the hazards, and such groups may find it inconceivable that the government would permit their entry into areas where there are substantial risks.

The findings that led to the adoption of the Interstate Land Sales Act might well be considered as evidence in support of the assumptions offered above. Numerous property transactions are made within our society by parties, who, imprudently and at great risk to themselves, place their faith in the integrity of land developers and property sales enterprises.

Of course, the interests of this stakeholder group involve more than the threat of loss from involuntary entry into high-hazard areas. Their interests extend also to choice-inhibiting public policies that might (a) prevent their vountary and knowledgeable entry into high-amenity areas in which greater than normal natural hazard risks may be present; or (b) require the use of building strengthening mitigations and thus escalate the cost of structures within such areas.

It seems likely that no single proposed criterion for trading off the risks of hazard exposure against either the cost of building mitigations or the amenity returns derived from entry into some areas will satisfy all members of this group.

3. Visitors or Workers in Natural Hazard Zones. Hospitals, nursing homes, and places of work or entertainment draw hundreds of thousands of transients into potentially hazardous areas during selected hours of the day or periods of the year.

Although the permanent resident population of U.S. central cities has been declining, it is likely that the daytime worker, shopper, and visitor populations of these same places have been increasing. Workers, shoppers, and tourists clog the streets of major central business districts during daytime hours throughout the nation, while conventioners, theatergoers, window-shoppers, and urban entertainment seekers may be found in great numbers in the same places during the evening hours.

Most of the deaths produced by the San Fernando earthquake of 1971 occurred within a Veterans Administration hospital whose occupants were drawn to the facility out of a concern for, rather than a rejection of, interest in their own longevity.

Scenarios depicting the plight of the above population subsets are not difficult to write. Surrounding the large high-rise structures of many cities in seismically active areas are literally thousands of individuals clogging the streets during the noon hour. Imagine a high magnitude earthquake that rocks and stresses the structures, releases parapets from their mountings, shatters the acres of glass enclosing the buildings, and sends tens of thousands of sharp and heavy missiles down on the unwary population below. Or, imagine the residents, as well as hikers and campers, in the inundation area below a large dam at the time of the dam failure there; or the vacationers jammed into a seashore hotel in a tsunami inundation area. There is no end to the examples one may cite.

Ignorance of their risk-taking, trust in the protecting influences of government, and a concern for the objectives and interests of the moment are more likely the concerns and perspectives of this group than the manipulation of some risk-taking calculus.

4. Owners and Users of Second Homes in Natural Hazard Zones. In recent years both the absolute numbers and proportions of second homes in the United States have been on the increase. Increased affluence among some segments of U.S. society and increased discretionary income have aided this trend. It would seem prudent to view second-home owners as a distinctly different group of stakeholders than those who are in permanent residence within a high-hazard area. Also, second-home owners consist of 2 major groups: those who frequently rent their properties, and those who reserve their second homes primarily for use by family members. The interests of both groups clearly will be influenced both by the risks associated with hazard exposures and the consequences associated with rigid, legally enforced risk mitigating requirements.

5. Non-Risk-Taking Bearers of Natural Hazards Costs. A significant fraction of the losses produced by exposure of people and property to natural hazards is now being transferred to nonexposed populations in the form of tax levies. These levies are destined to finance government relief, recovery, and insurance services to the voluntary and involuntary risk-taking residents and property owners in natural hazard zones. Similarly, the largest fraction of the costs of such federally provided area protection facilities as dams, seawalls, dikes, and channel improvements is being shifted to the general body of taxpayers rather than being imposed on those who are the primary recipients of the benefits.

By June 1975, the multidecade cost of flood control-projects completed under the supervision of the U.S. Corps of Engineers totaled $10.2 billion (in nonconstant dollars). The overwhelming fraction of these costs was incurred between 1936 and 1975. Direct federal expenditures for disaster assistance totaled more than $4 billion between 1953 and 1973. Currently, from $158 to $173 million per year are being expended from tax revenues to subsidize flood insurance programs and additional substantial sums are being expended for the annual operation and maintenance of existing area flood protection facilities, technology development and transfer operations, and administration of other hazard-related programs. As noted by Atkisson and Petak (1981), in terms of current federal tax deductions and disaster relief policies, the federal liability for annual expected natural hazard losses, based on 1970 exposures, can be set at between $2.7 and $5.8 billion per year (1970 dollars) (Table 5-28). Further, more than $1.0 billion represents transfers from donor to beneficiary states. The available evidence suggests that annual expenditures for this mix of purposes will continue to increase for several decades.

The many battles that have been fought at the local level over bond issues and property tax rates suggest the lively nature of taxpayer inquiry that may be directed to projects of general benefit to the community but which impose tax burdens on the members of that community. How many major hazard-mitigating area protection facilities would now be provided were the primary and intended local beneficiaries of such projects to bear the entire burden of costs associated with such works cannot now be quantitatively estimated. However, the California "Proposition 13" syndrome now seems to have become a growing national phenomenon; general taxpayer concern over mounting tax burdens is both a present and probable future fact of public policy within the United States. Of course, there are no major, popularly supported, and aggressive national constituencies concerned with this matter, although at numerous local sites throughout the United States, they can be identified and the consequences of their public activities more directly assessed.

Each of these major subsets of taxpaying citizen groups constitutes a theoretically distinct stakeholder group whose potential interests, perspectives, and activities may well constrain future policy making concerning natural hazards.

6. Financial Institutions and Mortgage Guarantors. Whether their seats of business activity are inside or outside hazard zones, financial institutions and mortgage guarantors clearly have a stake in the structural integrity and life-span of buildings located within hazard zones, and therefore have an interest in the costs and benefits associated with mitigations intended to protect and extend structural life.

Mitigation-Involved Parties

Included in this class are those stakeholders who must either make the basic public decisions to mitigate the effects of natural hazard exposures, or who must engage in the direct extension of mitigation-producing services.

1. State and Federal Policy Makers. Perceived public problems are the basic phenomena that motivate policy makers to take action. However, not all public problems are, or can be, acted on. Those that receive policy maker attention have produced dissatisfaction among significant constituencies and have led to the generation of demands or claims on the policy making system.

Whether dealing with natural hazards or other socially significant phenomena, policy makers prefer to adopt "distributive" rather than "regulatory" policies (Mayhew, 1974; Ripley and Franklin, 1976; Vogler, 1977; Mann, 1975) and exhibit a propensity to evaluate proposals in terms of what they will do for their own constituencies and their own chances for reelection (Mayhew, 1974). Proposals advanced by large constituencies command more attention than those advanced by small or virtually nonexistent ones (Schattschneider, 1960; Greenwald, 1977; Dexter, 1969). With respect to natural hazards and other potential problems, advocates of public policies must, therefore, consider the need for a constituency to be associated with their proposals (Olson, 1971;

Jones, 1977). Further, members of a potential risk-exposed constituency must perceive the risk associated with a hazard exposure if they are, subsequently, to be willing to organize to support risk-reduction proposals (Cobb and Elder, 1972; Schattschneider, 1960), and bring them to the attention of appropriate policy makers.

Finally, with respect to problems involving the public health and safety, policy makers seem to respond more vigorously to infrequently occurring but very visible "high consequence" events than to less visible but more frequently occurring "low consequence" events. Thus, federal safety research and development appropriations per death in total commercial air accidents are orders of magnitude greater than expenditures per death produced by motor vehicle accidents. In terms of total life loss per year, the latter is far more important than the former. However, fatal motor vehicle accidents are everyday events with few deaths per event. In contrast, air accidents occur rarely, but they are highly publicized and involve many deaths.

This perspective of death-producing phenomena is not unique to policy makers. There is an apparently pervasive social tendency to overstate the impacts of hazards that produce deaths (rather than just injuries), and that take multiple rather than single lives. Moreover, there appears to be a companion propensity to underestimate the risks from common, undramatic hazards that claim one person at a time.

2. Local Policy Makers. The local policy maker rides the horns of a special dilemma with respect to hazards management. On the one hand, the ethic of representational government decrees that he or she faithfully represent the views of the existing electorate, and on the other hand, the interests of countless other parties may be adversely impacted by the prevailing views of that electorate. Nonvoting second-home owners, the interests of potential future residents, the interests of outside investors, and the interests of vacationing and transient populations may be affected by local policy-making decisions; but the local policy maker is held accountable only to the resident electorate. Similarly, the policy maker in small jurisdictions usually is not a full-time or technically trained individual. He or she is beset by competing demands from other jurisdictions of government and constituents, and is asked to make decisions involving complex questions of fact and value. Limited by time and technical capacity, these individuals are greatly dependent on the findings, opinions, and recommendations of their local staffs and highly regarded "others" in their environments.

In a recent study, researchers found that local policy makers and opinion leaders in hazard-prone areas do not view natural hazards as being a high priority policy subject. To the extent that these individuals perceive a need for mitigating the effects of natural hazard exposures, they prefer "distributive" to "regulatory" policies (Rossi et al., 1982).

3. Local and State Planners and Building Officials. Over the past 2 decades several responsible authors, study groups, and commissions have suggested that local planners and building officials are less well trained and less informed on important matters than the real demands of their jobs require (Municipal Manpower Commission, 1962; National Commission on Urban Problems, 1968; Field and Rivkin, 1975; Working Group on Earthquake Hazard Reduction, 1978). Wedded to the interests of their local communities, this group of underpaid but extraordinarily important public officials is beset by numerous competing demands from adjoining and higher jurisdictions of government. Many of these demands are more procedural than substantive or technical; and this situation has all too frequently created adherence to procedural detail rather than to the substantive requirements of community problems. Limited by time, interest, constituency requirements, past training, and their current willingness and capacity to acquire knowledge concerning new subjects, a major burden of future natural hazard management activities will fall on this group. Far too little is known about the characteristics and development needs of these stakeholders, but it seems safe to suggest that their training and staff development needs are large and, as yet, essentially unsatisfied.

4. Code Writers and Criteria Developers. Numerous groups engage in the development of model building codes, suggested planning and zoning ordinances, and other documents widely used in offical policy-making efforts at local, state, and federal levels. In this effort the assumptions, perspectives, and motivations of many individuals play important roles in the quality, efficacy, and cost feasibility of the outcomes. Functioning in a world of imperfect knowledge, much emphasis is placed by such groups on past experience and informed guesswork. To the members of this group the need to empirically verify the assumptions on which such documents reside may be of far less importance than "getting the job done."

5. Insurers and Reinsurers. One potentially effective means for avoiding the severe personal, familial, and community distress produced by large hazard-induced losses involves schemes for spreading such losses among the relevant total group of local, regional, or national risk takers. However, tax law restrictions on maintenance of liquid reserves, problems associated with the non-liquidity of investments, and the sheer magnitude of the episodic losses that may be sustained during major hazard-induced catastrophes constitute severe impediments to the willingness and capacity of insurers and reinsurers to engage in the design and delivery of needed risk-distributing systems for occupants and investors in high-hazard areas (Atkisson et al., 1980).

Mitigation-Constraining Parties

Members of this final set of stakeholders are distinguishable, not because of the losses they sustain as a result of exposure to natural hazards, but because of their potentially important roles in generating external constraints on public policy makers involved in natural hazards management activities.

1. Land Speculators and Developers. Because of their natural scenic, climatologic, or recreational amenities, many hazard-prone land areas are attractive sites for future development. Accordingly, the interests of land speculators, other land owners, subdividers, and builders are affected both by any future imposition of governmental restrictions governing the use of the land, the disclosure of information concerning its hazard-proneness to prospective purchasers, and by future governmental investment policies concerning the provision of hazard-related area protection works.

Major opposition to hazard zone land planning and hazard management schemes may be expected from such parties. Moreover, to the extent that the interests of land speculators coincide with those of businessmen and other economically motivated parties whose future economic well-being may be tied to rates of development within hazard-prone areas, it can be expected that highly organized and vocal opposition will emanate from such quarters in respect to proposed hazard zone avoidance and risk reduction schemes.

2. Opponents of Governmental Regulation. In the United States, recent years have witnessed the development of widespread theoretical and political opposition to numerous types of governmental regulatory activities, including many that were once popularly supported and clearly addressed to public needs. Solid questions have been raised concerning the public benefit/cost ratios associated with many existing regulatory programs, and have fueled developing citizen interest in less overt forms of government intrusion into market operations. Both "conservatives" and "liberals" are increasingly discussing "informed personal choice" and decentralized forms of decision making as possible alternatives to traditional governmental regulatory programs and policies.

Individuals committed to these views as a matter of principle rather than personal self-interest, comprise the membership of this stakeholder group. They may be expected to demand that governmental approaches to natural hazard exposure losses consider the full range of loss-reducing alternatives with particular emphasis on those that stop short of comprehensive, nationwide regulatory schemes.

3. Advocates of Governmental Economy. Prompted by California's adoption of Proposition Thirteen (the local tax rate limitation amendment to the California constitution), a national campaign to reduce governmental expenditures and public tax burdens is under way. Supported by a broad base of public and policy maker opinion, the campaign is forcing governmental entities and jurisdictions to reevaluate once unquestioned programs and expenditures, to reorder old priorities, and to divert tax monies to expenditure targets that yield the greatest publicly desired payoffs per unit of investment.

In the future, members of this stakeholder group can probably be expected to question or oppose government expenditure policies that have the effect of transferring the cost of area protection facilities and/or the losses experienced by some types of natural hazard risk-takers to non-risk-taking or nonbenefiting parties.

4. Supporters of Natural Hazard Management Programs. No clearly identifiable constituency has emerged on the national scene to argue on behalf of comprehensively oriented natural hazards management policies and programs. Discussions of this topic have thus far been dominated by researchers committed to this subject as a field of scientific inquiry, by professionals involved in existing aspects of the field, by environmentalists, and by groups whose interests in the subject are of a short-term and "pork-barreling" nature. The development of such a constituency may well be a necessary condition to any future comprehensive revision of natural hazard management policies and programs.

Candidate Public Problems

In general, it may be said that the rational capacity of the public policy system may be increased if that system is provided with 2 types of inputs: public problem analyses and public policy analyses. "Problems" are as much a product of human values and perceptions as they are of empirical reality. Because problems require acts of human valuing, they must be viewed as potentially different from states of affairs identified objectively as "impacts," "consequences," or "effects." Objectively, an analyst may state that the effect of activity *A* will be to increase the cost of product *B* by an amount *C* to a consumer *D* and therefore consume *E* percent of *D*'s family income. Whether that increased cost produces a "problem" for *D* hinges on the values and reality perceptions of *D* or some other value-holding party who presumes to fix the values by which the affairs of others are to be judged.

Further difficulties are faced when one tries to distinguish between the class of events viewed as "problems" and those viewed as "public problems." In our society, not all problems are viewed as being public problems, and all public problems are not necessarily placed on the agenda for public policy action. Some are simply acknowledged as being extant but as being either too trivial or too politically infeasible to resolve to warrant the concern of the community and/or government. Accordingly, considerable peril is faced by the analyst who presumes to state what "public problems" are associated with a particular situation or set of situations under examination. For this reason, the following list is to be viewed as consisting of *candidate* problems each of which may be viewed by some significant stakeholder group as constituting a public problem.

Intrinsic Problems

The ten states of affairs discussed below have been identified as comprising the major candidate *"intrinsic" public problems* associated with natural hazard exposures or with efforts to mitigate such exposures. The first seven of these problems are intended to represent the publicly unwanted situations that comprise the essential justification for natural hazard management policy making, particularly at the national level. The remaining three problem statements are intended to represent the *primary* constraints to be met by decision making aimed at the resolution of the first seven problems. Considered as an interlocked whole, the suspected existence of these ten problems explains why the situations described as "instrumental problems" are viewed as "unwanted" situations or states of affairs.

The relationship of each "intrinsic" public problem to the several stakeholder groups identified above is depicted in Table 8-1. In each case, the matching of a stakeholder group with a problem through the note "P" is meant to suggest that the identified group is the "problem-impacted" party. The note "B+" is meant to suggest that the identified group is a "beneficiary" or "affected" party whose interests may positively be served by problem-solving activities, while a "B–" indicates that group interests will be negatively served by such problem-solving activities.

1. Large Numbers of Individuals, Families, and Businesses Are Now at Risk of Experiencing Significant Economic and Other Losses as a Result of Their Exposure to Natural Hazards in Their Home Communities. In 1970, annual expected total dollar losses arising from the exposure of buildings and their contents to 9 natural hazards totaled approximately $8.1 billion. These same exposures were expected to yield 979 deaths per year, a loss of 114,000 housing units, 129,850 person-years of homelessness, and nearly 80,000 person-years of unemployment. Approximately 33% of all building damage losses were sustained by flood-threatened parties who constituted less than 8.0% of the U.S. population. At the scale of whole counties and states, annual expected per capita losses from natural hazard exposures range from nearly $220 for the highest damage rate county to only $16.86 for the lowest.

Within the nation's 500 most damage-prone counties (for all natural hazards except riverine flooding), annual expected per capita losses ranged from $21.18 to approximately $200 in 1970. In terms of more finely drawn hazard zones, annual expected losses totaled slightly more than $402 per capita for occupants of riverine Flood Zones A to C in communities not protected from 50 year floods; $121 per capita per year in the storm surge floodplains of Georgia; $24 per capita per year for the three counties exhibiting the highest average annual number of tornado strikes; $28 per capita per year for occupants in states with the highest annual expected damage rate from earthquakes; approximately the same for individuals in the state exhibiting the highest hurricane wind hazard; and $26 per capita per year for occupants of intracounty areas that were within high expansive soils zones. Occupants of 500 year tsunami inundation plains in 1970 were found to be at risk of experiencing annual expected losses totaling $37 per capita. In single catastrophic events, such as major earthquakes or hurricanes, residents in the counties experiencing the highest magnitude of the occurrence were found to be at risk of experiencing per capita episodic losses in excess of $16,000.

2. Many Whole Communities Are at Risk of Experiencing Catastrophic Losses as a Result of Hazardous Natural Events. Although a recent study (Wright et al., 1979) has shown that few long-term economic impacts have been incurred by

Table 8-1. Relationship of Intrinsic Public Problems to Thirteen Stakeholder Groups

Stakeholder groups	Personal and familial losses	Community losses
Population currently at risk	P B+	P B+
Investors in hazard-prone land and structures	P B+	B+
Potential future occupants and investors	B+	B+
Visitors or workers in high-hazard zones	P B+	
Insurers and reinsurers	B+	
Second home owners and developers	P B+	P B+
Local policy makers		P B+
Financial institutions and mortgage guarantors	B+	B+
Taxpaying subsidizers of hazard mitigations	B+	
Local and state planners and building officials		B+
Advocates of environmental protection		B+
Researchers and professionals		B+
Code writers and criteria developers		B+

Note: P = problem-impacted party; B+ = party who stands to gain; B— = party who stands to lose.

communities as a result of their exposure to hazardous natural events, the recovery capacity of these communities has been greatly influenced by the infusion of disaster relief funds from outside areas and jurisdictions. Also, natural hazard exposures of the high magnitudes that can be expected to occur in this country in one or more major metropolitan areas have not yet been experienced during the current century.

In 1972, the U.S. Office of Emergency Preparedness reported to the Congress that increasing population density, inadequate evacuation routes, ineffective building codes, and insufficient provision of safe refuges are increasing the probability of a major hurricane catastrophe along the Atlantic and Gulf Coasts. The report noted that some states vulnerable to hurricanes do not have statutory authority to order evacuation on the scale and at the time a hurricane advisory might indicate the need for such action; that numerous local communities and

Intrinsic public problems							
Involuntary risk for mobile population	Excessive annual economic losses	Reduced probability of mortgage payoff	Expected increase in magnitude of problem	Costs transferred to non-risk-takers	Increase in cost of operational requirements	Policies incompatible with personal or community values	Impaired improvements in life quality
B−	P B+		P B+	P B+	P B+	P B+	P B+
B−	P B+	P B−	P B+	P B−	P B+	P B+	P B+
B+	B+	B+	P B+	B−	P B+	B+	B+
P B+			P B+				P B+
B+	P B+	P B+	P B+	B−	B+	B−	
P B+		P B+	B−	P B+	P B+	P B+	P B+
			P B+			B−	B+
	P B+	P B+	P B+	B−	B+	B−	
B+	B−	B+	B+	B+	B+	B+	B+
	B+	B+	B+		B−	B−	B+
				B+	B+	B−	P B+
	B+		B+	B−	B+		
	B+		B+		B+	B−	

large jurisdictions lack adequate knowledge concerning their vulnerability to natural disaster; that both private citizens and local public officials are not aware of the natural disaster hazards existing in the area in which they live, the likelihood of their occurrence, and the measures that property owners can take to avoid or mitigate them; and that comparatively few are sufficiently detailed or in a form that can be used in land use planning and in implementing hazards avoidance or mitigation plans.

3. Involuntary Risks Are Being Incurred by Significant Numbers of Migrants, Workers, Customers, and Tourists Who Enter Unknowingly into Natural Hazard Zones. When the general attitude of the public that government will protect them is combined with the high level of residential mobility, intercommunity migration, and away from home shopping, working, and vacationing habits, and when these factors are further associated with the fact that neither hazard zone occupants nor outsiders possess much knowledge concerning the risks associated with specific sites, then a substantial number of nonhazard zone residents are being exposed unknowingly and, therefore, involuntarily to risk taking within natural hazard zones. At present, those parties who sell, offer to sell, or participate in the transfer of property are not generally required to disclose to the possible new owner information concerning the hazard-proneness of the property.

4. Annual Economic Losses in Excess of the Probable Annual Amortized Costs of Mitigations Are Being Experienced by Parties Located within Many Natural Hazard Zones. In approximately 230 U.S. counties, the application of building-strengthening strategies is capable of reducing annual expected natural hazard losses by amounts greater than the annual amortized costs of the mitigation. In numerous riverine flood zones, coastal storm surge areas, and other hazard zones, combinations of building-strengthening, area protection, or avoidance strategies also may reduce annual expected natural hazard losses by amounts greater than the annual amortized costs of the mitigation. Therefore, to the extent that economic "efficiency" is itself a deeply rooted value and motivation of this society, this situation must be viewed as a public problem.

5. The Probability of Payoff of Some Loans and Mortgages to Private Financial Institutions, Public Entities, and Mortgage Guarantors Is Being Reduced by Continuing Imprudent Investments in High Hazard Zones. Although the quantitative magnitude of this problem cannot readily by identified, the data generated by this study suggest that the situation described in this problem statement is both real and probably expanding.

6. Unless Appropriate Corrective Measures Are Taken, the Above Problems Will Increase in Magnitude Between 1980 and 2000. Current patterns of interstate, intercommunity, and intracommunity residential change and capital investment are increasing the size of the nation's population at risk of exposure to natural hazards. The projected increase in total natural hazard economic losses between 1970 and 2000 is in excess of the annual expected loss in 1970.

7. A Significant Fraction of the Costs of Voluntary and Involuntary Natural Hazard Risk-Taking Is Being Transferred to the General Body of Non-Risk-Taking Taxpayers. Because of the numerous aids and subsidies contained in current federal natural hazard mitigation and relief legislation, a significant and growing fraction of the national annual cost of natural hazard exposures is being transferred to parties who do not reside within the hazard zones in which these losses are incurred. Although the 1936 Flood Control Act provided for cost sharing by state and local governments in federally initiated flood control projects, that provision was dropped shortly after the enactment of the statute, and the bulk of flood control costs subsequently have been borne by the general body of taxpayers. Similarly, numerous occurrences now qualify for disaster designation under the terms of the Federal Disaster Relief Act, and parties experiencing losses during such events are entitled to a wide variety of benefits from the Federal Government, including forgivable loans, interest-free loans, direct recovery services, and federally financed reconstruction of community lifelines and other infrastructure.

The availability of these benefits has sometimes been cited as constituting an incentive for individuals to enter high-hazard zones and as a deterrent to local and personal initiation of more economical hazard avoidance, hazard mitigation, or hazard adjustment activities.

8. Purchasers and Users of Property Face an Escalation in the Initial and Annual Amortized Cost of Such Property as a Result of Governmentally Required Building-Strengthening and Other Hazard-Mitigating Requirements. This analysis has made clear that overzealous, across-the-board strengthening of building codes may well increase by a substantial amount the net annual burden associated with natural hazard exposures. Other studies performed within the past decade or so also have suggested that building code requirements in many areas may exceed the objective needs of public health, safety, and building life, and be imposing unnecessary cost burdens on a housing-hungry population (National Commission on Urban Problems, 1968; Field and Rivkin, 1975). More recently, the Working Group on Earthquake Hazards Reduction (1978) noted that current seismic codes do not adequately balance the risks of quake-inducing damage against the cost of applying mitigations, and further observed that some building codes do not reflect the current state of the art in respect to earthquake-resistant design.

A more careful study of the contents, costs, and benefits of current building code requirements seems warranted. However, present evidence suggests that the problem statement, as composed, is a fully defensible interpretation of the current state of affairs.

9. The Ability of Individuals and Whole Communities to Engage in Courses of Personal and Community Action Compatible with Their Own Values May Be Threatened by Action-Forcing and Regulatory Policies Adopted by State and Federal Governmental Entities. From a national, regional, or state perspective, the data presented in this and other similar reports does not make it easy for any compassionate or rational analyst/observer to determine "what is right." Indeed, the data suggests that there may be many "rights," many possible courses of action that are consistent with the objective realities of the natural hazard situation and the prevailing values of specific communities and the nation as a whole. Decision making by higher units of government that require uniform or near uniform application of limited perspectives and value choices across the face of the country therefore run the risk of overlooking "acceptable alternatives" that might be applied in specific locales by men and women guided by intelligence and goodwill.

A leading nineteenth century policital and legal philosopher once observed that the "public interest requires that we do today what men of intelligence and good will would wish–five or ten years hence–had been done." From this perspective, natural hazard situations are so complex, the values contained within them so numerous, and the possibilities for action so various, that there may be many "rights," and many ways to fulfill the "public interest" as that interest is perceived in many specific places and at varying specific times.

10. An Overreaction to Natural Hazard Losses May Impair Our National Capacity to Realize Otherwise Achievable Improvements in Human Longevity and Life Quality. Like other nations, the United States is plagued by a long list of public problems and an expanding agenda of public wants and goals. Given limited knowledge concerning the costs and benefits associated with the expanding variety of policies and actions that may be taken in respect to these problems and goals, beset by competing claims from politically organized groups, and further limited by objective evidence and perceptions, the major actors within the American public policy system are in continuing danger of misallocating limited available public resources.

As compared with other problem situations, it is easy for researchers, activists, regulators, and technical advisers to overstate the importance of the phenomena that are the focal points of their professional lives.

Additional study and information synthesis is needed to identify and rank the numerous natural and man-made hazard zones within which both resident and transient populations are at risk of experiencing consequences of various magnitudes, frequencies, and importance to themselves and the nation. If our overall national objectives are to achieve the greatest extensions in human longevity, maximum improvements in life quality, and the large gains in the output of public benefits per million dollars of public expenditures, then the subject of natural hazard risk reduction would profit from inquiry conducted within larger contexts.

To the extent that the "Proposition 13 syndrome" is one that may beset most regions of the country and all levels of government, then it may well be that public expenditures at all levels will be constrained more severely in the future than they have in the past and that the need for such comprehensive problem assessment and policy planning will be even larger than intimated here.

Instrumental Problems

Situations or states of affairs whose existence contributes to the occurrence of, or impedes the mitigation of, "intrinsic" problem states are referred to here as "*instrumental*" public problems. In these terms, their solution is justified only in terms of the extent to which they contribute to the solution of "intrinsic" problems. The assumed relationship between the candidate "instrumental" problems identified below and the "intrinsic" problems identified above are depicted in Table 8-2.

The following candidate list of 13 instrumental problems was developed using both the data from this analysis and the published findings of recognized experts (see Advisory Commission on Intergovernmental Relations, 1966; National Commission on Urban Problems, 1968; U.S. Office of Emergency Preparedness, 1972; United Nations Office of the U.N. Disaster Relief Coordinator, 1976; Water Resources Council, 1976; Dacy and Kunreuther, 1969; Kunreuther, 1978; Field and Rivkin, 1975; White, 1974, 1975, 1976; Williams, et al., 1968; Wright et al., 1979).

1. Hazard Zone Identification, Mapping, and Classification: The past rate of progress in identifying, mapping, and classifying natural hazard zones has been inadequate; too few zones have been so mapped; and inappropriate or incomplete information has been provided map users concerning the frequency and magnitude of hazard occurrences within such zones. Timely, effective, and rational natural hazard management activities are dependent on the availability, accessibility, quality, understandability, and comprehensiveness of information concerning the boundaries of local natural hazard zones and their component sections, the frequency and magnitude of hazardous occurrences expected within the various sections of these zones, and the hazard-relevant geophysical characteristics (such as soil type) of elements of the several sections of the zone.

Thus, the Working Group on Earthquake Hazards Reduction (1978) has observed that current seismic risk maps are in conflict with each other, do not present alternative levels of risk, do not adequately incorporate "engineering considerations," and are not available for a sufficient number of areas within the nation. The group noted that existing information about earthquake hazards is generally neither sufficiently detailed nor in a form that can be used either in land use planning or in actions intended to implement plans for avoiding hazards and mitigating damages. Agreeing with the Office of Emergency Preparedness, the group noted that tsunami flood zones have not been adequately defined. Virtually all those who have examined natural hazard occurrences within the United States have noted the need for hazard zone and risk mapping and have commented on the general lack of such information. This information deficiency may well lie behind the generally unsatisfactory level of public and policy maker understanding of natural hazards within local communities and larger jurisdictions.

Cost requirements associated with high-quality and timely hazard zone mapping activities, together with the human and technical resource requirements associated with this function, suggest that central federal priority setting should take place.

Whatever the final priorities may be, it seems appropriate to suggest that they should be fixed in consonance with the annual expected losses associated with exposures within the various types of hazard zones, the sizes of the populations currently at risk, and with the growth in losses and population exposures expected between now and the year 2000. Moreover, it may be wise to reconsider the appropriateness of the past federal role in this process and to place greater

Table 8-2. Relationship Between Intrinsic and Instrumental Candidate Public Problems

Candidate intrinsic problems	Inadequate progress in hazard zone mapping	Inadequate data for developing mitigation criteria	Inadequate data to establish cost-loss reduction ratios for mitigation
1. Individuals residing in hazard zones	++	++	++
2. Whole communities at risk of loss	++	++	++
3. Visitors incur involuntary risk	++	++	++
4. Losses exceed annual mitigation costs	+	++	++
5. Reduced probability of mortgage payoff	+	++	++
6. Problems will continue to increase in magnitude	++	++	++
7. Costs transferred to non-risk-taking taxpayers	0	0	+
8. Policies increase property cost	+	+	+
9. Policies threaten individual and community values	++	++	++
10. Impaired improvements in human longevity	+	+	+

Note: Effect of resolution of instrumental problems on the resolution of intrinsic problems: ++ = very supportive; + = supportive; 0 = no effect; — = detrimental.

emphasis on federal establishment of criteria and standards to be employed in such mapping and zone-classifying activities, as contrasted to direct federal conduct or financing of these operations. Expanded roles for state and regional planning agencies are not difficult to conceive, nor are requirements for public hearings and technical review of map outputs prior to their official publication and endorsement.

2. Inadequate Methods and Data Are Available for Use in Operations Targeted on the Development of Empirically Defensible Mitigation Criteria. The term

Candidate instrumental problems									
Inadequate information to support model building code requirements	Local building codes of uncertain quality can be incompatible with natural hazards management	Limited effective local planning and implementation	Low level of local natural hazard perceptions	Weak constituency for natural hazard policy	Policy favors financial assistance and area protection works	Inadequate coordination of federal natural hazard programs	Limited opportunity to purchase hazard insurance	Low public interest in natural hazard insurance when available	Federal research expenditures not matched with annual expected losses
++	++	++	++	++	++	++	++	++	++
++	++	++	++	++	++		++	++	++
+	+	+	++	+	+	+	0	0	+
++	++	++	+	++	+	++	++	++	++
++	++	++	+	+	+/–	+	++	++	+
++	++	++	++	++	++	++	++	++	++
+	++	+	0	–	++	++	++	++	0
++	++	+	0	0	–	+	++	++	+
+	+	+	++	+	+	++	0	0	+
++	++	+	++	0	+	+	0	0	+

criteria in this problem statement refers to a predictive or descriptive statement describing the level of loss or damage reduction that may be expected at any given level of intensity or frequency of hazard occurrence or that may be derived from use of any specified mitigation. For many hazards, such as expansive soils, the current data base is woefully inadequate; for almost all hazards, some additional data are required.

3. Inadequate Procedures and Pools of Data Have Been Provided to Assist Technical and Regulatory Bodies in Their Establishment of Empirically Defens-

ible Statements Concerning the Cost/Loss Reduction Ratios Associated with the Use of Specific Mitigations in Specific Types and Sections of Natural Hazard Zones.

4. In Respect to Natural Hazard Mitigations, Significant Differences Can Be Noted in the Content of "Model Building Codes"; and Too Little Empirically Defensible Information Is Available to Support the Numerous Judgments that Have Entered into the Specification of "Model Code" Requirements.

5. Local Building Codes Are Generally of Uncertain Quality, Too Infrequently Reviewed and Revised, and Sometimes Based on Motivations and Purposes that May Not Clearly Be in the Public Interest or Compatible with the Ends of Rational Natural Hazards Management. The Working Group on Earthquake Hazards Reduction noted that some building codes do not reflect the current state of the art with respect to earthquake-resistant design and further observed that some of the model building codes may not be completely adaptable to all areas of the United States with their different degrees of seismic risk. The group also observed that these codes do not adequately balance the risks of quake-induced damage against the costs of mitigation; and noted that inadequate action has been taken to implement building code and other regulatory provisions intended to deal with the hazards posed by older, non-earthquake-resistant structures.

Similarly, the U.S. Office of Emergency Preparedness reported in 1972 that state and local lesiglation has not kept pace with the growing problems of natural disasters, and that "ineffective building codes" in communities along the Atlantic and Gulf Coasts are exacerbating the probability of a major hurricane catastrophe in that area. The National Commission on Urban Problems reported in 1968 that "building code jurisdictions are thousands of little kingdoms, each having its own way: what goes in one town won't go in another–and for no good reason." The Commission concurred in the view that "the provisions in codes are antiquated and outdated and that the procedures for modernizing and amending them are slow, laborious, and lacking in objective standards." The Commission was so concerned with the quality of local building codes that their recommendations called, in part, for "minimum standards below which no community might fall and maximum limits in order to prevent restrictive practices" (National Commission on Urban Problems, 1968).

6. Local Planning and Building Regulation Departments and Professional Staffs Exhibit Limited Capacity to Engage in Effective Natural Hazard Policy Planning and Implementing Activities. The staffs of building departments at the local level generally are drawn from local construction trades, are paid extraordinarily low salaries, and typically lack the time and training to engage in the technically demanding function of reviewing, revising, and developing hazard-related building codes. Similarly, land use and regional planners typically have not received academic training on subjects related to natural hazards; and even professionally trained architects and engineers frequently exhibit the same incapacity.

Although quantitative defense of this view is rarely offered, most groups that have examined this subject seem to be in agreement with the assumptions implicit in this problem statement. If true, the statement suggests the need for much technical assistance and training support to these staffs by state and federal entities. At a minimum, there is a need for a well-executed and unbiased study to determine the scope of the problem.

7. There Is a Comparatively Low Level of Public and Policy Maker Perception of the Natural Hazards that Exist within Their Communities and of the Consequences Associated with Continuing Exposure to Those Hazards.

8. There Is No Apparent "Political Constituency" That Has Emerged to Argue on Behalf of Comprehensive and Rationally Conceived Natural Hazard Management Policies at any Level of Government; but Limited Constituencies Have Developed that Argue Against Imposition of Natural Hazard Management Action-Forcing Policies on Local Governments by State and Federal Units of Government.

9. The Major Public and Policy Maker Demands in Respect to Natural Hazard Management Policies Are Tilted in Favor of (a) Financial and Other Forms of Assistance to Disaster-Impacted Parties, and (b) Area Protection Works Funded by the Nation as a Whole.

10. Inadequate Coordination Has Been Provided to Hazard Management Programs Conducted by the Federal Government. Several reports of recent vintage have noted the absence of central leadership and coordination of natural hazard management programs within the federal structure; the lack of coordination of floodplain-directed programs; and the lack of programs intended to develop public and policy maker understanding of the extent of the natural hazard problem and appropriate mitigations. According to the Working Group on Earthquake Hazards Reduction, there has been virtually no integrated coordination of federal land use planning and development programs with those more pointedly related to natural hazard exposures. Similarly, there seems to have been little integration of employment programs for the chronically or occasionally unemployed (such as CETA) with the aims of natural hazard management activities (such as removal of structures from high-hazard zones).

The Presidential announcement that accompanied the reorganization of federal disaster management functions contained an acknowledgment of the need for a more comprehensively and centrally oriented organizational structure to cope with the mix of functions related to the initiation of hazard mitigation programs and to the provision of relief and recovery services after the occurrence of hazard-induced disasters. However, it is not yet clear that the objectives have been achieved by establishment of the Federal Emergency Management Agency, or that the complexity and importance of natural hazard management policies and problems has been well understood by the top administrators of the Agency.

11. In General, Populations at Risk of Exposure to Natural Hazards Do Not Now Have the Opportunity to Purchase Insurance of Appropriate Coverage at Desirable Rates to Spread the Risk of Their Exposure-Induced Losses Among the Relevant Population of Risk Takers. Although federally subsidized flood insurance is now available and commercial insurance against selected other natural hazards is now offered by some companies in some parts of the United States, it is not generally possible at present for property owners to purchase all-purpose natural hazard insurance coverage and, therefore, to take advantage of this means for avoiding catastrophic losses arising from natural hazard exposures. The absence of this opportunity probably directs undue attention to other methods for mitigating potentially catastrophic natural hazard losses, including those related to provision of area protection facilities and to use of building-strengthening technologies. From a cost-benefit point of view, insurance coverage may well be a better solution to some aspects of the loss problems associated with natural hazard exposures than other approaches. At the very least, insurance can be an important partner in a comprehensively oriented loss-reducing strategy.

A variety of factors may be influencing the lack of these insurance opportunities, including some relating to tax code constraints on maintenance of liquid reserves, and the difficulties and risks faced by insurers and reinsurers in meeting the "unusual" financial requirements associated with possible catastrophic occurrences. There is a substantial need for a thorough, empirical study of this subject.

12. Even Where Natural Hazards Insurance Is Available, Public Interest in the Purchase of Such Insurance Has Not Been Particularly High. Kunreuther (1978) in an examination of individuals' propensity toward protecting themselves against flood and earthquake damage has shown that a relatively small portion of those exposed to these risks have personally protected themselves. This was due largely to limited knowledge of what insurance was available, limited awareness and knowledge of the hazard, continued expectation of federal aid, and, finally, a general lack of concern with the consequences of low probability type events.

13. Federal Expenditures for Hazard-Related Research and Technology Development Projects Have Not Exhibited an Appropriate Match with the Annual Expected Losses Associated with Exposure to Various Hazards or with the Size of the Populations at Risk of Exposure to Such Hazards. However laudable, natural hazards research and technology development efforts within the federal structure have tended to respond to the "natural disaster of the moment" rather than to the total mix of potential natural hazard exposure problems. However well-advised the research expenditures authorized under the Earthquake Hazards Reduction Act, it appears that far more attention is being given to this hazard than to other types of hazard exposures of equal or greater national importance. Thus, apparently large annual losses are being sustained as the result of property exposures to expansive soils, but comparatively little is known about this phenomenon. Similarly, too little research and technology development is being directed toward means of coping with coastal flooding, landslide, and wind hazards. The whole subject of cost-benefit relationships in building code standards has for too long been ignored; and the substantial technical assistance and training requirements of state and local planning, hazard management, and building code agencies and personnel have gone too long unattended. The reports by the Working Group on Earthquake Hazard Reduction (1978), the National Commission on Urban Problems (1968), and White and Haas (1975) provide a context for future deliberations on this subject.

However, none of the reports examined have clearly, explicitly, and rationally offered defensible criteria through which the desirability, feasibility, and priority of alternative research, technology development, and technology transfer opportunities could be judged. Certainly, such factors as the scope of the problem addressed, the need for the information, the loss-reducing potential of new knowledge, and the state of the current knowledge base are factors that should be used to judge such matters. Moreover, there has been too little effort made to comprehensively assess research and other needs related to human and property exposures in all types of hazard zones, including those of natural and man-made origin. All too little now is known about the comparative risks incurred by individuals and property exposed to conditions in a variety of natural and man-induced hazard zones, and therefore, an appropriate basis for policy making is lacking.

If the *candidate problems* identified above are those accepted by a policy maker, then two other questions must quickly be addressed: (1) What priority should be given to the solution of each problem, or any component thereof? and (2) What types of policies may be employed to deal with each priority problem?

Establishing Priorities

It is necessary to establish priorities because resources are scarce. When it is impossible to meet all needs, then a determination must be made concerning the appropriate allocation of resources among the problems being considered. Typically the priority question is a matter of relative emphasis, not a matter of which problem is treated first to the exclusion of others.

Generally, priorities should be based on the magnitude of the problem faced, or the ease with which the problem may be solved. More important problems should be addressed with more resources than should less important problems, and those problems most amenable to solution should be tackled before those believed to be "impossible" or difficult to solve.

The establishment of these priorities is as much an exercise in making "value choices" as it is one in responding to objective evidence. For example, if the total loss of life sustained by the nation is the primary criterion used to judge the relative priority of natural hazard exposures, then the exposures of greatest year to year importance are tornadoes, riverine, and coastal flooding.

If, however, the greatest value is to be placed on the phenomena that involve the largest life loss in a single event, regardless of frequency, then priority would have to be given to major earthquakes, tsunamis, and hurricane wind/storm surge occurrences. If total economic loss is to be the guiding criterion, then emphasis would have to be placed on wind hazards, coastal and riverine flooding, earthquakes, and expansive soils. If the immediate technical and cost feasibility of solutions is to be emphasized, then priorities should be given to mitigation of landslides and to the future avoidance of flood zones.

If the annual magnitude of *future* losses is to be given greater weight than those of the present, then priority would be given to those phenomena and hazards that are associated with the largest fraction of the projected increases in natural hazard costs of all types.

In general, problems can be operationalized and their relative priorities determined by addressing four questions:

1. How many people or how much area is adversely affected?
2. How intensely are these people and areas being affected?
3. Will the problem situation get better or worse if nothing is done?
4. Is the greatest gain to be derived from dealing with existing exposures or from preventing future ones?

However, these priority questions may be resolved, the policy maker will face the further task of determining the applicability of each *general* type of policy to each specific problem, whether of an intrinsic or instrumental character.

This analysis identifies nine types of policies that offer a range of potential approaches to decision makers: (1) action-forcing policies, (2) attention-focusing policies, (3) disaster recovery policies, (4) technology development policies, (5) technology transfer policies, (6) regulatory policies, (7) system management policies, (8) system optimization policies, and (9) investment and cost allocation policies.

The applicability of each of these types of policies to each of the candidate public problems is depicted in Table 8-3. As shown in the table, and in terms of the time frame immediately surrounding the policy, a "0" suggests little or no direct impact on the problem state of affairs, a "+" suggests a high degree of problem resolution, and a "—" suggests that a worsening of the problem might result from use of that type of policy. Of course, "avoidance" policies do nothing for those now exposed to hazards: the construction of a dam may do much to reduce losses for a currently exposed population group, but serve to entice others into situations of higher risks than necessary; a federal regulatory policy may produce immediate impacts on a hazardous natural situation, but weaken the quality of intergovernmental relations and depress the capacity of local governments to deal with such situations in the future.

It seems clear that a mix of policies is probably required, both short- and long-term consequences should govern policy designs, both system improvements and direct hazard reduction policies should be executed concurrently, and no single value should govern all policy choices.

Federal Role

Even if agreement can be reached concerning the "problems" implicit in natural hazard management, even if agreement can be reached as to which of those problems are "public" in character, and even if agreement can be reached concerning the best mix of policies to deal with those problems, the question then remains: To what level of government falls the primary responsibility for dealing with each problem? What is the role of the federal government with respect to the entire mix of problems and to any specific problem within the set?

Throughout U.S. history, there has been a broad continuing shift in public opinion with respect to the proper role of the federal government. In the post-revolutionary period of U.S. history, the checks and balances, the pluralistic federal system of government, and the safeguards against governmental intrusion on personal choice that were built into the national Constitution and those of the states were responses to the deep-rooted philosophic, ethical, political, moral, and economic values of society at that time. That a limited, but very important, role for the federal government was imagined by the framers of those documents and the publics they represented cannot be denied.

Neither, however, can it be denied that later events in U.S. history recurrently altered the public mood and buttressed arguments in favor of expanding federal roles. When states and local governments were either unable or unwilling to cope with problems affronting the conscience and problem perspectives of the national public, general willingness to alter the federal role was expressed. Although the general social commitment to "states' rights," "grass-roots government," "pluralism," and "freedom of choice" remained intact throughout social confrontation with the events altering the federal role, there was nevertheless a willingness to get on with the tasks of both a substantively and procedurally democratic society. Frequently, other underlying values, such as those imbedded in the issue of "states' rights," became the slogans of campaigns conducted by stakeholder groups whose interests were less oriented toward the expressed value itself than toward some "selfish" cause that could be served if that value were protected.

This expanding federal role has been the case, through an almost unbroken chain of developments, since the days of the Depression and the presidency of Franklin Delano Roosevelt. Now, however, it appears that circumstances are leading to a public reappraisal of the federal system. In spite of such evident problems as those associated with eroding environmental quality, social inequity, inequality of opportunity, consumer abuse, and impediments to the personal pursuit of longevity, security, and self-actualization, there seems to be a growing

Table 8-3. Applicability of Public Policy Types to Candidate Public Problems

Candidate public problem	Action forcing	Attention focusing	Disaster recovery
Intrinsic			
1. Individuals residing in hazard zones	+	+	+
2. Whole communities at risk of loss	+	+	+
3. Visitors incur involuntary risk	+	+	0
4. Losses will exceed annual mitigation costs	+	+	+
5. Reduced probability of mortgage payoff	+	+	+
6. Problems will continue to increase in magnitude	+	+	+
7. Costs transferred to non-risk-taking taxpayers	–	+	–
8. Policies increase property cost	–	0	0
9. Policies threaten individual and community values	–	+	+
10. Impaired improvements in human longevity	–	+	0
Instrumental			
1. Inadequate progress in hazard zone mapping	+	+	0
2. Inadequate data for developing mitigation criteria	–	+	0
3. Inadequate data to establish cost-loss reduction ratios for mitigation	–	+	0
4. Inadequate information to support model building code requirements	–	+	0
5. Local building codes of uncertain quality can be incompatible with natural hazards management	–	+	0
6. Limited effective local planning and implementation	–	+	0
7. Low level of local natural hazard perceptions	–	+	–
8. Weak constituency for natural hazard policy	–	–	+
9. Policy favors financial assistance and area protection works	+	+	–
10. Inadequate coordination of federal natural hazard programs	–	+	–
11. Limited opportunity to purchase hazard insurance	–	–	+
12. Low public interest in natural hazard insurance when available	+	+	–
13. Federal research expenditures not matched with annual expected losses	–	+	–

Note: + = policy can assist in problem resolution; 0 = policy will have no effect on problem resolution; – = policy can be detrimental to problem resolution.

Policy type					
Technology development	Technology transfer	Regulatory	System management	System optimization	Investment & cost allocation
+	+	+	+	+	+
+	+	+	+	+	+
0	+	+	+	+	+
0	+	+	+	+	+
0	+	+	+	+	+
+	+	+	+	+	+
−	0	+	+	0	0
+	+	0	+	+	+
+	+	−	−	+	+
+	+	+	+	+	+
+	0	+	+	+	+
+	0	−	−	−	−
+	0	−	−	−	−
+	+	+	+	+	−
+	−	+	+	+	−
+	+	+	+	+	+
+	−	+	+	+	+
+	+	+	+	+	+
0	0	+	+	+	+
+	−	+	+	+	+
+	+	+	+	+	+
0	0	+	+	+	+
+	+	+	+	+	+

public disenchantment with the effectiveness and cost of the Federal government's current role in public choice-making operations.

Federally imposed policy and program priorities have obfuscated the agenda of local governments throughout the country, have produced a wave of local policy maker involvement in procedural detail rather than the substantive merit of program content, and have frequently produced program mixes and expenditure balances inconsistent with the priorities of local inhabitants. As a result, we may now be entering into a period of growing opposition to an extension of federal action-forcing policies. So complex and numerous are the facts and values surrounding recognized "public problems" that many cannot be, and perhaps should not be, resolved at a level that imposes homogeneity on people nurtured in the tradition of pluralism. It is no easy task to determine "what is right." The answer to this question may require that facts unique to a particular time and place be integrated with the society-consistent but slightly different values operative within that situation, in accordance with a calculus in which there is no national agreement.

If this nonquantitative assessment of a public mood is correct, then the architects of natural hazard management policy may pointedly consider the question "What is the proper role for the federal government in natural hazards mangement"?

The raising of this issue is prompted not by any philosophic or political bias, but rather out of a sense that the changing attitudes of the national public will exert powerful constraints on whatever decisions are made concerning the role of the federal government in natural hazards management. Within this context, it may be appropriate to reexamine some old assumptions concerning that role and to reappraise their contemporary validity:

1. State and Local Communities Lack the Numbers and Quality of Human Resources to Innovate Solutions to Important Problems. This assumption may well have been correct in the early days of the Republic. For example, the 13 colonies contained only 3.5 million persons when the Constitution was written, and the human resources of the entire nation had to be tapped in order to identify and recruit into government service such talents as those exhibited by Washington, Jefferson, Hamilton, and Madison. Today, however, few states are so sparsely populated that large reservoirs of outstanding talent cannot be discovered within their borders.

2. State and Local Communities Lack the Fiscal Resources Necessary to Deal with Such Major Problems as Those Posed by Natural Hazard Exposure. In an earlier period of our history the onset of a major flood was viewed virtually as an "act of God," as an unpredictable occurrence that could not be planned for, and which sometimes had a devastating impact on the economy of the affected area. Today, we understand that such occurrences can be predicted, if not in terms of the precise hour and exact place of their occurrence, then, at least, in terms of the areas that will be affected with some magnitude of occurrence, at some predictable level of frequency, and within reasonably definable geographic boundaries. If it was impossible in an earlier era for states and local governments to prepare themselves in advance for such occurrences, this charge can no longer be made. In terms of the individual states, the annual expected building damage and total damage rates presented in Chapter 5 suggest that any state could well establish actuarially sound reserves of sufficient magnitude to cope with the consequences of any natural disaster occurring within its borders.

Although it cannot be denied that the buildup of such reserves might involve difficult and unpleasant challenges, including new and continuing cost burdens, states are not lacking in the capacity to carry such burdens. In today's world, it may well be that state and local planning of this type is not engaged in for the simple reason that such action is unnecessary; the field has been preempted by the Federal Government, and it is far easier to rely on that institution and the resources of the entire nation than to tackle one's own problems with one's own resources. As a consequence, there may be a propensity on the part of state and local officials to favor "disaster" declarations for numerous happenings of all types, regardless of whether or not they fit past definitions of truly "disastrous" occurrences.

3. If the Federal Government Doesn't Exercise Leadership in Solving a Problem, Then the Problem Will Not Be Solved at All. The implication of this assumption is that most major problems were first identified by the federal government and that the primary means for solving those problems were first invented by that jurisdiction. The facts, of course, are to the contrary. For example, federal intrusions into the field of air quality management were accomplished only after long years of campaigning by western, primarily Californian, legislators, local officials, and community leaders. The broad outlines of the national attack on air quality problems were drafted outside of any federal building and were "sold" to major actors within the federal system. Approaches to problems of the coastal zone were pioneered by one or two states, and the major elements in the nation's attack on urban problems were derived from the pioneering experiences of numerous local governments. That it may be more convenient for the federal government to draw together the ideas generated by numerous institutions within our society than for any single local or state unit of government to do this job cannot be denied. However, it cannot be charged that all the major problems will go unaddressed if the federal government fails to consider them.

However, even if some of the oldest assumptions concerning the need for federal activity are in error, it is also clear that the federal government can handle some problems more effectively, economically, and efficiently than can the individual states or local units of government. Technology development and transfer programs are illustrative. Fifty natural hazards research institutes in the United States do not appear to be an efficient alternative to a more centralized hazards research operation supervised and/or conducted by such a distinguished central institution as the National Science Foundation. Also, few would deny that the Federal government must continue to stand in readiness to act because of the oversight, imprudence, or caprice of state or local policy makers to deal with problems of concern to more than their own constituency. Equal opportunity laws are illustrative here. Thus, it is important to seek *the most appropriate balance* among governmental jurisdictions for the management of natural hazards and other policies.

One criterion that might be utilized to determine the "federalness" or "localness" of a problem is this: To what extent is the problem infused with value questions and conflicts? The higher the value content, the more numerous the value conflicts between problem-impacted parties, and the more intense the feelings of those parties concerning these questions, then the more "local" is the problem. By its very nature, public policy making is concerned with reconciling competing claims and value conflicts within a relevant community. The greater the number of value conflicts and the larger the number of value-holding groups, the more difficult is the task of reaching effective decisions. Localization of decision-making responsibility and authority in theory permits holders of value premises to more directly make their views known to decision makers and, because of the sheer number of local points at which such decisions can be made, permits much pluralism within the system and much opportunity, therefore, for citizens to seek and find a community whose values conform to the taste of the individual seeker.

In short, it appears that the range of information which is necessary for rational decision making, from the objective to the subjective, from the value free to the value laden, may well influence the "correctness" of any decision concerning the locus within the federal system at which primary responsibility for decision making should be assigned. The importance of where to assign responsibility and authority within the federal system has been recognized by the Congressional Office of Technology Assessment which has observed that:

> The federal government has the overall responsibility for conducting the Implementation Plan (for the Earthquake Hazards Reduction Act of 1977) because earthquakes and other natural hazards are national problems. However, when all decisions are made at the federal level, local and state governments tend to adopt a resigned or desperate attitude of "letting George do it," and gradually abdicate responsibility at their own levels. This tension between paternal federal government and maturing local governments must be resolved.

Policy maker choices with respect to the resolution of jurisdictional questions, and those relating to the relative importance of problems and problem-solving strategies, are resolved, in part, through policy maker definition of the purposes and goals that are to comprise the targets of any designed set of public policies.

Conclusion

Like other phenomena of our time, hazardous natural occurrences present public policymakers with difficult, sensitive, and complex decision choices. Whether to act, when, how, and in respect to what set of impacts are central questions in these decision processes. Believing that these decision operations can be aided through explicit definition of *candidate problems* and affected *stakeholder groups*, this chapter has targeted on these subjects. It therefore provides the framework for consideration in following chapters of the alternative goals, policies, and action possibilities that might be considered by policymakers and those who influence policy at federal, state, and local levels.

References

Advisory Commission on Intergovernmental Relations. 1966. Building Codes: A Program for Intergovernmental Reform. ACIR, Washington, D.C.

Atkisson, A. A., and W. J. Petak. 1981. Earthquake Insurance: A Public Policy Analysis. Draft of a report to the Federal Insurance Administration. J. H. Wiggins Company, Redondo Beach, California.

Atkisson, A. A., and W. J. Petak, and D. R. Anderson. 1980. Earthquake Issues Workshop Report. Report to the Federal Insurance Administration. J. H. Wiggins Company, Redondo Beach, California.

Cobb, R. W., and C. D. Elder. 1972. Participation in American Politics: The Dynamics of Agenda Building. Allyn & Bacon, Boston, Massachusetts.

Dacy, D. C., and H. Kunreuther. 1969. The Economics of Natural Disasters, Implications for Federal Policy. Free Press, New York.

Dexter, L. A. 1969. How Organizations are Represented in Washington. Bobbs-Merrill, Indianapolis, Indiana.

Field, C. G., and S. R. Rivkin. 1975. The Building Code Burden. D. C. Heath, Lexington, Massachusetts.

Greenwald, C. S. 1977. Group Power: Lobbying and Public Policy. Praeger Publishers, New York.

Industrial Union Department. 1980. AFL-CIO vs. American Petroleum Institute et al., U.S. Supreme Court 78-911, 2 July 1980. U.S. Govt. Printing Office, Washington, D.C.

Jones, C. O. 1977. An Introduction to the Study of Public Policy. Duxbury Press, North Scituate, Massachusetts.

Kates, R. W. 1971. Natural hazards in human ecologic perspective: Hypotheses and models. Econ. Geogr. 47:438–451.

Kunreuther, H. 1978. Disaster Insurance Protection: Public Policy Lessons. John Wiley and Sons, New York.

Mann, D. E. 1975. Political incentives in U.S. water policy: Relationships between distributive and regulatory politics. *In*: M. Holden, Jr., and D. L. Dresang (Eds.), What Government Does. Sage Publications, Beverly Hills, California, pp. 94–123.

Mayhew, D. R. 1974. Congress: The Electoral Connection. Yale Univ. Press, New Haven, Connecticut.

Mileti, D. S., T. E. Drabek, and E. J. Haas. 1975. Human Systems in Extreme Environments: A Sociological Perspective. Institute of Behavioral Science, University of Colorado, Boulder.

Municipal Manpower Commission. 1962. Governmental Manpower for Tomorrow's Cities: A Report of the Municipal Manpower Commission. McGraw-Hill, New York.

National Commission on Urban Problems. 1968. Building the American City. U.S. Govt. Printing Office, Washington, D.C.

Olson, M., Jr. 1971. The Logic of Collective Action: Public Goods and the Theory of Groups. Schocken Books, New York.
Ripley, R. B., and G. A. Franklin. 1976. Congress, The Bureaucracy and Public Policy. Dorsey Press, Homewood, Illinois.
Rossi, P. H., J. Wright, and E. Webber-Burden. 1982. Natural Hazards and Public Choice. Academic Press, New York.
Schattscheneider, E. E. 1960. The Semisovereign People. Holt, New York.
United Nations Office of the U.N. Disaster Relief Coordinator. 1976. Guidelines for Disaster Prevention. 3 vols. U.N., Geneva.
U.S. Office of Emergency Preparedness, Executive Office of the President. 1972. Disaster Preparedness. U.S. Govt. Printing Office, Washington, D.C.
U.S. Water Resources Council. 1976. A Unified National Program for Flood Plain Management. U.S. Govt. Printing Office, Washington, D.C.
Vogler, D. J. 1977. The Politics of Congress. 2nd edition. Allyn & Bacon, Boston, Massachusetts.
White, G. F. 1958. Changes in Urban Occupance of Flood Plains in the United States. Research Paper No. 57. Department of Geography, University of Chicago, Chicago, Illinois.
White, G. F., Ed. 1974. Natural Hazards: Local, National, Global. Oxford Univ. Press, New York.
White, G. F. 1975. Flood Hazard in the United States: A Research Assessment. Monograph NSF-RA-E-006. University of Colorado, Boulder.
White, G. F. 1976. Natural Hazard Management in Coastal Areas. Prepared for the U.S. Department of Commerce, Office of Coastal Zone Management, Washington, D.C.
White, G. F., and E. Haas, 1975. Assessment of Research on Natural Hazards. MIT Press, Cambridge, Massachusetts.
Williams, L. A., E. M. Young, and M. A. Fischetti. 1968. Survey of the Administration of Construction Codes in Selected Metropolitan Areas. Prepared for the National Commission on Urban Problems. Clearinghouse for Scientific and Technical Information, Springfield, Virginia.
Working Group on Earthquake Hazards Reduction. 1978. Earthquake Hazards Reduction: Issues for an Implementation Plan. Office of Science and Technology Policy, Executive Office of the President, Washington, D.C.
Wright, J. D., P. H. Rossi, S. R. Wright, and E. Weber-Burden. 1978. After the Clean-up, Long-Range Effects of Natural Disasters. Sage Publications, Beverly Hills, California.

9

Constraints on Public Hazards Management Policy Making

Introduction

In theory, a wide variety of legislative strategies can be used at federal, state, and local levels to manage the risks associated with the exposure of people and property to natural hazards. In practice, however, the ability of any jurisdiction to use these strategies may be constrained by legal, sociopolitical, administrative, economic, and other factors, including the normative commitments of its citizens and political influentials.

Legal Constraints

Public efforts to reduce losses associated with the exposure of people and property to natural hazards tend to involve the use of comparatively few types of policy enactments, each designed to implement one or more *technical strategies* for reducing potential hazard exposure losses. The primary technical strategies for reducing such losses involve combinations of actions targeted on (1) reducing the intensity, magnitude, or frequency of hazardous occurrences through use of area structural mitigation systems (e.g., dams, levees, retaining walls, seawalls); (2) strengthening the resistance of buildings and other structures to the forces associated with hazardous occurrences, thereby reducing their vulnerability to hazard-induced damage; and (3) reducing the consequences of hazardous occurrences by limiting the exposure of populations and structures to such occurrences. The latter technical strategy can be thought of as embracing 2 types of activities: those calculated to avoid the location of people and structures in areas where high intensity hazardous occurrences are expected, and those that, although not preventing the location of structures within such zones, nonetheless limit the total exposure of people to such hazards. This latter technical loss-reducing strategy can be implemented by such measures as controlling the occupancy or use of hazard-exposed structures.

Each of these technical strategies for securing reductions in hazard exposure losses can be employed only if appropriate policy action is taken by the relevant governmental jurisdiction(s). Such policy action takes the form of one or more types of legislative or administrative enactments, as depicted in Table 9-1. These enactments include: land use zoning and subdivision control ordinances, regulations, or statutes; governmental condemnation or purchase of land and improvements on the land; governmental budgets and appropriations providing authorization for public investments in area structural mitigation systems; and building and housing codes, statutes, and regulations. Typically, building codes are enacted to control, in part, the structural quality of new structures and existing structures that are substantially modified, while housing codes are directed at assuring that existing structures used for purposes of human occupancy meet minimum health and safety structural requirements.

Table 9-1. Governmental Strategies and Policy Enactments for Reduction of Natural Hazards Losses

Strategy	Implementing policy enactment
1. Hazard avoidance	Land use zoning legislation; subdivision control legislation, title transfer enactment; governmental condemnation or purchase of lands
2. Area structural mitigation	Governmental investments (budgets, appropriations); subdivision control legislation
3. Structural strengthening (vulnerability-reducing mitigations)	Building and housing codes, statutes, and regulations
4. Control of population exposures	Building and housing codes, statutes, and regulations

As noted in Chapter 3, the first legal constraint that determines whether these several types of enactments may be used is concerned with the basic legislative and regulatory powers of the jurisdiction in question. For example, in the United States the general legal view is that the states of the union are the primary sources of all governmental authority. Local units of government, such as cities, counties, and special districts, possess only such authority as may be given to them by the constitutions and legislatures of their parent states. The federal government, of course, possesses only that authority which has been *delegated* to it by the several states of the union through the instrument of the Constitution, as well as such powers as may be reasonably inferred from those delegations.

With respect to local units of government, one simplified explanation of the scope of authority possessed by such units is that offered by *Dillon's rule:*

> It is a general and undisputed proposition of law that a municipal corporation possesses and can exercise the following powers and no others: First, those granted in express words; second, those *necessarily* or *fairly implied* in, or *incident* to, the powers expressly granted; third, those *essential* to the declared objects and purposes of the corporation—not simply convenient, but indispensable. Any fair, reasonable doubt concerning the existence of power is resolved by the courts against the corporation, and the power is denied. . . . (Mandelker, 1963)

Assuming that a governmental jurisdiction seeking to apply one or more of the technical strategies for hazard risk reduction is judged to possess the authority to utilize one or more of the types of policy enactments specified in Table 9-1, other legal constraints then are faced by the enacting jurisdiction. Although these constraints typically present few insurmountable problems to the enacting jurisdiction, careless approaches to their use may lead to court action and to the possible voiding of the enactment. Although courts in the United States are extremely reluctant to substitute their judgment for that of legislative and rule-making bodies, they sometimes do so if enactments are inappropriately phrased, developed through inappropriate processes, or are unsubstantiated by an appropriate evidentiary base.

On the latter point, U.S. courts have been extremely reluctant to weigh the factual merits of legislative enactments. For example, a California court has held:

> A legislative body, in the exercise of its police power, has a broad discretion to determine both what public interests are and the measures necessary for the protection of such interests. The determination of the need for a mode of exercising the power is primarily for the legislative body and the courts will not hold enactments invalid unless they are palpably unreasonable, arbitrary or capricious, having no tendency to promote the public welfare, safety, morals, or general welfare. Every presumption is in favor of the reasonableness of the law and its validity. A court is not concerned with the wisdom or policy of the law and cannot substitute its judgment for that of the legislative body. If reasonable minds might differ as to the reasonableness of the regulation, the law must be upheld. (Justesen's Food Stores vs. City of Tulare, 1941)

Similarly, a New York Court has stated:

> This court is unable to say that the city legislative authorities are wholly unreasonable to the point of irrationality in believing that the police and fire problems will be somewhat alleviated by not permitting laundromats . . . to be open between midnight and 6 a.m., and by requiring that between 6 p.m. and midnight laundromats have someone in attendance. Even if the municipal legislative authorities are mistaken in

> this view, that does not invalidate the legislation. Courts do not sit as an additional house of either state or municipal legislature. (Schact vs. City of New York, 1963)

All regulatory legislation at state and local levels of government may be traced ultimately to the "police power" of the states. Although this phrase is incapable of precise definition, the term has been "used and given shape by the courts in the course of thousands upon thousands of adjudications. The term identifies valid governmental actions undertaken to protect the health, safety, and welfare of the community" (Krier, 1970). At the federal level, the "working equivalent" of the "police power" is the grant of power contained in Article III, Section 8, of the U.S. Constitution.

At both state and federal levels, regulatory policy making is constrained by specific Constitutional prohibitions, such as those contained in the Fifth and Fourteenth Amendments to the U.S. Constitution. These amendments, respectively, state that: "no person shall be . . . deprived of . . . property without due process of law; nor shall private property be taken for public use without just compensation." The Fourteenth Amendment goes on to state that "no state shall . . . deny to any person within its jurisdiction the equal protection of the laws."

With respect to "just compensation," the courts have not been reluctant to uphold the constitutionality of statutes and regulations that seriously abridge the power of property owners to use their property for the purposes for which such property was acquired. However, the constitutional validity of such abridgement of property rights depends on the purposes of the regulation. Where such purposes are related to public health and safety, the courts tend to uphold the enactments, but where such purposes are related to other public ends, the courts are less willing to support the enactments unless they are accompanied by just compensation of the property owners. For example, Pollack (1968) has observed that enactment based on legislative objectives that are unrelated to the protection of the public from the harm, injury, and health disturbances occasioned by hazard exposures may require the payment of "just compensation" to persons whose property rights have been impaired:

> At least one of the frequently used tests of whether police power regulations have gone so far as to be compensable government "taking" is whether the legislation simply restrains conduct harmful to others or whether its purpose is positive enrichment of the public at the expense of private property.

However, when hazard mitigating statutes and regulations are based on concern for public health and safety, they may impose considerable economic burdens on regulated parties and even lead to the closure of otherwise lawful businesses. One leading case of this sort was decided by the United States Supreme Court in 1916.

> So far as the federal constitution is concerned, we have no doubt that . . . the harshness of such legislation, or its effect upon business inter-

> ests, short of a merely arbitrary enactment, are not valid constitutional objectives. Nor is there any valid federal constitutional objection in the fact that the regulation may require the discontinuance of the use of property, or subject the occupant to large expense in complying with the terms of the law or ordinance. (Northwestern Laundry vs. Des Moines, 1916)

Under the police power of government, governmental entities and jurisdictions may engage in a broad variety of activities. According to some members of the United States Supreme Court, attempts to define the reach of the police power or

> its outer limits is fruitless, for each case must turn on its own facts . . . public safety, public health, morality, peace and quiet, law and order—these are some of the more conspicuous examples of the traditional application of the police power . . . yet they merely illustrate the scope of the power and do not delimit it . . . it is within the power of the legislature to determine that the community should be beautiful as well as healthy, spacious as well as clean, well-balanced as well as carefully controlled. (Berman vs. Parker, 1954)

Thus, it seems clear that governmental entities and jurisdictions are comparatively free to deal with the public problems posed by natural hazard exposures so long as they are attentive to a comparatively few, but important, legal criteria. These criteria require that the public purposes behind the enactment be specified, that an evidentiary base that is not "wholly unreasonable" be offered in support of the enactment, that constitutionally sound procedures be used in the enactment of the policy, and that just compensation be provided to property owners when the public purposes to be accomplished are not related essentially to the public health and safety. In general, then, the constitutional validity of any natural hazard statute or regulation depends on showing that the enactment is responsive to a public purpose, that the required procedures are reasonable, and that the regulation is not unduly burdensome. Although an ambiguous phrase, it seems clear that "the undue burden" criterion decrees that the abridgement of private property rights be somewhat limited. If a regulation leads to physical occupation of a landowner's premises, or to his being denied all use of them, the regulation almost certainly will be struck down. However, if it precludes uses that have traditionally been characterized as nuisances, or that will produce threats to the public health and safety, it will almost certainly be upheld. Typically, the courts approach a challenge to the constitutionality of a hazard mitigating regulation with the presumption that such enactments are constitutionally valid, that they accomplish the public purpose, and that they impose no undue burden on property owners. As a practical matter, this means that to prevail, the affected property owner must establish the contrary of one of these presumptions. Unless he does this by a preponderance of the evidence, the state will carry the day. Evidence to this end introduced by the owner may of course be rejected or subjected to another interpretation as a result of evidence introduced by the state.

The existence of this validity presumption has made some legislative bodies careless. The facts that a truly conservative, or even prudent, legislative body would require to be verified have been skimped. In part this is a product of economics. Verification of the facts is a time consuming and expensive process. It is easier to follow precedents, neglecting the local peculiarities that may have shaped them. Fortunately (or in some cases unfortunately) for the regulations, this legislative carelessness is more than matched by property owner laziness. The reported cases contain relatively few instances of well-prepared factual attacks on police power regulations.

When legislative bodies attempt to legislate with care, the facts requiring verification are myriad. This is particularly true in the case of natural hazards. Most important would be a risk assessment that identifies or estimates hazard occurrences of different geographic extents and physical intensities together with estimates of the property damage and personal injury associated with each occurrence and of the costs of mitigating that damage and injury. Evidence bearing on these facts may be culled from a variety of sources, most generally from the history of past occurrences, expert opinion and field investigations.

Legislatures that seek to verify such facts and then go on to reach findings concerning the extent and intensity of the hazard occurrence to be abated or mitigated and the regulations thought appropriate to that end will provide themselves with the maximum possible protection against constitutional challenge. Against some property owners this may of course not be sufficient. The degree of risk abatement or mitigation sought by the regulation may be so great as to effectively deprive them of all ownership benefits. In these situations it may be better to proceed by condemnation rather than tolerate the uncertainties of owner action. In most situations, however, the outlined procedure will protect the state. A court may find the evidence relied on by the legislature to be weak, or it may disagree with the factual and policy conclusions based theron, but will be extremely reluctant to substitute its own judgment for that of the legislature. It will be helped to this conclusion if, in adopting regulations, the legislature has given notice to all affected property owners and provided them with an opportunity to be heard.

In some cases, of course, a legislature will not be able, or would find it wasteful, to verify all the facts required for some policy enactments. For example, the legislature may not find it economical, or possible, to trace with exactitude the boundaries of the hazard zones in which certain uses will be permitted or abridged. Instead, it may generally describe the characteristics of these areas. In these cases, instead of substituting a guess for the missing information, another procedure is available, a procedure that makes the enactments less vulnerable to constitutional challenge. Given that the legislature can with reasonable specificity provide criteria for decision, then it can specify by elevation or by soil or slope characteristics, or in some other manner the areas in which specified uses will be permitted, and can delegate to a permit-issuing body the task of making the impact of its criteria concrete. Such a procedure is obviously less time consuming and costly than one in which the legislature makes all the required determinations, may also be fairer, and may further provide protection against constitutional attack. However, the legislature may not delegate the whole authority to legislate to administrative entities. It must provide reasonably specific criteria for decision making by such entities.

Finally, the constitutional vulnerability of a hazard management policy may be reduced if the enacting entity is able to demonstrate a clear linkage between the contested policy and some comprehensive set of plans or programs for dealing with the hazard. If the policy action deals only with a single flood-prone parcel of land, and all other generally comparable parcels are ignored, the enactment will probably be struck down. If an entire repertoire of mitigation actions are included in a program, not simply abridgements of land use, but also building strengthening and occupancy control measures, then any single segment of the total policy or program is likely to be viewed more favorably by the courts. If the approach to risk reduction in a specific hazard zone reflects appropriate multijurisdictional planning and coordinative activity, further support will be provided.

Sociopolitical Constraints

However large the legal capacity of a governmental jurisdiction or entity to deal with a natural hazard impact, the willingness of governmental decision makers to deal with these matters may be influenced by a variety of sociopolitical factors. These factors relate to the values, perceptions, motivations, priorities, and preferences of hazard-exposed populations, other involved parties, public opinion formers and leaders, political activists, whole political constituencies, professionals and administrators, decision influencers within government, and the policy makers themselves. Additionally, the representational foci of community and political leaders, policy-influencing parties, and the policy makers themselves may be expected to influence the outcome of the policy-making process.

The gross outline of the linkages between these many factors is depicted in Figure 1-1. The activators of the public policy system are the set of phenomena referred to as "public problems," which, under appropriate circumstances, may become the focus of "political demands" on the policy system and, thereafter, may be escalated to the status of "public policy problems" and placed on the "public policy agenda" for action (Anderson, 1975).

The model, of course, denies the validity of "folk wisdom" concerning the American legislative process: i.e., that public policy springs full-blown from the wisdom, compassion, public problem-sensing, and problem-solving skills of legislators. Without denying that legislator-initiated and engineered public policy sometimes, perhaps frequently, emerges from the system, the model tends to accept the perspectives of Anderson (1975), Davies (1975), and Keefe and Ogul (1977).

As observed by Keefe and Ogul,

> the fact that the legislature is empowered to make laws does not mean that it initiates the ideas for legislation. Indeed, for the infusion of ideas

> and the origination of most legislation, the legislature is dependent upon familiar "outsiders". . . . Neither the wide perception of a social problem by legislators nor their recognition of a group's particular claims for governmental action is certain to lead to legislation. The chances . . . increase when (1) influential pressure groups mobilize their members and seek a governmental solution; (2) the unorganized public becomes intensely concerned with the matter or, conversely, is indifferent to the special measures sought by a pressure group; (3) the parties and powerful legislators take up the cudgels; and (4) the formation of strong counter-pressures to defend the *status quo* fails to materialize.

Both a law professor and a state legislator, Davies (1975) asserts that the success of a legislative proposal depends on action by outsiders, actions that result in the formulation of a specific policy proposal and make it possible for legislators to support the proposal. Within the context of the Davies perspective, legislators are seen as functioning more as "judges" of proposal efficacy than as discoverers of public problems and initiators of solutions. To legislative outsiders, according to Davies, falls the task of generating public and legislator interest in problems and problem solutions and the further function of creating a sociopolitical environment within which it becomes possible to design and enact public problem-focused policies.

Even if the above perspectives of the American legislative process are overstated, it is clear that public policies do not emerge full-blown from the head of Zeus, that policy compatible sociopolitical environments must be fashioned, and that much time must be devoted by policy-sponsoring legislators to the guiding of proposals through the legislative process. In regard to the latter, Clapp (1968) quotes a U.S. congressional leader as having stated, "I don't believe there is any man in the House who can do a really effective legislative job on more than two bills in a year. If you do that, you have a full-time job." Commenting on the necessity for such activity, Keefe and Ogul (1977) observe that

> the overriding strategy in the advancement of legislation, from introduction to final vote, is to fashion a bill that can attract and consolidate the necessary support, preferably with a minimum of concessions to opponents. The process of winning support calls for tapering demands from the optimal down to the acceptable—ranging from what is most desirable to what, if necessary, will do. . . .

Within the limits of these perspectives and the general model presented in Figure 1-1, it, therefore, is important to dispose of the following questions: (1) What factors influence the social definition of a phenomenon as being a public problem? (2) What factors influence whether or not public problems become the focus of constituency development and political demand-generating activities? (3) What factors influence the placement of any such problem on the public policy agenda? (4) What factors influence policy maker ranking of items on the agenda and the nature of the action which is taken to dispose of agenda items?

The development of answers to these questions seems to be particularly important with respect to natural hazard management policies. As noted before, the impact of natural hazards is not felt by the public on either a continuous or near continuous basis. Neither are such effects necessarily visible to all segments of society, nor are the causes and solutions to these impacts clearly perceived. In this regard, catastrophic occurrences such as earthquakes, or great hurricanes and storm surges do not occur very often. To potential impacted parties they are "unexpected" and "unfortunate" occurrences that are the result of remote chance and bad luck. Even when visited by such a calamity, impacted parties tend to shrink from the conclusion that the phenomenon may occur again in the future and, instead, seek to continue with their lives as before (Mileti et al., 1975).

Illustratively, the most devastating civil disaster in American history was the Galveston hurricane and storm surge of September 8, 1900. The tropical disturbance from which the storm evolved was first detected on September 1 and struck Galveston, full force, on September 8 and then took another week to disappear over the Great Lakes. During the single night of its strike upon Galveston Island, hurricane winds and storm surge flooding claimed approximately 6000 lives, about 10% of the Galveston population, and left only a few undamaged structures standing on the island (Halstead, 1900). In the days following the strike of the hurricane upon the island, on-the-scene observers described the situation in words such as these:

> Fever is appearing among the survivors and the medical staff is unable to cope with it. What little water there is is polluted with disease germs, and the pestilence is spreading. Nothing can stop it but the arrival of doctors and supplies. Parents are warned to keep away from the dead wagons in which their children have been tumbled, and long-barreled rifles back up the warning. . . . The safety of the . . . living necessitates this course, for the hot sun is already breeding disease and pestilence from the hundreds of decaying bodies. The city is a vast carnal house. The dead wagons hurry from place to place, filling with the awful cargo, and then speed away to the docks, where the bodies are dumped into scows and towed out into the Gulf. The waters of the Gulf float the bodies for a few minutes, then swallow them up. There are no services, no prayers, no tears—just a small "splash", "splash", as the corpses are thrown to their last resting place. Then the funeral cortege of scows returns to the docks for another load.

One week after the storm passed through Galveston, the community's Mayor Jones reported that:

> We are broke, the majority of us. I hesitate to say how much it will take to put Galveston where her people can care for themselves . . . There is not a building but is damaged, not a house of those left standing but will have to be re-roofed, and few that will not need to be straightened on their foundations. . . . It is true Galveston is represented as being one of the wealthiest cities of the country. But our rich people had every-

> thing here and are crippled. The people of moderate means, who had homes and worked on salaries, are scarcely an exception, ruined. . . . The island in the sparsely settled parts seems to have been swept clean. (Halstead, 1900)

Two weeks after the storm strike, a report from Galveston indicated that

> many families refuse to leave. Scores of persons are living in their wrecked homes here. Many of these houses are without floors and devoid of all sanitary provisions . . . The foul stench from the carcasses makes sleep almost impossible at night, and strangers who come here do not remain long on account of the terrible odor. The lime which was ordered a few days ago for disinfection purposes has not yet arrived. . . . The work of burning the decaying human bodies and other carcasses which are to be found under almost every pile of wreckage continues . . . The waters of the Gulf are giving up dead bodies continuously and the shore of the mainland and the beach of the island are strewn with them . . . There are only ten houses in a habitable condition south of High Island . . . (Halstead, 1900)

On September 18, 1900, a correspondent from a Chicago newspaper reported from Galveston that the city

> has been struck three times with floods and hurricanes, but even this experience is not enough to convince the residents that it will ever happen again. Only a few more cautious have any idea of taking steps to prevent a repetition of the recent disaster. Asked if there will be anything done to make future floods impossible, they will quote the old saw: "Lightning never strikes in the same place twice." (Halstead, 1900)

A representative of the U.S. Coast and Geodetic Survey reported from the disaster scene that some of the damage from the storm surge was due, in part, to removal of sand dunes that originally had protected the eastern end of the low, flat island on which the city was located. As the city expanded, these dunes had been leveled off and, thus, there was no break against a violent wind or sea moving from the east. Much of the postdisaster discussion focused on providing for the construction of a seawall to protect the city from future occurrences and on means for advancing the reconstruction of the entirety of the area.

Amid the scenes of the immediate postdisaster period, one of Galveston's best known citizens and merchants stood near the center of the devastation and said, "We are going to straighten out everything. We are going to stay here and work it out." A day later, the local newspaper stated:

> Galveston must rise again. . . . The loss of life and property is appalling . . . but with resolute faces, the sentiment [is] . . . that out of the awful chaos of wrecked homes and wrecked business Galveston must rise again. The sentiment [is] . . . not that of bury the dead and give up the ship; but, rather, bury the dead, succor the needy, appeal for aid from a charitable world, and then start resolutely to work to mend the broken chains. . . . Galveston shall rise again. . . . The blight and ruin which have destroyed Galveston are not beyond repair; we must not for a moment

> think Galveston is to be abandoned because of one disaster, however horrible that disaster has been. . . . We will bravely undertake the vast work of restoration and recuperation. . . . (Halstead, 1900)

Today, Galveston has, of course, been fully rebuilt and has, apparently, nearly forgotten this disaster. From end to end, the 35 mile long island, which rises only a few feet above sea level, is marked by subdivisions and commercial establishments that reach out nearly to the mean high tide line. The devastation of the sand dunes appears to have continued unabated throughout the years since the hurricane and storm surge of 1900. Stronger buildings have been constructed than were in place in 1900 and most beach-lot residential structures have been elevated several feet above the ground level. The area is an attractive site for second-home developments and for permanent residents. There is little evidence that would lead one to conclude that any of the island's current inhabitants seriously believe that another disaster could strike.

As near as one can tell, the "problems" that Galvestonians and others read into the great catastrophe of 1900 were phenomena that related to the availability of aid from outside sources during the postdisaster period, to the need for capital to rebuild a disaster-struck city, and to the need for engineered systems that might reduce life loss during the actual hours when hurricane winds and storm surge waters sweep through a hazard-prone area. The diversion of capital, manpower, and other resources from non-hazard-prone areas was apparently never viewed as a problem itself, nor was the building of a city in such a hazardous area so viewed. Thus, all the public policy efforts in the postdisaster period were directed to these phenomena, rather than to the underlying causes of the disaster.

The response of Galveston to its disaster is not unique. After hurricane Camille struck the Gulf Coast in 1969, the slogan "We shall rise again" was emblazoned on automobile bumper stickers and store windows and was repeated in various mass media reports (Quarentelli and Dynes, 1973). Apparently, the victims of community disasters usually see the postdisaster future in optimistic terms and do not seek to remove themselves from the hazard zone. With respect to victims in 2 tornado-impacted communities (Waco and San Angelo, Texas), for example, 52 to 74% thought their neighborhoods would be better in the long run (Quarentelli and Dynes, 1973). Thus, a sense of optimism about the future and a willingness to rebuild seems characteristic of disaster victims.

Within the context of the available evidence, therefore, it is appropriate to suggest that the socially disvalued aspects of major hazard-induced events are those that impede efforts to rebuild and to get on with life as it was before. The notion that the area might better be left in an undeveloped or less-developed state does not seem typical, and the possible need to avoid the hazard in the future does not seem to motivate the impacted population to make a social commitment to future hazard avoidance. Indeed, regarding physical hazards, avoidance motivations are strongest among those individuals who have not previously experienced the hazard (Golant and Burton, 1970). Apparently, the acceptability of any risk, including those associated with exposure to natural hazards, is a function of such factors as the extent to which the risk is familiar, known, voluntary, controllable, catastrophic, and dreaded (Fischoff et al., 1976). Even when a risk meets the last two criteria, those who actually have experienced the consequences of the risk taking view the possibility of future exposures with less apprehension than those who have not and, in regard to natural hazards, typically continue the pattern of prior risk taking. However, the accuracy of hazard perception is influenced by whether one has "more," "recent," or "intense" experience with the specific hazard (Kates, 1971; Burton and Kates, 1964; Saarinen, 1966). Also, the preparedness level of individuals in communities seems to be higher among those with prior experience with the hazard (Kates, 1971; Russell et al., 1970; Fritz, 1961; Burton and Kates, 1964). Rates of adoption of natural hazard adjustments increase immediately after the extreme event (Saarinen, 1966; White, 1964) but decline thereafter. Hazards characterized by high frequency of impact are associated with the use of more adjustments (Kates, 1971), and low probability occurrences motivate fewer adjustments than do high probability occurrences. Burton (1962) and Mileti et al. (1975) summarize the available evidence as follows:

> a general positive relation [exists] . . . between perception of hazard and adoption of adjustments. . . . [It] is exceedingly rare for relevant adjustments to be adopted . . . insofar as they are designed to cope with highly unusual events such as damaging earthquakes. But immediately following particularly dramatic disasters there appear to be . . . attempts at improving adjustments to the hazard. . . . [A] large-scale disaster in one locale stimulates for a time the serious consideration of adoption of relevant adjustments in nearby similar hazard locales.

This tendency is clearly illustrated in the response to earthquakes in California. Slosson (1975) suggests that a strong "emotion reaction" factor stimulated natural hazard mitigation actions in California. In a study of legislation in that state for three years before and after the San Fernando earthquake of 1971, he notes that during 1969 and 1970 10 earthquake-related bills were introduced, but only one passed. Immediately after the 1971 earthquake 47 seismic bills were introduced, of which 23 passed; and the following year, 24 bills were introduced, 12 of which passed. During the next 2 years, 50 bills were introduced and 16 passed; of these, the majority were amendments or corrections to bills that were passed in 1971 and 1972. In the 7 years examined, approximately 160 bills were introduced, whereas approximately 60 were adopted and signed into law (Figure 9-1). On the basis of this evidence, Slosson suggests that California legislators are more prone to respond to problems than to anticipate them; that they are more willing to conform to public passions than to shape public opinion. Between disasters, there is generally a lack of legislative action; but, during the emotional period following a disaster, many hurriedly prepared and ill-conceived legislative bills are introduced, requiring corrective legislation. As indicated by the 1973-1974 legislative results, he found that good, well-prepared, and technically sound legislation generally fails. According to Slosson, this sequence strongly suggests that it is the responsibility of concerned professionals to have

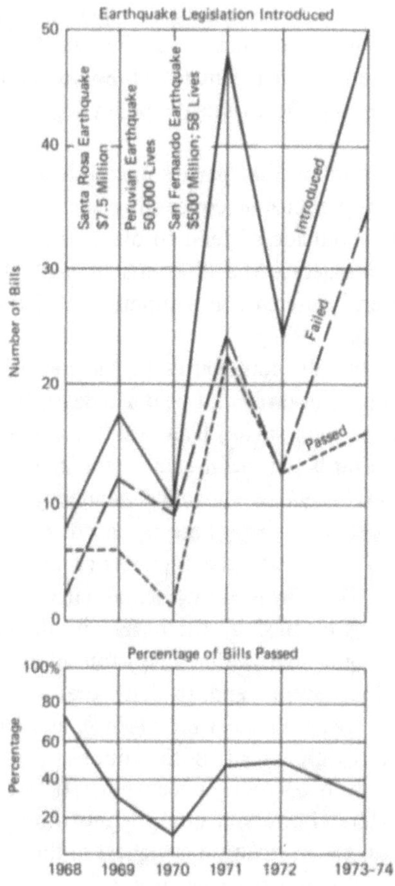

Figure 9-1. State of California earthquake legislation, 1968–1974. From Slosson, 1975, p. 37.

technically sound legislation prepared prior to a disaster and then be willing to volunteer time and effort to assist the legislators when the emotional reaction runs high.

However, even in the wake of disastrous occurrences it may be difficult for legislative bodies to adopt appropriate hazard mitigation measures. Thus, Solomon and Okrent (1977) have examined the reaction of public bodies in Los Angeles to the 1971 San Fernando earthquake, and have documented the failure of the Los Angeles City Council to quickly adopt stringent protective aseismic building code measures for hazardous existing buildings. In Los Angeles, there are approximately 14,000 unreinforced masonry structures built prior to October 6, 1933. These structures are considered to be extremely vulnerable to destruction during an earthquake. However, it wasn't until 1981 and after 10 years of debate, that the Los Angeles City Council was able to approve a program of rehabilitation or demolition of these buildings. The new law requires pre-1934 buildings to be strengthened to levels roughly equal to 60 percent of the current code requirements for new construction, if they are to be rehabilitated. Thus, even in Los Angeles, a city with recent (1971) major seismic activity, as well as in California as a whole, significant seismic policy legislation has not easily been adopted.

Consistent with the finding that parties inexperienced with natural disasters tend to exhibit more apprehension concerning such events is a further set of findings suggesting that "outsiders" tend to overestimate the recovery needs of disaster areas and to exaggerate the disaster's impact. Media exaggerations of disaster impacts apparently increase as a function of distance from the disaster site (Quarentelli and Dynes, 1973).

This situation undoubtedly contributes to the pervasive social tendency to (1) overstate the impacts of hazards that produce death (rather than just injuries) and take multiple (rather than single) lives and (2) underestimate the risks from common, undramatic hazards that claim one person at a time.

Thus, there is a general social or "outsider" tendency to overstate the impacts of hazardous natural events, to exaggerate the needs of those who are impacted by such events, and to understate the importance of less dramatic hazardous incidents and activities. Those who do not experience the disaster tend to deny that they, too, may be threatened by the hazard by denying its existence ("we don't have floods here; only high water") (Burton, et al., 1968). Those actually impacted by a hazardous event tend to deny that a similar event will recur ("lightning never strikes twice in the same place"), and to seek a restoration of things as they were before the event. Both groups apparently prefer to assume that others will take action to prevent recurrences ("the government is taking care of it"). To the extent that either group is motivated to take action to prevent future occurrences, the motivation is highest in the immediate aftermath of a self-experienced or nearby disastrous occurrence.

Whether dealing with natural hazards or with other socially significant phenomena, policy makers exhibit a distinct preference for adoption of "distributive" rather than "regulatory policies" (Mayhew, 1974; Vogler, 1977; Mann, 1975) and a propensity to evaluate proposals in terms of what they will do for their own constituencies and their own chances for reelection (Mayhew, 1974). Proposals advanced by large constituencies command more attention than those advanced by small or virtually nonexistent ones (Schattschneider, 1960; Greenwald, 1977; Dexter, 1969). With respect to natural hazards, and other potential problems, advocates of public policies, therefore, must consider the need for a constituency to be associated with their proposals (Olson, 1971). Further, of course, it is clear that members of a potential risk-exposed constituency must perceive the risk associated with a hazard exposure if they are, subsequently, to be willing to organize so as to support risk-reduction proposals (Cobb and Elder, 1972; Schattschneider, 1960).

Since almost none of the above evidence provides a very glowing picture concerning the readiness of groups and individuals to initiate and/or support economically optimal, "cost-internalized," and loss-avoiding natural hazard management policies, it was assumed that, even in high-hazard areas, local communities probably had done little that was not required by the federal government in order to mitigate the risks associated with natural hazard exposure.

To test this hypothesis and to gain a better understanding of the attitudes of local officials toward a variety of possible hazard management issues, a systematic survey was conducted of local officials in a selected sample of 200 counties chosen from the 500 counties with the highest expected damage rates from natural hazards (see Chapter 5). Slightly more than 50% of the communities surveyed responded to the questionnaire, but fewer than half of these respondents indicated that their local building codes included specific provisions aimed at mitigation of natural hazards losses. In those cases where local building codes were declared to contain such provisions, the natural hazard most frequently addressed was flood, but other hazards mentioned by respondents included hurricane, storm surge, wind, earthquake, landslide, and expansive soil.

Except for such frequently occurring and obvious natural hazards as hurricane, tornado, severe wind, and storm surge, few respondents were aware of the range of natural hazards potentially capable of producing losses within their communities (Table 9-2). The risk model utilized in this study identified 101 of the respondent counties as being subject to significant losses from expansive soils, 100 as being subject to landslide, and 66 as being subject to earthquake losses. As indicated in Table 9-2, however, only 4% of officials in the landslide communities were aware of this hazard, only 18% were aware of their expansive soils problems, and only 21% were aware of the earthquake threat. Granted, the distribution of expansive soil and landslide zones within the surveyed counties may be such that the precise communities which we surveyed lack sufficient continuing damage to produce awareness of these potential problems but the more widespread threat posed by periodic earthquakes suggests that these possible local considerations do not fully explain the pattern of responses. Instead, it seems much more likely that local officials are comparatively unaware of the threats presented to their communities by potentially high consequence hazardous natural events which exhibit low annual rates of occurrence. Lacking an understanding of the threats presented by natural hazards, it is understandable why so few of the surveyed communities have adopted hazard-mitigating provisions within their local building codes. However, this does not explain why so few of the communities which exhibited a high level of understanding of the hurricane, tornado, severe wind, and storm surge hazards nonetheless failed to take direct building code action to mitigate the effects of these hazards. It appears that locally accurate perceptions of natural hazard risk levels are not enough to prompt enactment of hazard-mitigating policies; that other conditions must also be satisfied.

Of the entire group of survey respondents, 8% indicated that no building codes of any nature were utilized within their jurisdictions, 28% had adopted the Uniform Building Code, and an additional 23% had adopted the Southern Building Code. In addition, several respondents, such as those in North Carolina, indicated that state-level codes were enforced and that such codes represented

Table 9-2. Recognition of the Threat of Natural Hazards by Local Officials, in Comparison to the Risk Model (Excluding Riverine Flood)

Hazard	Recognized by risk model	Recognized by risk model and by respondent	% of municipalities recognizing hazards established in the risk model	Recognized by respondent but not by risk model
1. Hurricane	34	31	91	4
2. Tornado	98	78	80	0
3. Severe wind	75	56	75	10
4. Storm surge	31	27	87	0
5. Tsunami	4	1	25	3
6. Earthquake	66	14	21	3
7. Expansive soil	101	18	18	0
8. Landslide	100	4	4	0

modification of national model codes. In those communities in which building codes were being enforced, approximately 75% of the respondents believed that the codes provided at least a medium degree of protection from natural hazard losses. Over one-half of the respondents believed that such code provisions had produced no particular effects on conventional building practices within their communities, but about one-third said that the provisions either had hindered standard building practices or rendered them more expensive.

Although 80% of the respondents indicated their communities had adopted general land use plans, only 66% of this set indicated that their land use plans contained specific language addressing natural hazards, and few of these indicated that their plans addressed hazards other than flood. However, expansive soils, earthquake, severe wind, tsunami, and landslide hazards were included in the general plan of one of the surveyed communities. Only 38% of the respondents indicated that their local zoning ordinances specifically addressed the subject of natural hazards, and in virtually all of these cases the hazard addressed was flood. Those respondents who indicated that no special natural hazard provisions were included in their comprehensive plans, zoning regulations, or subdivision ordinances tended to deny that any such provisions were needed. Nearly two-thirds of this group stated that the level of natural hazard risk within their communities was too low to warrant special consideration in these documents.

Typically, also, respondents exhibited negative attitudes toward a possible requirement that "natural hazard impact assessments" be obtained from parties engaged in major building or subdivision projects. Those communities in which some form of environmental impact assessment is now required expressed the view that any necessary natural hazard assessment should be included in these documents, but most expressed concern that hazard assessment requirements would complicate local construction procedures.

All respondents were also asked to identify the societal sector and/or governmental level at which responsibility should be fixed for the policy setting, program planning, financing, and program implementation concerned with the management of natural hazards. Shown graphically in Figure 9-2, respondents overwhelmingly indicated that local governments should dominate policy setting, program planning, and program implementation but that federal and state governments should pay the bills. Local governments and regional planning entities were seen as the proper organizations to plan hazard-management programs and local governments were overwhelmingly viewed as the proper groups to implement such programs. The possible role of federal and state governments in policy setting was acknowledged, but most respondents believed that local governments should dominate these operations as well. Considered as a whole, the findings from this survey are consistent with those derived from a study conducted during approximately the same period by a research team from the University of Massachusetts. The latter study involved interviews with approximately 2300 politically influential individuals in a sample of 20 hazard-prone states and 100 hazard-prone local communities. It revealed that natural hazards are not viewed as a very serious or high priority policy subject by local

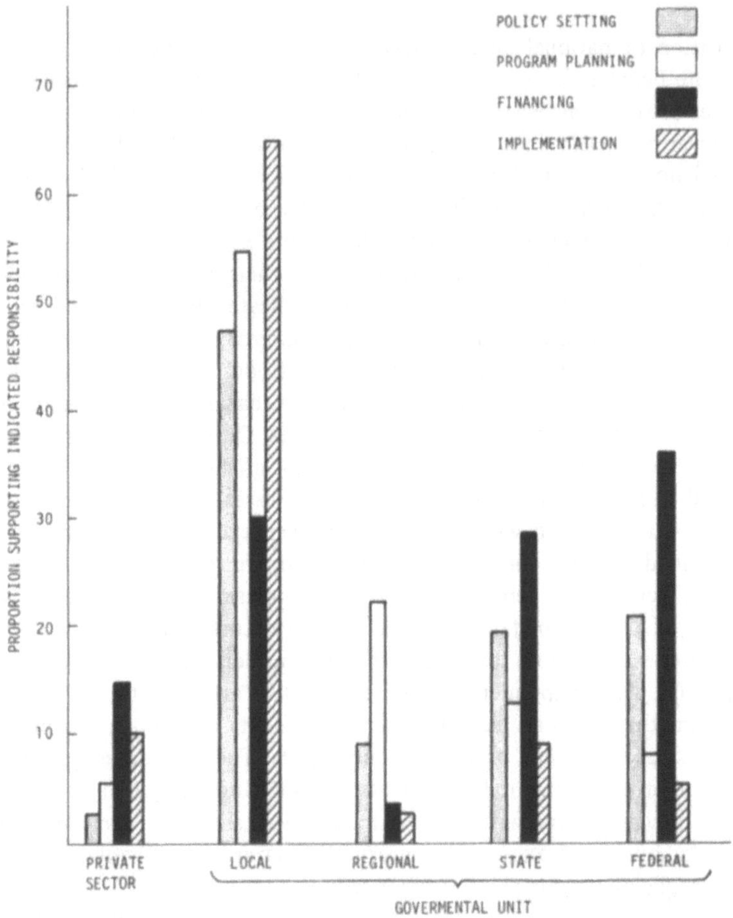

Figure 9-2. Proportion of local officials in hazard-prone areas who support indicated hazard management responsibility for specified level of government and private sector.

policy makers and political influentials; and that, when the need for risk mitigation is recognized, preference is expressed for "distributive" rather than "regulatory" policies (Rossi et al., 1982). In a completely separate study, in which we examined earthquake hazard mitigation policies and practices in 3 U.S. metropolitan areas and their parent states, we reached similar conclusions (Atkisson and Petak, 1981).

Value Constraints

Underlying the legal dicta and the personal, group, political, and policy maker preferences that constrain natural hazards policy making are the value propositions and systems to which individuals and groups within our society are committed. A value proposition, of course, is both normative and prescriptive and defines an end that *ought to be achieved* or a course of action that *ought* to be followed. Although the *correctness* of value commitments cannot be determined through empirical methods, the consequences associated with the operationalization of values may be examined. The holders of values may, of course, be commited to more than one value proposition or set of such propositions; and these commitments may exhibit some hierarchical or preferential ordering.

Great risk is incurred by anyone who presumes to assess the prevailing value commitments of a society. Values held by any one individual may be in conflict with those held by another and, of course, there is a lot of difference between groups in terms of their ascendant value commitments.

As used here, an "issue" is a question so framed that it may be answered in either the affirmative or the negative and is, further, a question to which different parties will respond differently. Issues of fact are those which are capable of empirical resolution or verification and deal with what *is* or what will be. On the other hand, issues of value are not subject to empirical resolution and can only be examined in terms of the predictable consequences that may result from their operationalization.

Although numerous issues of value beset natural hazards policy makers, the following propositions may be viewed as "givens" within our society and suggest policy perspectives which are rarely contested.

1. Special Protection Should Be Given to the Young, the Disabled, and the Infirm, Who Are Unable to Care for Themselves in the Event of a Hazardous Natural Occurrence. Commitment to this value proposition underlies such enactments as the California Field Act, which requires that both old and new schools and hospitals, meet earthquake resistant standards of a high order. Throughout our history, there seems to have been little social or political conflict surrounding the operationalization of this proposition.

2. Individuals, Groups, and Governmental Entities Should so Guide Their Decision Making and Behavior so as to Avoid Inflicting Serious Damage and Injury on Other Parties, Unless Some Overwhelming Public Purpose Requires Such a Course of Action. In its oldest form, this proposition is, of course, embedded in the tort liability dicta we have inherited from English common law. Even without any statutory prohibition against such conduct, one who willfully and knowingly engages in a course of activity that will cause injury to others invites a legal action by the injured parties that will be viewed with sympathy by our courts.

If anything, our social commitment to this proposition has grown stronger, rather than weaker, in recent decades. Thus, the nineteenth century slogan, "Let the buyer beware," has now essentially given way to the slogan, "Let the seller beware." Product liability decisions made by the courts and product liability statutes enacted by legislative bodies throughout the country provide vivid evidence of this shift in social commitment.

Implications for builders, developers, and governmental units, with respect to natural hazards management, may be larger than the literature now reveals. Thus, there may be real questions as to the liability of a local government entity that, knowing that development within a hazard zone will result in future harm and/or dollar loss to those who occupy the developed area, nonetheless permits such development to take place. Similarly, a builder or developer who knowingly constructs dwelling units within such a hazardous zone and then withholds information from potential purchasers concerning those hazards may also well be inviting future tort liability actions.

3. Opportunities for Personal Choice and Decision Making Should Be Preserved, and Governments Should Intervene in Such Processes Only When Required by Some Ascendant Public Purpose. Pervading our society is a strong and long-lasting social commitment to pluralism, to the right of individuals to "do their own thing," as long as that "thing" does not do harm to others. Although specific applications of this proposition sometimes provoke social controversy, as in the case of "crimes" involving acts between consenting adults, there is a general social willingness to gauge the efficacy of proposed policies against this standard. Thus, however hazardous to its participants, we are reluctant to deny individuals the opportunity to jump out of airplanes with parachutes strapped to their backs unless such individuals choose to do so over populated areas where such activities may result in injury to innocent non-risk-takers on the ground below.

4. The Right of Individuals to Hold and Use Property Is to Be Respected by Their Fellow Citizens and Enforced by Government. Social commitment to this proposition predates the founding of this republic, and is, of course, one of the cornerstones of modern capitalistic and democratic societies. When conjoined with a prevailing social commitment to the view that individuals have a right to engage in activities that do not cause injury to others, this value provides a major constraint to the governmental execution of policies that might otherwise prevent future injury and death arising from exposure to natural hazards.

5. Society Should Protect the Rights of Individual States and Local Communities to Adopt Rules and Regulations That Are Considered in the Best Interests of Their Publics. The constitutional tradition of the United States has been to leave as much decision-making power as possible with the individual states. Thus, there has been a consistent value commitment to decision making based on grass roots politics and pluralistic democracy. Natural hazard emergency preparedness and mitigation practices have therefore traditionally been the responsibility of individual states, with county and local governments performing the actual work.

When the federal government attempts to exercise its authority to engage in direct regulatory activity in this field or to establish action-forcing conditions on the availability of benefits, loud complaints are voiced by local and state governments. The complainants view these acts as interference with their traditional rights and powers.

Value Issues

Although other value commitments of our society might be listed above, this brief recitation underscores some of the major value constraints within which natural hazards policy making will take place. Also of great potential influence on such policy making are the following *value issues*, which, predictably, will comprise the focus of local, state, and national debates concerning natural hazards and the policies framed to deal with exposures to those hazards:

1. Shall the Death or Injury of an Individual Exposed to a Catastrophic Event Be Given a Higher Value Than the Death or Injury of an Individual Exposed to a Noncatastrophic Occurrence? It seems undeniable that the American policy-making system responds more acutely and vigorously to catastrophic occurrences than it does to the less visible, less publicized occurrences that inflict injury or death on many more persons, but in events that produce these consequences on only one or a few persons at a time. If our ascendant value commitments are to prevent the greatest number of premature deaths and unnecessary injuries per million dollars of available resources, then catastrophic occurrences would receive far less policy attention than is now the case. Whether our social and policy reaction to catastrophic and noncatastrophic events is due to our ignorance, to our misperceptions of reality, or to some underlying set of values that give greater weight to catastrophic than to noncatastrophic occurrences cannot be answered by these researchers at this time. Instead, the value issue can only be identified.

2. Shall Local Governments Be Permitted to Take Action That Is Consistent With the Interests of Their Current Constituencies But Antagonistic to the Interests of Their "Extended" and Probable Future Constituencies? This issue is one that assumes that the failure to decide or the failure to act may, itself, be viewed as a decision or action. The question, also, takes into account the phenomenon of the "shifting constituency" and the further phenomenon of the "extended constituency." By the former term we mean, simply, that the constituency of any geographically constrained jurisdiction of government may change drastically with the passage of time. At one point in time, the majority of the electors within such a jurisdiction may be owners of large plots of land devoted to the growing of citrus crops; but, only a few years later, the majority may be residents of single-family homes. In a "representative" system of government, it is easy to suggest that elected policy makers must be responsive to the demands of those who have elected them and who constitute their contemporary constituency. But what about the constituents who have, at any point in time, yet to arrive on the scene? Similarly, actions taken by a local elected board in response to the clear interests and demands of citizens within the area may do damage to others beyond the jurisdictional boundary lines of that unit of government. Illustrative is the mix of actions that might be taken by a local unit to protect its citizens from the hazards posed by riverine flooding. The actions selected might well result in the mitigation of the local community risk but might increase stream flow within the floodways of downstream areas and, therefore, exacerbate the risk faced by members of the community's "extended constituency." As posed, therefore, this value issue is one that pervades discussions of federalism and of intergovernmental relations.

3. Should Voluntary Risk Taking That Has a Low Probability of Causing Injury to Others Be Prevented by the Actions of Local, State, and Federal Governmental Entities? This issue presents markedly different implications, depending on the level of government. Consistent with the prevailing U.S. social commitment to the notion of "pluralism," our segmented system of federal government, in theory, provides numerous opportunities for individuals with common interests and value commitments to come together in communities of their own choice. Some communities may opt for one style of life and one level of risk taking, while others may reject such an orientation. When higher levels of government choose to act in certain areas, an inevitable consequence of such action is to push our system toward homogeneity and, therefore, away from pluralism. The point is important whenever the question of *voluntary risk taking* is considered. Without denying that some risk taking is almost purely "suicidal" in its root impulse and apparent consequences, it must also be remembered that there is much voluntary risk taking that is associated with the receipt of compensating benefits. Thus, it would be folly to suggest that recreational parachutists are simply suicidal in impulse. Clearly, there is compensation to such individuals in terms of the sheer joy they derive from their hazardous sport. Similarly, the retired individual who builds a home on a beautiful, outcropping cliff above the blue waters of the Pacific may derive enormous joy from the panorama, while at the same time be running the risk of losing life and property through a slippage of the land into the sea below. A Midwestern couple that flees the hardships of the northern winters for the balmy climes of the Gulf Coast may be running the risk of death from storm surge but be experiencing the rewards of a more salubrious climate and the joys of walks on the beach. If the risk is voluntary, which, if any, unit of government is the better to deal with constraints on that risk-taking?

4. Shall Government Policy Making in Respect to Natural Hazards Seek to Serve "Optimizing" (Effective Long-Term) or "Satisficing" (Short-Term Partial Solutions) Goals?

5. Should One Type of Natural Hazard (i.e., Earthquakes) Be Given Priority Over a More Comprehensive Approach to Natural Hazard Mitigation?

6. Should Cost Effectiveness and/or Cost-Benefit Criteria Be the Principal Basis for Natural Hazard Policy Selection?

7. If Disaster Assistance Is to Continue, Should Disaster Assistance Be Given to "Individual" Victims Who Have Suffered Natural Disaster Losses on the Same

Basis as Now Provided to Whole Communities That Have Suffered Equivalent Per Capita Losses? Where Does Equity Lie in the Current Policy?

8. Once Made Aware of the Risks From Natural Hazards, Should the Individual Citizen Be Required to Internalize Any Losses Incurred as a Result of a Natural Event?

9. Should Local and State Legislative Bodies Act as the "Consciences" for Their Communities and Select a Policy Risk Level Based on Informed Analysis and Testimony, as the Basis for the Adoption of Natural Hazard Mitigation Rules and Regulations?

10. Should Natural Hazard Exposures Be Considered a National Problem and Thus Subject to Regulation by the Federal Government?

11. Should the Citizens of States Who Experience Low Damage Rates, by Virtue of the State's Geography or Management Practices, Be Responsible Through a Federal Taxing Mechanism for Bearing the Cost Burdens Created by the Poor or Improper Land Use and Building Code Policies of Another State?

12. Should States and Local Governments Bear Primary Responsibility for the Mitigation of Natural Hazards and for Dealing With the Aftermath of a Natural Disaster?

Administrative Constraints

Whatever direction may be taken in respect to natural hazards management policy making, the objective outputs of the policy-making process will also be influenced by constraints having to do with the characteristics, configurations, internal quality, authority, and funding capacities of individual jurisdictions of government and of their organizational components. Thus, administrative constraints relate both to institutions and organizations and to the quality and quantity of the human and other resources that comprise their elemental building blocks.

With respect to natural hazards management, the organizations of principal interest will be those concerned with land use planning and zoning, development and enforcement of subdivision and building regulations, hazard zone mapping, and policy analysis. Although there have been no major recent studies focused on the development of quantitative descriptors of the excellence and capacities of these institutions and organizations in the several levels of U.S. government, three studies have been made of this general subject within the past two decades: the report of the Municipal Manpower Commission (1963), the research reports underpinning the activities of the National Commission on Urban Problems (1968), and the study conducted by Field and Rivkin (1975). Of these, the latter two seem particularly relevant to this discussion.

In 1968, the staff of the National Commission on Urban Problems conducted a survey of a stratified sample of the nation's 17,993 local governments. The sample consisted of 3104 governments, of which 81.7% responded to the survey. The principal findings of the survey by the commission's staff were as follows:

1. Land use planning and building quality regulatory activities are widespread, directly affecting a high proportion of the nation's population and involving many thousands of local governments.
2. Most of the regulating governments are relatively small, apparently too small in most instances to engage any full-time employees for such work. This, of course, is a reflection of the prevailing atomized pattern of local government under which, for example, one-third of all the incorporated municipalities in metropolitan areas have less than 2500 inhabitants and one-half are less than one square mile in area.
3. Even among the regulating governments that do have full-time employees for such work, pay rates are generally low, and only the largest governments have top-ranking jobs paying enough to attract and hold well-trained professional or technical people.
4. Local expenditure for these planning and regulatory activities is not insignificant—some $300 million annually. However, this sum is far less than one percent of all urban government expenditure and is even more strikingly dwarfed by the property values that are affected by such activities—more than $1000 billion worth of urban real estate and over $50 billion annually of new construction.
5. Similarly, local government employees engaged in these activities number only 33,000 (full-time equivalent) persons, compared with some 3 million persons employed in the construction activities affected by their work.
6. Control of land use through local zoning ordinances and subdivision regulation is widespread and expanding. Of all zoning ordinances, a large proportion originated since 1950, and many have been considerably revised in recent years. Also, most zoning governments have reportedly prepared "master plans" of prospective land use.
7. Zoning governments deal with large numbers of requests for rezoning and "zoning variances" and, on the average, reject less than one-fourth of such requests.
8. Nearly all municipalities in metropolitan areas and a majority elsewhere have a local building code, but a considerable fraction of these codes has not been materially changed in recent years.
9. Of the cities and towns of 5000-plus that have building codes, about two-thirds report that their local provisions are based on a national or regional "model" code. However, only about one-fourth of these have recently adopted at least 90% of the updating changes recommended by the model code organizations.
10. There is great diversity in local code regulation of particular residential building practices. The survey asked about 14 specific building practices,

including 13 approved by all applicable "model" codes, and one practice accepted by some, but not all, the "model" codes. Of these 14 practices, one is prohibited by over half the municipalities of 5000-plus that have building codes, 4 others are prohibited by more than one-third, 3 by about one-fourth, and each of the remaining surveyed practices is rejected by some of these governments. Similar proportions of rejection appear for the municipalities whose local codes are reportedly based on some national or regional "model" code.

The Municipal Manpower Commission also examined procedures for enforcement of zoning and building laws and the processes for appeal from local findings. In respect to this subject, the Commission (1963) commented as follows:

> The construction industry and related interest groups are extremely influential in the operation of appeal procedures. Representatives of the industry frequently are requested to recommend individuals for appointment to appeal boards, and codes and ordinances frequently require that members of appeal boards be architects, engineers, and contractors. Such practices would not appear to provide adequate protection to the public.

In the Field and Rivkin (1975) study, the authors comment:

> Most cities offer a salary and budget that fail to attract good people into code-enforcement jobs. According to 1970 data, in only the largest cities do salary ranges for building officials provide monetary incentives for efficiency and fairness in code enforcement. The potential for salary growth is limited. In most cities surveyed, there is a narrow range between beginning and maximum salaries. For all cities, $7,490 is the median for starting salaries and $9,600 for maximum salaries. Only in cities with populations of more than 500,000 did salary schedules offer much chance of job advancement. The overall median salary for chief building officials in 1970 was only $10,586, surely no incentive for aggressive leadership. . . . Building department budgets are generally too small to support adequate in-house training programs or to undertake ongoing code revision. . . . To date the federal government has committed few of its resources to improving enforcement.

The authors of this same study also observed that one chief building official in seven is over 60, and over half of all chiefs are past 50.

> Limited budget, limited salaries, and limited prestige add up to uninviting job opportunities for bright young professionals eager to do a job. Even those willing to accept the unattractive entry inducements are handcuffed by lack of funds for professional training, comprehensive code revisions, and evaluations of their own enforcement activities.

Field and Rivkin further observed that:

> Most officials come out of the local construction industry trained as members of building trades . . . or as general contractors. . . . It is not

> surprising that inspectors by work and social habit are tied to their localities. The officials themselves acknowledge that their most frequent work contacts are with local builders, local building material suppliers, local architects, and local engineers. Their social relationships follow the same pattern. . . . Thus, the parochial view is reinforced; it is seldom modified by contacts with people outside the community who might hold more cosmopolitan views on construction technology. . . . The evidence indicates that building officials are sensitive to both the political structure and the interests of the construction industry in their communities.

In the recent report of the Working Group on Earthquake Hazards Reduction (1978), the group concluded that even professionally trained individuals are not fully prepared to deal with natural hazards. Thus, with respect to the earthquake problem, the group concluded that

> Many universities and professional schools simply are not preparing their students to deal with the earthquake problem. Many universities, even some in "earthquake country", are graduating structural engineers without providing them with course work on seismic-resistant design. Similarly, schools of architecture generally do not provide courses on seismic-resistant design for their students. This problem is sometimes further compounded by the fact that in many communities building officials, who are expected to enforce seismic regulations, often have less training than building designers. . . . Professionals frequently have insufficient backgrounds to incorporate the available information into their decision-making. In all of the relevant areas–seismology, geology, earthquake engineering, the social sciences–knowledge is expected to expand even more rapidly in the future.

The working group went on to recommend an aggressive, federally sponsored program of personnel upgrading and a federally financed program of educational services to meet these needs. The working group also focused separately on the professional preparation of land use planners to deal with natural hazards and concluded that "professional land use planners do not have the training or experience to understand and apply earthquake hazards information." The group noted that "few academic curricula introduce urban and regional planning students to, or train them in, avoiding or mitigating natural hazards." This suggests that such action is warranted as an adjunct to the ongoing federally sponsored natural hazards management program. Further justification is provided by a recent examination of the seismic safety policies and practices in three U.S. metropolitan areas (Atkisson and Petak, 1981). How much of the nation's past failure to develop means for curbing losses from natural hazard exposures may be charged to oversights, limited capacities, organizational disabilities, and staffing problems of land use planning, program development, and building regulatory agencies at local and state levels cannot now be judged; but certainly these disabilities may well have played a part in bringing about the current national situation.

Economic Constraints

Another set of factors that will constrain the capacity of national, state, and local bodies to implement loss-reducing natural hazard mitigations is economic in character. Thus, it seems clear from the evidence presented in Chapter 5 that many natural hazard mitigations have the potential to substantially increase both the initial and the annual amortized cost of new housing and other structures. Yet, as a result of other factors within our economy and society, the median cost of new housing starts has been increasing more rapidly than has median family income. Without belaboring this point, it seems undeniable that the nation rapidly is approaching a point where we may be pricing our housing stock beyond the reach of most. To the extent that our concern for natural hazards losses triggers further escalations in the cost of housing and other structures, the existing problem may well be exacerbated.

Moreover, since logic would suggest that both the sale and rental value of existing structures may be influenced by the total supply of such structures, as well as by the cost of new and replacement facilities, increases in the cost of new structures may well exert substantial "ripple" effects throughout the entire economy, thus fueling inflationary pressures already all too apparent within the economy. This highly speculative perspective is offered, not as a definitive judgment concerning the potential impacts of natural hazard mitigations on the U.S. economy, but, rather, as a cautionary note. This analysis did not include sophisticated study of this question, and yet, the question clearly is of importance to the making of public policy in this field.

Federal Constraints on State and Local Decision Making

Through a variety of "carrot and stick" provisions in current federal statutes, the federal government possesses considerable capacity to influence the nature and timing of natural hazard policy making and program implementation by state and local governments.

One set of possible constraints is that contained in the Disaster Relief Act of 1974. This Act delivered to the Executive Branch the principal legislative authorization necessary for the design and partial implementation of a comprehensive, well integrated, balanced, and rational national-level natural hazard management program. The Act defines "major disaster" as "any hurricane, tornado, storm, flood, high-water, wind-driven water, tidal wave, tsunami, earthquake, volcanic eruption, landslide, mudslide, snow storm, drought, fire, explosion, or other catastrophe in any part of the United States . . ." The Act authorized the President [Section 201 (A)] to establish "a program of disaster preparedness," utilizing "services of all appropriate agencies;" such program to include "plans for mitigation, warning, emergency operations, rehabilitation, and recovery" as well as plans for "training . . . annual review of programs . . . coordination of federal, state, and local preparedness programs . . . application of science and technology . . . research." The President was required by the Act to "provide technical assistance to the states in developing comprehensive plans and practicable programs for preparation against disasters, including *hazard reduction, avoidance, and mitigation* . . . [emphasis added]. The Act further required the President to "conduct annual reviews of the activities of federal agencies and state and local governments providing disaster preparedness and assistance, in order to insure maximum coordination and effectiveness of such programs and . . . from time to time report thereon to the Congress" [Section 316].

Section 406 of the Act requires that, as a

> condition of any loan or grant made under the provisions of this Act, the state or local government shall agree that the natural hazards in the areas in which the proceeds of the grants or loans are to be used shall be evaluated and appropriate action shall be taken to mitigate such hazards, including safe land-use and construction practices in accordance with standards prescribed or approved by the President after adequate consultation with the appropriate elected officials of general purpose local governments, and the State shall furnish such evidence of compliance with the Section as may be required by regulation.

Section 201(C) of the Act states that Presidentially-authorized planning grants to the states shall be used for the "development of plans, programs, and capabilities for disaster preparedness and prevention" and requires states to submit, to the President, plans which shall "set forth a comprehensive and detailed state program for preparation against and assistance following emergencies and major disasters." The President is authorized by Section 601 of the Act to "prescribe such rules and regulations as may be necessary and proper to carry out the provisions of this Act," and further authorizes the President to "exercise any power or authority conferred on him by any Section of this Act either directly or through such federal agency or agencies as he may designate."

Thus, the Disaster Relief Act of 1974 provided the Executive Branch with several specific requirements which it was to meet and further authorized and encouraged action by executive agencies to develop plans for mitigating the risks associated with exposure to potentially disastrous natural hazards, including all of those examined in this study with the single exception of expansive soils. Interestingly, the United States Water Resource Council Report on a "Unified National Program for Floodplain Management," which was submitted to the President in response to Section 1302(C) of the National Flood Insurance Act of 1968 (PL 90-448), does not indicate that the proposals in any way relate to the Congressional charge upon the President under the terms of the Disaster Relief Act of 1974. Beyond the general authorities contained in the Disaster Relief Act of 1974 are, of course, the specific mandates and authorizations contained in the National Flood Insurance Act of 1969 (PL 90-448), the Flood Disaster Protection Act of 1973 (PL 93-234), the National Dam Inspection Program (PL 92-367), the Coastal Zone Management Act of 1972 (PL 92-583), and the Earthquake Hazards Reduction Act of 1977 (PL 95-124). Statutory authorizations which underpin the programs of the National Oceanic and Atmospheric Administration, the National Science Foundation, the U.S. Soil Conservation Service, the U.S. Corps of Engineers, and other agencies also buttress the existing legal capacities for hazard-related program planning and development operations within the federal establishment. Also, of course, the specific authorizations contained in the Environmental Education Act and other similar statutes provide a basis for extensive implementation of technology transfer and attention-focusing policies in respect to natural hazards management.

The Coastal Zone Management Act was adopted by Congress to facilitate more effective management of the nation's coastal zones. Section 302(H) of the Act states:

> The key to more effective protection and use of the land and water resources of the coastal zone is to encourage the states to exercise their full authority over the lands and waters of the coastal zone by assisting the states, in cooperation with federal and local governments and other vitally-affected interests, in developing land and water use programs for the coastal zone, including unified policies, criteria, standards, methods, and processes for dealing with land and water use decisions of more than local significance.

The Act also permits states that qualify under its terms to participate more directly in license, permit, and other regulatory activities affecting the coastal zone, which are otherwise assigned by federal law to such agencies as the U.S. Environmental Protection Agency, the U.S. Army Corps of Engineers, the Bureau of Land Management, the U.S. Coast Guard, and the Department of Energy.

Regulations issued pursuant to the Coastal Zone Management Act specify that states seeking to qualify for participation under the Act must develop comprehensive programs of coastal zone management, parts of which are to focus on such phenomena as floods, erosion, land stability, climatology, and meteorology. The regulations further require that states identify coastal zone areas of particular concern, including hazard areas due to storms, slides, floods, and erosion, and require states to identify the hazard potential of such areas insofar as such potential "affects reasonable and safe use of resources."

Under the terms of the National Flood Insurance Act, states and communities are required to develop land use and control measures for the management of their riverine and coastal floodplains. Even with its more recent amendments, that Act contains many provisions intended to stimulate more effective floodplain management by states and local communities. Amendments made in May 1977 removed from the earlier provisions of the Act a prohibition against loans by any federally insured lending institution to property owners in HUD-designated flood-prone areas who were not participants in the flood insurance program, i.e., occupants in a "managed" floodplain. The amendment, nonetheless, required lending institutions to notify loan recipients that they would not be eligible for federal disaster relief and that the federal government could not assist individuals who undertook development in flood-prone areas without adequate flood-proofing measures.

Any sober examination of the current mix of hazard-related federal statutes makes clear that there are numerous federal points of leverage which may be employed to support future action-forcing policies at the federal level in respect to the management of natural hazards. The United States Congress is comparatively free to establish conditions that must be met by state and local governments in order to qualify for the financial and other benefits otherwise authorized by statute. These qualifications and conditions may require states or local units of government to present plans or programs that demonstrate they have initiated and will continue to enforce appropriate regulations over natural hazard exposures. Adoption of appropriate hazard mitigation laws, regulations, and plans by state and local governments can be made a condition for award of federal disaster relief funds, federal flood insurance subsidies, subsidies for hospitals and other health care facilities enumerated in the National Health Planning and Resources Development Act (PL 93-641), and could further be made a condition for award of federally-supported or guaranteed mortgages under the programs administered by such agencies as the Veterans Administration, the Farmers Home Administration, and the Federal Housing Administration. Any comprehensively conceived set of federal action-forcing policies could focus on land use planning and zoning, other forms of hazard zone avoidance, building codes, notification of hazard exposure to buyers at time of transfer of property title, and on other similar risk reducing methods.

Conclusion

The following conclusions seem supported by the discussions presented above:

1. Constitutional Constraints on the Exercise of the Police Power by State Governments Pose No Serious Barriers to the Regulation of Natural Hazard Exposures by These Entities. In law, state governments are vested with the "police power" of government which authorizes them to adopt regulatory and other policies to protect the health, safety, and welfare of the community. So long as such enactments are directed to the accomplishment of a public purpose, observe the constitutional requirements for due process of law, do not deny to any person the equal protection of the laws, and observe the admonition that private property shall not be taken for public use without just compensation, these enactments will generally be found to be valid by the court.

Although the constitutional validity of a regulatory statute depends both on it being "reasonable" and not "unduly burdensome," these constraints are comparatively easy to meet if statutes are appropriately drafted and the facts bearing on the focal situation are clearly established. In respect to factual findings, the courts are reluctant to substitute their judgments for those of the legislature; in respect to legal questions, the courts begin their inquiry with the presumption that the statute is constitutionally valid. On the whole, the burden of establishing the unconstitutionality of a regulatory statute falls on the aggrieved party rather than the enacting legislating body.

2. No Serious Legal Obstacles are Posed to the Design and Implementation, by the Federal Government, of Policies Intended to "Force" State and Local Units of Government to Engage in Natural Hazard Management and Regulatory Programs. The United States Congress is comparatively free to establish qualifications and conditions that must be met by state and local governments in order to qualify for the financial and other benefits otherwise authorized by statute. These qualifications and conditions may require states or local units of government to present plans or programs that demonstrate that they have initiated and will continue to enforce appropriate regulations over natural hazard exposures.

3. Legally, the Federal Government May, Itself, Adopt and Enforce a Limited but Important Set of Natural Hazard Regulatory Policies. Under its constitutional power to regulate the flow of interstate commerce, to engage in treaty making, and to accomplish the purposes of Article III, Section 8 of the U.S. Constitution, the federal government has engaged in a wide range of regulatory activity, including, for example, the establishment and enforcement of air pollution emission standards governing discharges from fixed sources of air pollution. Within reasonable limits, the Federal Government may also engage in direct regulatory activity in the field of natural hazard management. Of course, it may not deal with matters of a purely local concern, and may not engage in activities alien to the constitutional charges and authorizations placed on the federal level of government.

Also, powerful political and other constraints operate to prevent the U.S. Congress from regulating matters that, by tradition, are viewed to be matters of peculiarly local or state concern.

4. Typically, Local Governments Can Engage in Effective Regulatory Activities to Control Natural Hazard Exposures Only When Authorized to Do So by (a) State Constitutional Provisions, (b) Enabling Legislation Enacted by Their State Legislatures, or (c) "Home Rule" Grants of Power From the State Government. Although municipalities are comparatively freer to enact and enforce regulatory policies than are counties, both units of government are subject to the constraints in the above statement. Both units of government may, if authorized by state government, engage in the enactment and enforcement of building codes, land use planning and zoning regulations, housing codes, and subdivision regulations. Typically, however, these regulations are applicable only to land areas within the geographic jurisdiction of the concerned local government. Thus, the applicability of a county's regulations typically would be limited to the unincorporated territory of the county, while the applicability of a municipality's regulations would be limited to lands within its corporate boundary lines. Some states have, however, vested municipalities with the power of "extraterritorial jurisdiction" which permits such cities to enforce their building, housing and land use regulations on unincorporated territory bordering the boundaries of the city.

5. Although the Legal Barriers to the Adoption of Effective Natural Hazard Regulatory Policies are Few and May Be Readily Overcome, Carelessly Drafted, Poorly Supported, and Inappropriately Focused Enactments May Be Found Invalid by the Courts. Although the courts begin their inquiry into the validity of an enactment on the presumption that it meets all constitutional tests, and although the courts are reluctant to substitute their own judgments for the factual findings of a legislative body, the courts will not hesitate to strike down an enactment if it is so carelessly drawn as to suggest that constitutional prohibitions may have been violated. Thus, regulatory statutes may not be used for the "taking of private property without just compensation" and must further meet such constitutional tests as those related to equal protection under the laws, reasonableness, and the prohibition against the imposition of "undue burdens."

6. Natural Hazard Regulatory Policies May Be Directed at a Large Number of Potentially Desired Ends, Including (a) the Strengthening of New and Old Buildings, (b) the Avoidance of Certain Kinds of Uses of Specified Land Areas, (c) the Legally Mandatory Purchase or Offering for Sale of Natural Hazards Insurance, (d) Mandatory Notification of Potential Property Purchasers or Users of the Hazard-Proneness of a Land Parcel or Structure on the Land; and (e) Restrictions on the Conduct of Hazard Increasing Activities by Individuals, Corporations, Business, and Governmental Entities.

7. The Primary Constraints on the Adoption of Effective Natural Hazard Regulatory Policies by Any Level of Government are Those of a Political and Social Nature. The primary impediment to the adoption and enforcement of effective natural hazard regulatory policies has to do with the "willingness" rather than the "capacity" of governmental law-making bodies to act. Typically, some political constituency must be present to advance the need for a proposed regulatory enactment, the enactment must clearly be directed at resolution of a state of affairs generally recognized to constitute a problem to some significant population subset, and the cause must rank comparatively high on the political agenda of the community and its lawmakers. The probability of enactment is lessened if other political constituencies are opposed to the measure. Also, lawmakers, and their constituents, seem to prefer problem-solving strategies involving "distributive" as contrasted to "regulatory" policies.

8. Among Populations At Risk of Exposure to Natural and Other Physical Hazards, Avoidance Motivations are Strongest Among Those Individuals Who Have Not Previously Experienced the Hazard, and the Primary Motivation of Those Who Have Recently Experienced the Hazard is to Support Efforts Aimed at Rebuilding the Damaged Area and Enabling Hazard Zone Occupants to Get on With Life as it Was Before.

9. Hazardous Events That Produce Multiple, Rather Than Single, Deaths and/or That Inflict Property Damage on Many Rather Than Few Persons Tend to Evoke

More Immediate and Vigorous Responses From Policy Makers and Concerned Publics Than do the Cumulatively Larger Losses That May Be Sustained by Individuals One or Two at a Time as the Result of Less Dramatic Hazardous Occurrences.

10. Individuals and Groups Inexperienced With Natural Hazards Tend to Exhibit More Apprehension Concerning Them Than do Those Who Have Experienced Such Occurrences; "Outsiders" Tend to Overestimate the Recovery Needs of Disaster Areas and to Exaggerate the Impact of a Disaster on a Community; and Media Exaggerations of Disaster Impacts on a Community Apparently Increase as a Function of Distance From the Disaster Site.

11. Local Officials Tend to be Unaware of the Degree to Which Their Communities are Threatened by Natural Hazards, Particularly Nonflood Hazards. Local officials tend to resist the imposition of natural hazard management and other policies by federal and state authorities and to support the proposition that local governments should be responsible for natural hazard policy making.

12. The Contents of Building Codes, Zoning, and Other Community Development and Hazard Abatement Regulations at Local Levels Appear to Respond More Sensitively to Economic and Political Pressures Than to Objective Standards of Safety or Health. This conclusion, though not independently framed by the study team, has been advanced by such researchers as Field and Rivkin (1975) and Siegen (1976). This conclusion also is one that may reasonably be viewed as having been advanced by the 1968 National Commission on Urban Problems.

13. There is a Pressing Need for a Contemporary Study to Determine the Capacity of State and Local Units of Government to Undertake the Drafting and Implementation of Effective Natural Hazard Management and Regulatory Policies. Various studies conducted over the past 15 years have suggested that the staffs of local building departments and planning agencies are of such limited size, so poorly paid, and so overburdened with a range of functions as to limit their capacities to engage in fully comprehensive and technically sound regulatory and planning activities. Also, some observers have suggested that the prior education and training of such staff members may be inappropriate to contemporary natural hazard management and environmental quality improvement needs.

14. Additional Study is Needed to Determine the Possible Broad-Scale Economic Effects That Might be Produced by Hazard Mitigation-Induced Escalations in the Costs of New Housing and Buildings. Project constraints prevented this study from examining the possible "ripple" effects that might be induced on the economy as a result of the broad-scale application of costly building strengthening strategies to mitigate the risks associated with natural hazard exposures. However, the subject is important and further study is warranted.

15. State and Local Action to Promote the Avoidance of Development in Natural Hazard Zones, to Reduce the Vulnerability of Persons and Property Within Such Zones, and to Otherwise Mitigate the Risks Associated With Natural Hazard Exposures Have Been Spotty, of Widely Varying Quality, and Inconsistent With the Dimensions of Natural Hazard Exposure Losses.

References

Anderson, J. E. 1975. Public Policy Making. Praeger Publishers, New York.

Atkisson, A. A., and W. J. Petak. 1981. Seismic Safety Policies and Practices in U.S. Metropolitan Areas: A Three City Case Study. Report to the Federal Emergency Management Agency. J. H. Wiggins Company, Redondo Beach, California.

Berman vs. Parker. 1954. 348 U.S. 26, 99 L. ed. 27, 75 Sup. Ct. 98. U.S. Govt. Printing Office, Washington, D.C.

Burton, I. 1962. Types of Agricultural Occupance of Flood Plains in the United States. Dissertation, Research Paper No. 75. Department of Geography, University of Chicago, Chicago, Illinois.

Burton, I. 1965. The shores of megalopolis: Coastal occupance and human adjustment to flood hazard. Climatology 18, No. 3.

Burton, I., and R. W. Kates. 1964. The perception of natural hazards in resource management. Nat. Resour. J. 3:412–441.

Burton, I., R. W. Kates, and G. F. White. 1968. The Human Ecology of Extreme Geophysical Events. Natural Hazards Research Working Paper No. 1. Department of Geography, University of Toronto, Toronto.

Clapp, C. L. 1968. The Congressman as a legislature. *In*: S. C. Patterson (Ed.), American Legislative Behavior: A Reader. Van Nostrand-Reinhold, Princeton, New Jersey, pp. 98–129.

Cobb, R. W., and C. D. Elder. 1972. Participation in American Politics: The Dynamics of Agenda Building. Allyn & Bacon, Boston, Massachusetts.

Davies, J. 1975. Legislative Law and Process in a Nutshell. West Publishing Company, St. Paul, Minnesota.

Dexter, L. A. 1969. How Organizations are Represented in Washington. Bobbs-Merrill, Indianapolis, Indiana.

Field, C. G. and S. R. Rivkin. 1975. The Building Code Burden. D. C. Heath, Lexington, Massachusetts.

Fischoff, B., P. Slovic, S. Lichtenstein, B. Combs, and S. Read. 1976. How Safe is Safe Enough? A Psychometric Study of Attitudes Toward Technological Risks and Benefits. Decision-Research Report 79-9. Decision Research, A Branch of Perceptronics, Eugene, Oregon.

Fritz, C. E. 1961. Disaster. *In*: R. K. Merton and R. Nisbet (Eds.), Contemporary Social Problems. Harcourt, New York, pp. 651–694.

Golant, S., and I. Burton. 1970. A semantic differential experiment in the interpretation and grouping of environmental hazards. Geogr. Anal. 2:120–134.

Greenwald, C. S. 1977. Group Power: Lobbying and Public Policy. Praeger Publishers, New York.

Halstead, M. 1900. Galveston: The Horrors of a Stricken City. American Publishers' Association, Chicago, Illinois.

Justesen's Food Stores vs. City of Tulare. 1941. 43 Cal. App. 2d 616, 621, III P. 2 d 424. U.S. Govt. Printing Office, Washington, D.C.

Kates, R. W. 1971. Natural hazards in human ecologic perspective: Hypotheses and models. Econ. Geogr. 47:438–451.

Keefe, W. J., and M. S. Ogul. 1977. The American Legislative Process. 4th edition. Prentice-Hall, Englewood Cliffs, New Jersey.

Mandelker, D. R. 1963. Managing Our Urban Environment: Cases, Text and Problems. Bobbs-Merrill, New York.

Mann, D. E. 1975. Political incentives in U.S. water policy: Relationships between distributive and regulatory politics. *In*: M. Holden, Jr., and D. L. Dresang (Eds.), What Government Does. Sage Publications, Beverly Hills, California, pp. 94–123.

Mayhew, D. R. 1974. Congress: The Electoral Connection. Yale Univ. Press, New Haven, Connecticut.

Mileti, D. S., T. E. Drabek, and E. J. Haas. 1975. Human Systems in Extreme Environments: A Sociological Perspective. Institute of Behavioral Science, University of Colorado, Boulder.

Municipal Manpower Commission. 1962. Governmental Manpower for Tomorrow's Cities: A Report of the Municipal Manpower Commission. McGraw-Hill, New York.

National Commission on Urban Problems. 1968. Local Land and Building Regulation. Research Report No. 6. Washington, D.C.

Northwestern Laundry vs. Des Moines. 1916. 239 U.S. 486, 491–492. U.S. Govt. Printing Office, Washington, D.C.

Olson, M., Jr. 1971. The Logic of Collective Action: Public Goods and the Theory of Groups. Schocken Books, New York.

Pollack, L. W. 1968. Legal boundaries of air pollution control—state and legal legislative purposes and techniques. Law Contemp. Probl. November, No. 33.

Quarantelli, E. L., and R. R. Dynes. 1973. Images of Disaster Behavior: Myths and Consequences. Disaster Research Center, Ohio State University, Columbus.

Rossi, P. H., J. Wright, and E. Webber-Burden, 1982. Natural Hazards and Public Choice. Academic Press, New York.

Russell, C. S., D. G. Arey, and R. W. Kates. 1970. Drought and Water Supply: Implications of the Massachusetts Experience for Municipal Planning. Johns Hopkins Press, Baltimore, Maryland.

Saarinen, T. F. 1966. Perception of Drought Hazard on the Great Plains. Research Paper No. 106. Department of Geography, University of Chicago, Chicago, Illinois.

Schact vs. City of New York. 1963. 40 N.Y., Misc, 2d 303, 243 N.Y.S. 2d 272, 279.

Schattscheneider, E. E. 1960. The Semisovereign People. Holt, New York.

Siegen, B. H. 1976. Other People's Property. D. C. Heath, Lexington, Massachusetts.

Slosson, J. E. 1975. Legislation related to earthquake. Calif. Geol. February, p. 37.

Solomon, K. A., and D. Okrent. 1977. Seismic Building Codes for the City of Los Angeles, California. Brief Case Study, Rand Report No. P-6018. Rand Corporation, Santa Monica, California.

Vogler, D. J. 1977. The Politics of Congress. 2nd edition. Allyn & Bacon, Boston, Massachusetts.

White, G. F. 1964. Choice of Adjustment to Floods. Research Paper No. 93. Department of Geography, University of Chicago, Chicago, Illinois.

Working Group on Earthquake Hazards Reduction. 1978. Earthquake Hazards Reduction: Issues for An Implementation Plan. Office of Science and Technology Policy, Executive Office of the President, Washington, D.C.

Wright, J., P. H. Rossi, S. R. Wright, and E. Weber-Burdin. 1979. After the Clean-Up, Long-Range Effects of Natural Disasters. Sage Publications, Beverly Hills, California.

10
Public Policy Alternatives

The Policy Options

In theory, public policy makers at all levels of government currently face at least three major policy options in the management of natural hazards: (1) adopt no new legislative policies or major administrative policies but continue to implement present federal, state, and local policies; (2) adopt no new public regulatory policies and make no major changes in public distributive policies, but concentrate instead on improving the implementation of existing policies and programs through a focus on natural hazard-related system management, system optimization, attention focusing, technology development, and technology transfer policies, programs, and operations; (3) initiate proposals for major changes in current public regulatory and distributive policies in the field of natural hazards management.

Option 1: "Do Nothing." Continue Current Practices and Policies as They Are. Any effort to modify existing federal, state, or local policies in the field of natural hazards management must inevitably be based on the answers to two simple questions: Why? Why, now?

Based on the conclusions presented in the earlier chapters and the lists of candidate problems presented in Chapter 8, the answers to these questions are not difficult to find. Moreover, a recent report by the Comptroller General of the United States (1980) has found that current federal disaster relief policies are characterized by much potential inequity and inefficiency. The report observed that:

> We are able to present guiding principles for the formulation of such a policy. Comparing the current design and operation of disaster assistance policy with such principles is important in designing alternatives that will deliver assistance to disaster victims more equitably and efficiently . . . Equity is the consistency of assistance awarded among victims of the same disasters, and over time. Efficiency is the influence upon a person's decision to live or do business in a particular place . . . i.e., to discourage location or production in hazard-prone areas, or, on the other hand, encourage mitigation to avoid hazards . . . achieve better control over budgeting for natural disasters; and improve the cost-effectiveness of programs.

The Comptroller General further noted that:

> The extent of the government's role in providing relief from natural disasters is defined by the proportion of financial risk it is willing to bear on behalf of society. *What* proportion is appropriate, however, is unclear. The level of risk-sharing borne by the federal government on behalf of society is a political decision based on value judgments. We take no position on what level is appropriate.
>
> We believe that the larger the share of total disaster losses paid for by taxpayers, the less efficient will be the individual decisions in locating housing, business, and agricultural activity. On the other hand, the smaller the role paid by taxpayers, the more difficult it will be for disaster victims to recoup the losses or avoid financial ruin.

On the other hand, given other high priority matters which now clog the policy agendas of legislative bodies at all levels of government, some policy makers may well deny that any of the situations described in the earlier chapters provide compelling evidence of the need for action, now.

Since the end of World War II the federal government has dramatically extended and modified prior disaster relief policies. An impressive set of services to victims of disasters is now in place; research efforts to extend the base of knowledge concerning natural hazard mitigations and effects are being conducted by a number of federal entities and contractors. A variety of technical assistance services are being extended to state and local governments.

If considerable inequity is associated with federal efforts to aid the victims of natural disasters, it is nonetheless true that all victims are aided in more than a trivial fashion, whether through income tax relief or other more direct forms of assistance. If the interstate tax transfers occasioned by current federal policies are large and growing, they are nonetheless smaller than other income and wealth redistribution policies being implemented by federal authorities, including those which effect transfers between the young and the old, the rich and the poor, the defense industry states and the non-defense industry states. If the costs of a "probable maximum loss" hazardous natural event are as large as now estimated ($60 to $90 billion), and if the national economic and other impacts created by such an event would strain resources, it is nonetheless possible that such an event might prove to be of no greater impact (in 1980 constant dollars) than a single peak year of expenditures during the Vietnam War.

Why, then, consider any change at all in disaster relief and hazard mitigation policies? Why not leave things pretty much as they are? This policy alternative is one which may be appealing to many, at least for the moment. Other national policy matters, it might be argued, are of greater importance and more worthy of immediate policy attention.

Any rational decision process which leads to this policy outcome would have to be based on one or more of the following findings in respect to the list of candidate public problems which were inventoried in Chapter 8 and to the other factual circumstances and situations described elsewhere in earlier chapters.

1. The factual descriptions implicit in the problem statements are incorrect.
2. The factual components of the problem statements are correct, but the described situations are not inconsistent with prevailing policy maker and social values, are tolerable, or are less important than other possible targets of public policy.
3. The "candidate" descriptions of public problems are correct, as measured both by factual and value criteria, but this is not the time to inaugurate problem solutions. Problem-solving solutions should be implemented at a later date when other more pressing or important items on the governmental policy agendas have been disposed of.
4. The candidate descriptions of problems are correct and consistent with appropriate value criteria, but they represent matters which are more appropriately within the domain of state and local governmental entities, or private sector enterprises, than that of the federal government.

Option 2: Initiate "Dramatic, New" Initiatives and Changes in Current Policy. If one rejects the wisdom of "doing nothing" to change existing natural hazards management policies, one then must dispose of the question as to whether or not comprehensive new initiatives should be taken by governments at all levels to cope with the full set of situations and candidate problems described earlier.

However compelling one may view the need for such comprehensive activity, current political realities might well dampen one's enthusiasm for such efforts. The political and social climate of the nation appears to be such that dramatic new federally initiated approaches to natural hazards management problems would not receive appropriate support by broadly based political constituencies or by important policy makers at any level of government. Convincing evidence suggests that community-level opinion leaders and policy makers do not currently assign a very high priority to natural hazards management activities and that their understanding of the dimensions of natural hazards threats to their communities is far too limited. When this fact is considered together with the economy-in-government campaigns that are sweeping the nation, with the fact that a sufficiently precise and broad-based pool of data is not yet available to assist rational and comprehensive natural hazards decision making at all levels of government, and with the current wave of professional and public concern over the burdens produced by governmental regulatory programs, the overall infeasibility of this policy option seems clear. However, this conclusion is not meant to imply that it would be imprudent at present to establish the general outlines of possible future major changes in natural hazards management policies and programs. Indeed, activities in the near future might well be directed to the development of such outlines, and to the establishment of a base of policy-maker understanding and support necessary for the adoption and implementation of any substantial new proposals.

Option 3: Concentrate Current Activities on "Fine-tuning" the Current System. Given the above observations, it may appear to some that the most feasible option for the present is to concentrate on design and implementation of changes in system management, system optimization, attention focusing, technology development, and technology transfer policies and operations to the end that a more effective, efficient, equitable, publicly supported, and internally consistent approach to natural hazard problems may be taken by all levels of government at some future time.

The Policy Plan

As the term is used here, a policy plan is a set of statements describing the goals to be achieved through implementation of one or more specific policy proposals. Whether Option 2 or 3 is selected, a policy plan is required in order to define both the central and subsidiary purposes to be served by changes in existing policies. The definition of the goals and purposes to be specified in the policy plan will require a series of choices between competing future states of affairs and between the preferences of competing stakeholders. The goal statements will indicate which stakeholder group interests are to be given preference over others, which candidate public problems are to be accepted as being both correct factual statements and as inconsistent with the public interest, and how the policy maker has resolved the central issues of value and fact. In the course of designing the policy plan, potential conflicts between alternative goals will become evident, but the multiple goal-accomplishing attributes expected of specific action possibilities also may be clarified. In any event, it is on the basis of the desired future states of affairs which are specified in the policy plan that criteria may be developed to assist policy makers in making rational choices between the numerous decision alternatives they will face during any effort to change existing natural hazard management policies. In this spirit, the following candidate goal statements are those which seem to represent the principal decision options for policy makers at federal and lower levels of government. In this document they represent, not recommendations from one party to another, but the major choices which our analysis suggests are to be confronted, and disposed of, by policy makers and their constituents.

Readers will observe that an array of candidate goal statements has been presented which is sufficient, in itself, to prompt the initiation of those activities necessary to the solution of each of the intrinsic and instrumental candidate problems identified in Chapter 8. Thus, it is from the identification of candidate problems that one is able to compose candidate goal statements. Without a referent candidate problem, there can be no candidate goal statement. The latter are justified by the former, and the former are justified on the basis of guiding value propositions related to specific factual situations believed to be occurring in the present, or believed to be descriptive of probable future states of affairs.

It is the candidate goal statements which then guide the development of specific *action possibilities*, which are the final alternatives from which policy makers must choose. *Action possibilities*, then, represent the specific policy-setting and implementing activities which are deemed to be necessary to the accomplishment of specified goals. Their design is influenced both by the nature of the defined goals and by an understanding of the possible targets and methods available for intervention in the causal sequences which were earlier referred to as *problem-causing systems*.

Candidate Goal Statements

Goal 1: Future Involuntary Exposures to Natural Hazards Should Be Reduced to an Effective Level of Zero. One of the oldest and least controversial function of any government is to protect its citizens from threats to their health, longevity, safety, and general welfare that are of such a nature that the individual citizen can only imperfectly take action to protect himself or herself. Situations involving involuntary natural hazard risk-taking, borne out of the ignorance of the risk taker or the inability of the risk taker to adopt a risk-avoidance strategy, or out of the deliberate withholding of hazard exposure and risk-taking information from the parties who are or might be exposed to the hazard, or out of the organizational and technical disabilities of specific community governments, clearly qualify as targets for governmental activity under this traditional role and responsibility of federal and state governments.

Goal 2: Future Possible Increases in the Risk of Life Loss From Natural Hazard Exposures Should Be Avoided.

Goal 3: Current Levels in the Risk of Life Loss From Natural Hazards Should Be Reduced. Although goal statements 2 and 3 are targeted for reduction in life loss, the first accepts the view that current levels of risk may be tolerated but that future increases in those levels should be avoided. The second of these two statements rejects the view that current levels of risk of life loss are acceptable and suggests that these levels also should be reduced.

The two goal statements probably describe situations in which there will be quite different mixes in the levels of costs and benefits associated with relevant goal-accomplishing action policies and strategies. If current levels of risk are to be reduced, then attention will have to be directed at retrofitting existing buildings and other structures, and at the identification, modification, or retirement from service of those hazardous buildings and other community structures (such as dams and bridges) which increase levels of natural hazard risk beyond those which are technologically avoidable. As indicated in earlier chapters, however, structural retrofit policies are far more expensive to implement than strategies targeted on completely new structures. Similarly, except for any possible local distributional costs, hazard avoidance strategies involve less cost than reducing the vulnerability of new buildings and structures to selected hazards.

Acceptance of either goal statement would further imply that policy makers have placed a higher value on avoidance of deaths resulting from natural hazard exposures than on reductions in death rate levels induced by other causes. Yet, suggestive evidence provided earlier indicates that, if the national budget for reduction of premature death is limited, then more cost-effective death avoidance policies may be identified than those associated with the management of natural hazards.

Consideration of these two goal statements therefore will require policy makers to address a number of complex questions and to consider the system-scale tradeoffs associated with the several possible choices which they must confront.

Goal 4: Substantial Reduction Should Be Made in the Level of the Projected Increases in Annual Expected Losses That Are Anticipated to Arise From Natural Hazard Exposures in the Year 2000. The projected 1970 to 2000 increase in annual expected natural hazard losses is in excess of the level actually tallied in 1970. However important the current level of loss may be, wisdom suggests the importance of a new and higher level of concern for the projected increases in the sizes of the population at risk of such exposure and the level of vulnerability of their exposed structures.

Goal 5: Expanded Opportunities Should Be Provided to Populations Currently at Risk of Exposure to Hazardous Natural Events to Reduce the Current Level of Risk Associated with Such Exposures; but Knowledgeably, Economically, and in Full Consideration of Relevant Costs and Benefits. Reductions in current, in contrast to projected future, levels of losses from natural hazard exposures are the targets of policies that deal with the decrease of danger in existing hazardous buildings, the operation of disaster warning and prediction systems, and the construction of flood control projects. Hazardous structure abatement programs may involve varying mixes of structure strengthening, modification, use, or abandonment/retirement measures.

This statement contemplates that the general goal of reducing existing risk levels may be of appeal to many policy makers and their constituents, but further assumes that parties desiring to implement such activities will wish to be further guided by an appropriate balancing of the costs and benefits entailed in such efforts. Numerous tradeoffs are necessary in the process of decision-making targeted on reduction of the risks incurred by current populations exposed to the conditions within major natural hazard zones. Rarely is there any single "right" level of risk reduction or any single "most feasible" strategy to be applied to achieve that risk level. Some parties whose interests are involved in particular risk-taking situations may wish to pursue natural hazard mitigations which involve a higher annual level of expected cost than an annual level of expected benefit. Such a judgment cannot, on the face of it, be viewed as illogical since that is precisely the situation that obtains whenever a party purchases property insurance. The annual premium payment for any insurance policy must always exceed the annual expected economic benefits to be derived from the policy; otherwise, sales of such policies could not possibly be made by any commercial enterprise. What prompts individuals to purchase property insurance is the higher practical worth they perceive in avoiding incurring a large loss in any given year, as contrasted to the much smaller annual cost associated with insuring against such an event. Similar considerations may prompt residents in one natural hazard zone to favor the community use of mitigations which produce somewhat higher annual expected costs than annual expected benefits, while the values of residents in still another zone may lead them to make a completely different type of decision.

In either case, however, rational decison making requires that adequate information be made available to facilitate the final measurement of benefits and costs and the final selection of the rule or principle that will govern the ultimate decision. When non-risk-taking parties external to the hazard zone are involved in the situation, as is the case under current federal disaster relief and tax policies, then an appropriate weighting of their interests also becomes important and the kinds of decision-supporting information that is required may need to be somewhat expanded.

Goal 6: The Federal Government Should Expand the Capacity and Readiness of Local, State, and Federal Governmental Entities to Curb the Present and Future Risks Associated with Natural Hazard Exposures.

Goal 7: Government Should Expand the Capacity and Readiness of Individual Citizens, Groups of Citizens, and Nongovernmental Organizations and Associations to Knowledgeably Participate in Decision-Making Activities Targeted on Determination of the Levels of Risks and Types of Mitigation Appropriate to Deal with Natural Hazard Exposures. Current evidence suggests that few entities at local, state, or federal levels now possess the range of personnel, legal and technical knowledge, and jurisdictional and financial capacity to engage in the broad range of functions necessary to support any fully rational natural hazards management program of policy setting and policy implementation.

At the federal level the creation of the Federal Emergency Management Agency has not resolved the prior problems caused by the segmentation of natural hazards research and management responsibilities among the several departments and agencies of the federal government. Disaster relief policies implemented by the Small Business Administration are still made in isolation from those associated with the implementation of FEMA's own direct disaster relief obligations under the terms of the Disaster Relief Act of 1974; policies toward the funding of hazard-related research are still influenced more by the authorizations provided by specific statutes (such as the Earthquake Hazards Reduction Act) and by the interests of specific departments and agencies than by any comprehensive perspective concerning the information needs associated with a fully integrated federal hazards management program. A coherent, comprehensive, and internally consistent policy-shaping and program-implementing perspective has yet to emerge as the result of interactions between the several committees of Congress, the Office of Management and Budget, the Federal Emergency Management Agency, the National Science Foundation, and other agencies. Although matrix forms for organizing federal responsibilities are certainly well known to federal administrative and policy staffs, nothing resembling a formal matrix management system has yet been invoked in respect to the development and implementation of federal natural hazards management policies.

Without such a scheme, it is difficult to understand how responsibilities can be fixed for internal leadership in respect to the risks associated with specific hazards (such as earthquakes, flood, hurricanes, etc.), much less in respect to the several distinguishable types of operations associated with the coherent management of such hazards (research, disaster planning, mitigation planning, disaster relief, postdisaster economic recovery, etc.). In the few cases where responsibilities seem to have been assigned, and where various critical actors within the system recognize the reality of such assignments, it is not at all clear that appropriate financial, technical, and staffing support has been provided to those who bear the responsibility, nor is it clear that the critical actors within the cells of the emerging matrix really possess the power to bring about change and to engage in the kind of intraagency and interagency activities that are necessary to the shaping of a more coherent set of federal policies.

At state and local levels, the problem is even more acute. Within major metropolitan areas much of the growth which is occurring is taking place beyond the boundaries of those municipalities which are best equipped with staffs able to engage in the kind of land use, new building control, and hazardous structures abatement activities which are essential to effective mitigation of natural hazards losses. Moreover, the jurisdictional balkanization of U.S. metropolitan areas is matched by the organizational segmentations which characterize individual jurisdiction within such areas. This segmentation is signaled by the separation of land use planning, subdivision control, hazardous structure abatement, and new building control functions. In few local or state units of government is anyone in particular responsible for natural hazards management. Instead, many agencies are responsible for specific elements of the whole.

Even if these capacity problems are solved at federal, state, and local levels, it is not at all clear that many of the major actors within the current system are now ready to act to deal more effectively with the management of natural hazards. Public policy agendas at all levels are crowded with other issues, the economizing pressures which have been exerted at all levels and jursidictions of government have lowered staffing levels within existing operating agencies and made it difficult to keep up with routine work, much less to meet new challenges. Public understanding of natural hazards risks and public support for measures to cope with those risks are at comparatively low levels in contrast to other issues.

Within the bounds of this situation, some might argue that the federal government could well play the role of a change agent and system designer. It could well initiate activities intended to expand the capacity of the existing interjurisdictional system to better cope with these subjects and to undertake the molding of a climate of public opinion in which a variety of problem-solving actions would be more understandable and acceptable to the public.

Goal 8: Distribution of Disaster Relief Benefits by the Federal Government Should Better Conform to the Criteria of Equity. Numerous critics have charged that federal disaster relief policies do not equitably distribute benefits and do not otherwise equitably meet the needs of parties who experience "disastrous" or "catastrophic" losses. A large fraction of the benefits available to disaster-impacted parties from the federal government is limited to individuals, businesses, organizations, and governmental jurisdictions whose losses are sustained within areas which have been described by federal authorities as "disaster" or "emergency" areas. Thus, a Houston, Texas homeowner who experiences a personally catastrophic loss as a result of expansive soils damage to his or her home is not eligible for some forms of federal disaster relief (not including the casualty loss provisions in the Internal Revenue Code), while a homeowner who experiences "minor" cosmetic damage to a home in an earthquake or hurricane disaster area may receive such benefits.

Moreover, criticism has also been voiced concerning the distribution of federal benefits within designated disaster areas. One of the most recent of these is contained in the report of the Comptroller General of the United States (1980), which approvingly cites the results of a study by Rettger and Boisvert (1979). Noting the more favorable impact of tax reductions, loans, and insurance on the losses sustained by higher income groups, the Comptroller General commented that:

> The most striking characteristic of these data is the effective burden of cost-sharing imposed on the less well-to-do under the various programs. Clearly, the proportion of losses borne by the federal government is directly related to a victim's income. This relationship violates one concept of equity—the degree of assistance should not vary across income classes. . . . Loan programs, whether they are used to cover property or income losses have, in practice, involved value judgments which have changed over time. Thus, similarly affected individuals are treated differently at different times. One result is that the increased generosity of SBA and FmHA loan programs usually associated with widespread disasters is not effective in adequately covering individual property losses that are the most severe, e.g., losses from an

isolated tornado. Furthermore, inherent in the loan device is the impact of the tax code on non-insured losses. This causes the effective burden of the rich to be less than that of the poor, which seemingly contradicts broad federal income redistribution goals and provides greater incentives for poor locational decisions by both rich and poor than does insurance.

Under the current mix of policies, it seems clear that: (1) victims of one disaster may be treated differently than victims of another disaster; (2) proportionately higher reimbursement of natural hazard-induced losses is provided to the affluent than to the poor; and (3) preferential treatment is provided to individuals experiencing catastrophic losses in "disaster areas" as contrasted to those experiencing such losses in other areas and at other times.

Independent of all other candidate goals, this statement is intended to guide policy-changing activities so as to produce more equitable treatment, under federal disaster relief and tax policies, of those parties who experience losses as a result of natural hazard exposures.

Goal 9: Reduce the Current Level of Federal Liability for Natural Hazard Losses and Assure That Federal Disaster Relief Policies Provide Only for Assistance to Those Individuals, Groups, Business Organizations, Nonprofit Entities, and Governmental Jurisdictions Which Have Experienced Catastrophic Losses as a Result of a Natural Hazardous Occurrence. As a result of federal tax, disaster relief, and disaster loan policies, the federal government currently bears a liability for natural hazard-induced catastrophic losses which is approximately equal to the toal value of all *building damage losses* (see Chapter 5). Under current statutes and administrative regulations, for example, the federal government reimburses governmental entities and selected nonprofit organizations in presidentially declared disaster areas for 75% of the losses which they sustain as a result of damage to their buildings, building contents, and other structures. (federal reimbursement was reduced from 100 to 75% in 1981 in an attempt to motivate states to implement more effective hazard mitigation policies.) Additional substantial federal funds are expended in disaster areas for debris clearance, for emergency restoration of services, for emergency medical care and housing, and for direct loans and grants to private parties who have sustained losses.

Although one of the oldest and highest purposes of government is to protect individuals, organizations, and institutions from events or consequences which threaten their viability, the future quality of their lives or organizational activities, or which might produce an unusual measure of pain and suffering, it is not at all clear that current federal policies are limited to the accomplishment of these objectives.

For example, the San Fernando Earthquake of 1971 resulted in a Presidential declaration naming Los Angeles County as a disaster area. Although the earthquake produced the second highest dollar value of losses of any earthquake which has struck the United States in the last century, caused damage to approximately 48,000 dwelling units, and caused a number of deaths, the economic value of the earthquake-induced losses to structures was equal to only 2% of annual personal family income within the County, and to substantially less than 1% of personal annual family income within the State. In terms of the ratio between dollar losses sustained and the size of its annual operating budget, only one of the local jurisdictions within Los Angeles County was heavily damaged by the earthquake—the City of San Fernando, which lost virtually its entire water distribution system. Although large dollar losses were sustained by the County of Los Angeles (through the destruction of Olive View Hospital and the damages to Juvenile Hall) and the City of Los Angeles (resulting from the damage to Van Norman Dam and other municipally-owned structures), in both jurisdictions the sum of these damages was a comparatively small fraction of their annual operating budgets, when appropriately amortized.

Was the earthquake a disaster? As defined by Webster, the term refers to " . . . a sudden calamitous event bringing great damage, loss, or destruction . . . a sudden or great misfortune . . . an event or situation that is regarded as a terrible misfortune." As defined by federal disaster relief statutes, regulations, and policies, it was! That the quake produced "painful" losses to local governments and to the community-at-large is clear. But whether it approached or exceeded the capacity of local governments alone to deal with the damages, the subsequent recovery operations, and the relief of the more severe losses sustained by individual families and businesses is open to question. Whether, for the many, it constituted a "terrible misfortune" may also be questioned.

In this and other comparable disaster areas in recent years, much federal relief (in the form of direct cash payments, low interest or forgivable loans, and other benefits) has been distributed to private parties whose losses totaled a minor fraction of the value of their net assets or their annual income. Existing policies might therefore fairly be described as being more oriented to the minimization of direct losses to exposed parties within disaster areas than to the relief of such parties from catastrophic loss burdens.

Consistent with this perspective, this goal statement is grounded on the assumption that the function of disaster relief is not to protect individuals and groups from all, or even the major portion, of the losses they experience as a result of hazardous occurrences, but only those losses which are *catastrophic.*

Goal 10: Current and Future Levels of Natural Hazard Costs to Non-Risk-Taking Taxpayers Should Be Substantially Reduced. Public policies may, and do, redistribute the economic burdens of natural hazard losses to parties who were not exposed to the hazard. Such redistribution of costs may occur between the members of a local community, between a local community and other areas within its parent state, and between the loss-experiencing community or state and other states of the union.

The first two types of loss redistribution are, to a considerable extent, controllable by both the donor and the beneficiary parties. State governments are vested with police power, and local governments obtain their governing authority from state constitutions or state legislative bodies. A political "community of interest" exists within the state, and donors are able, if they choose, to enforce their will upon the beneficiaries. Within injured communities, the same situation exists. If profligate risk-taking is engaged in by some citizens, perhaps, for example, as a result of the activities of land speculators, subdividers and builders, these acts are within the direct political and legal power of the community to control.

A somewhat different situation exists when sharing of natural hazard loss burdens occurs across state lines. Although the federal government, empowered by the United States Congress, clearly has the power to dangle the carrot in one hand and wield the stick in the other, it has–generally–not chosen to do so with respect to the implementation of disaster relief policies. Thus, the benefits which the federal government distributes to those who have experienced natural hazard losses are not conditioned upon appropriate modification in the risk-taking behavior of the beneficiary parties.

It is true that the Disaster Relief Act of 1974 contains a provision (Section 406) which vests the administrators of the Disaster Relief Program with the power to require that recipients of program benefits evaluate and mitigate the hazards within the area in which the federal funds are to be utilized. However, Section 406 lacks a legislative history, and much doubt has been expressed by the responsible administrators of the program concerning the real intent behind the Section's phraseology. Although Section 406 has been on the statute books for approximately 7 years, no effort was made to implement this Section until approximately two years ago. Even now, there is considerable concern as to the extent to which FEMA may utilize Section 406 to require state and local government recipients of Disaster Relief funds to engage in appropriate natural hazard evaluation and mitigation activities. Because of the lack of a legislative history, there is more than reasonable doubt as to what Congress intended when it included the Section within the Act, and concerning the extent to which Congress would support zealous enforcement of its plain words. In any event, it is clear that the provisions of Section 406 may not be invoked prior to the occurrence of a Presidentially-declared disaster (thus giving each community at least "one disaster on the house") and that its provisions do not apply as conditions for the award of loans by the Small Business Administration and other federal entities. As a result of this situation, federal tax and disaster relief policies provide a means for distributing, across state lines, the cost burdens associated with natural hazard losses, but do not clearly provide federal entities with a means for assuring that profligate risk-taking behavior does not occur among the beneficiaries of those services.

Similarly, under the fiscal pressures produced by Proposition 13, which limited the property taxing powers of local governments in the State of California, some were surprised to learn the extent to which the general body of taxpayers in Los Angeles County had been supporting water runoff and flood debris control projects serving isolated subdivisions in mountainous and other high hazard areas. Before these fiscal pressures were produced, community flood control costs were usually distributed in the form of general property tax rates to all property owners within the County. However, as a result of the enactment of the tax-limiting measure, the costs of local flood control projects are now being distributed directly to the property owner beneficiary of the project. This adjustment of internal costs to internal benefits in local flood and debris control projects seems to be producing a much clearer sense of the aggregate efficacy of developing alternative sites within the community. Retrospectively, it is making clearer the extent to which prior policies had assigned the costs of some public investment programs to parties excluded from all or most of the benefits resulting from those policies.

User fees, insurance systems, and other mechanisms provide a variety of opportunities for policy makers to internally connect the costs and benefits associated with natural hazard risk taking behavior by investors and other parties, and thereby provide incentives for minimizing such risks.

Goal 11: Governmental Entities and Other Organizations Should Be Protected From the Adverse Impacts Associated with the Large Single-Year Expenditure Demands Produced by Natural Hazard Induced Probable Maximum Loss Events. As suggested by the single event scenarios presented in Chapter 7, hurricanes, earthquakes, and other natural hazards have the capacity to produce extraordinarily high losses during the occurrence of single events. Elsewhere (Atkisson and Petak, 1981a) we have estimated that a high magnitude earthquake in California could produce single event losses having an economic value in the range of 60-90 billion dollars and that the federal government would bear financial liability for approximately one-half of these losses. Prime insurance carriers and reinsurers would experience losses which might well exceed the value of their accumulated reserves and such financial institutions as savings and loan associations might experience institution-threatening losses of a significant fraction of their assets.

The sudden large payout demands produced by rare catastrophic occurrences of hazardous natural events engender problems which are obscured by the annual expected loss estimates presented in earlier chapters. Under circumstances where governments and private entities have failed to accumulate reserves sufficiently large to meet all or a significant part of these probable maximum loss event requirements, massive borrowing, deficit spending, and strains on capital markets may be produced. Interest payments on borrowed funds may substantially escalate the real annual expected loss burdens associated with the hazard.

Priorities and Tradeoffs

This array of goal statements outlines the boundaries of the intellectual and political turf on which natural hazards policy issues will be contested in the future. Some policy makers and their constituents may choose to embrace only a single candidate goal, while others may support more comprehensively-conceived changes in existing policies and approaches to the set of problems associated with current and probable future natural hazard exposures. Whatever the choice of goals, however, contestants will find it necessary to relate their goal preferences to an appropriate referent set of candidate problems, to justify those problems and problem-solving purposes on the basis of a relevant body of fact, and to orient their problem-solving activities to the interests and values of a relevant set of stakeholder groups. Different parties may assign different priorities and utilities to the several goals within the set. The extent to which candidate goals are to be accomplished, and by what future date, at what cost, and through what methods, will further complicate the deliberations and the debate.

For each goal to which an advocate is committed, appropriate goal implementing action possibilities must be identified. Finally, of course, appropriate choices must be made between competing action possibilities on the basis of appropriate decision-aiding criteria.

Action Possibilities for the Federal Government

Efforts to accomplish all or a part of the candidate goals discussed above could well include a variety of specific actions by one or more of the several elements of the federal government. Based on our analysis, the following set of action possibilities has been identified. As in the case of the candidate goal statements, the action possibilities should be viewed not as recommendations, but rather as a specification of the alternative actions that would be appropriate for accomplishing the goals included in the proposed policy plan. Each, of course, may well involve its own unique set of costs and benefits and its own level of political/technological feasibility.

Action Possibility 1. Under the Terms of the Disaster Relief Act of 1974 (P.L. 93-288, Section 201, 302, 406, and 601) Appropriate Executive Orders and/or Administrative Regulations Should be Issued to Require Initiation of the Following Activities by Federal Agencies.

a. An integrated, centrally coordinated, appropriately budgeted, and appropriately phased program of hazards zone mapping and classification should be designed and implemented. The program should be undertaken as a joint venture of the several federal agencies responsible for this function under the terms of the Disaster Relief Act of 1974, the Earthquake Hazards Reduction Act, and the various statutes concerned with flood insurance and with the mitigation of flood hazards in coastal and riverine flood plains. Rationally conceived hazard management activities are dependent on precise understandings of the metes and bounds within which specified hazard occurrences of particular intensities and frequencies may be expected. Virtually all who have examined natural hazard exposures have recommended an expanded, more timely, more comprehensive program of hazard zone mapping on the part of the federal government. Such activities should go substantially beyond the mere mapping of the turf on which a particular hazard, such as damage-producing expansive soils, may be expected to occur. In addition, such activities should include a scheme for classifying hazard zones by type, frequency, and expected intensity or magnitude of hazard occurrences. Thus, in respect to such a well-studied phenomenon as riverine or coastal flooding, this recommendation is meant to suggest that riverine and coastal floodplains should be precisely identified as to their metes and bounds; classified as to the frequency with which flooding is expected in the various sections of the plains (Flood Zones A, B, C, D, E, F); and that each such section should further be classified and coded by height or damage-producing magnitude of the floods expected therein during any flood of a specified frequency. Thus, such a mapping and classification system would require use of a standard and easily understandable coding system that would be consistent across the range of types of hazard zones.

This recommendation is further intended to suggest that: (1) a priority set of areas should be identified for federal designation as hazard zones; (2) such areas should include zones in which both natural and man-made hazards are expected to occur; (3) the classification and coding system for such zones should conform with technical criteria to be developed through consultation with experts from research, mapping, planning, engineering, and user enterprises and fields; and (4) the classification system should extend, where appropriate, to such subjects as the soil characteristics of subsections of each hazard zone. Some hazard zones would encompass the whole of existing counties or major sections thereof (as in the case of major wind hazards) while other zones would encompass microareas internal to single communities (as in the case of 10 year flood return zones of specified magnitudes). Of course, any mapping and classification system should be designed to permit identification of parcels of land that fall within more than one type of hazard zone. The classification methods developed through an earlier grant by the National Science Foundation to the Southwest Center for Urban Research (Houston, Texas) may well be adaptable.

In view of the loss estimates developed in this analysis, priority in establishment of areas to be subject to the proposed system as well as priority for completion of the mapping and classification program should be assigned so as to deal, first, with: (1) those hazards exhibiting the largest current and projected losses; (2) those areas believed to be within hazard zones now highly populated or expected to receive a substantial number of migrants between 1980 and 2000; (3) those areas exhibiting the highest annual expected per capita losses of either life or property.

b. Hazard exposure criteria should be developed for use in association with the hazard zone mapping and classifications system discussed above, and should specify the effects predicted from exposure to various hazard intensities and/or magnitudes. The recommendation contemplates that such criteria would be quantitative in form, that the use of a variety of expert panels would be necessary to their preparation, and that each criteria statement would be based on empirical evidence or subject to empirical verification. The recommendation further contemplates that the criteria statements would be composed without regard to normative criteria or influences, that they be composed as statements of probable relationships rather than expressions of what "ought to be."

The process of developing such criteria should resolve many of the issues of fact now surrounding natural hazards management and should also identify gaps in our understanding of the relationships between natural hazard exposures and specified sets of consequences.

c. Hazard mitigation criteria should be developed specifying the reductions in damages or other consequences that may be expected from the application of a specified building code or other mitigation to a specified set of objects exposed to a natural hazard.

As in the case of hazard exposure criteria, these statements are viewed as the product of scientific inquiry, technical panel activities, and empirically defensible information, rather than a reflection of political, social or economic biases.

d. Cost feasibility criteria should be developed expressing the cost-damage reduction ratios expected from application of a specified hazard mitigation criteria to a set of hazard-exposed objects.

e. The National Science Foundation should be authorized, funded, and directed to undertake the preparation of such critical literature reviews and the conduct of such research and loss analyses as necessary to assure the timely, objective, and effective accomplishment of the above projects with respect to the full range of natural hazards to which the U.S. population is exposed.

Action Possibility 2. Hazard Zone Occupancy and Risk Analysis Studies Should Be Undertaken as the Basis for Establishment of a National Hazard Zone Model that Can Be Used on a Continuing Basis to Present Information Concerning the Size of the Human and Building Populations Exposed to the Major Types of Hazards, by Frequency and Magnitude Level. A comprehensive, balanced, and cost-effective approach to continuing decision making is necessary to maximize the benefits gained per unit of resources expended on hazards zone research, technology development, and management programs. The proposed national hazards zone model is intended to facilitate this end by providing quantitative information concerning the sizes of the populations at risk of exposure to the full range of important natural and man-induced hazards in the numerous hazard zones of the U.S. The recommendation contemplates that (a) the model would be capable of revealing the sizes of the populations at risk of exposure in such zones, by type, magnitude, and frequency of hazard occurrence; (b) the level of risk associated with such exposures would be expressed in terms of annual expected dollar loss, injury, and morbidity; and (c) the model would be capable of expressing the above in respect to populations and/or land areas located within more than one type of hazard zone.

We recommend an early review of current federal efforts and information systems and literature relevant to this proposal, and the preparation of an extended document that would examine the utility, feasibility, and possible cost of such a model.

Action Possibility 3. Determine the Capacity of State and Local Units of Government and Their Professional Staffs to Engage in the Development and Implemen-

tation of Natural Hazards Management Policies and Programs, with Particular Reference to Those Having Implications for Building Codes, Subdivision Regulations, Housing Codes, and Land Use Planning or Zoning Standards. A decade has passed since the report of the National Commission on Urban Problems was developed. With the exception of the study by Field and Rivkin (1975), the recommendations more recently developed by other groups are not generally linked to objective and quantitatively oriented study findings concerning the current administrative, organizational, and staff capacities of local and state governments. This action possibility assumes the need for such a study and further assumes that: (a) such a study would be directed at cities, counties, and states exhibiting the highest potential for natural hazard losses; and (b) the study would include an examination of plans and programs developed under the terms of the Coastal Zone Management Act, the Disaster Relief Act of 1974, and similar statutes. It further assumes that such studies would focus on explication of user agency perceptions of the quality, thoroughness, utility, availability, and consistency of current federal regulations, guidelines, technical assistance documents, and similar materials.

Action Possibility 4. A Comprehensive Analysis Should be Made to Determine the Substantive Content and Applicability to Natural Hazards Management of (a) the Several Model Building Codes; (b) the Several "Model" or "Recommended" State and Local Hazard-Related Building, Zoning, and Subdivision Ordinances and State Hazards Management Acts; (c) Regulations and Standards of Federal Agencies; and (d) State, County, and Local Building Codes, Subdivision Regulations, Housing Codes, and Planning Ordinances in Areas Most Impacted by Potential Natural Hazard Problems. This study would provide the information base necessary for development of agency and area-specific recommendations concerning possible modifications in natural hazard management policies and structures. The recommendation contemplates that exemplary enactments by local, state, and other bodies would be assembled; that objective criteria would be developed for classifying the sum of the regulations examined; and that the output of the study would be useful for a mix of policy planning, technical assistance, and staff development purposes.

Action Possibility 5. Current Federally Sponsored and/or Financed Training and Technical Assistance Programs and Activities in Natural Hazards Management and Related Fields Should be Identified, Their Respective Contents and Clientele Determined, and Their Compatibility with Current and Future Needs Assessed; and a Comprehensive, Well Integrated Program of Training and Technical Assistance in Support of Natural Hazard Programs at Local and State Governments Thereafter Should be Developed. The action contemplates that activities funded under the terms of the Environmental Education Act, the Intergovernmental Personnel Act, and the statutes relating to hazards management and urban or community development would be included within the scope of this recommendation. The suggested study should focus, in part, on identification of possible revisions in the guidelines governing these existing activities to increase their utility to the field of natural hazards management.

The action further contemplates that training and technical assistance activities should be targeted on such groups as (a) elected and appointed policy-making officials of local and state governments; and (b) technical personnel of land use planning, regional planning, resource development, building regulation, and subdivision control agencies and organizations.

Action Possibility 6. A Comprehensive, Well-Coordinated, and Continuing Program of Public Information and Education Should be Conducted to Acquaint the Public and Policy Makers with the Essential Facts Concerning Natural Hazard Exposures, Consequences, and Alternative Mitigation Strategies. It seems clear that responsible decision making in the field of natural hazards management will require the active participation of an informed public. Yet, the evidence is overwhelming that even the residents of high hazard areas do not fully comprehend the extent of their risk, the consequences associated with exposures, and the types of adjustments that may be made to mitigate the effects of those exposures. Of course, the design of such a program would require the identification and review of the numerous segmented public information and education efforts in this field that are sponsored by the panoply of relevant federal agencies.

Some specific possibilities for such a program are as follows:

1. A "White House" or "National Conference" on Management of Environmental Hazard Zones could be conducted.
2. Conferences, workshops, and symposia might be conducted for community leaders, influentials, and media representatives of high-damage-rate states and counties.
3. Highly readable, objective brochures on each hazard could be distributed to the public in areas exhibiting moderate to heavy annual expected losses from exposures.
4. Objective and illustrated feature articles could be prepared for distribution to community newspapers in moderate to high hazard areas.
5. A series of documentary films could be prepared for broadcast by the Public Broadcasting System and commercial stations.

Action Possibility 7. In Cooperation With Elements of the American Insurance Industry, a Comprehensive, Well-Budgeted Study Should Be Undertaken to (a) Determine the Possible Utility of Both Comprehensive and Categorical Hazard Insurance Systems in the Mitigation of the Consequences Associated With Natural Hazard Exposures; (b) Identify the Present Public Policies that Constrain the Development and Extension of Such Insurance Services; (c) Review, and Possibly Develop, New Data Concerning the Factors That Influence Consumer Attitudes Toward Such Insurance, (d) Identify the Possible Functional and Dysfunctional Consequences Associated With Legal Requirements Calling for Mandatory Purchase of Such Policies; (e) Identify the Conditions Under

Which Mandatory Insurance Purchase Schemes Would Be Feasible for Both Industry and the Public. For a single natural hazard, earthquake, it has been suggested that some form of federally-supported all-hazard or hazard-specific insurance system may be technically feasible to cover catastrophic losses experienced by various parties as a result of the occurrence of hazardous natural events; but that a complex mix of supporting federal legislative enactments would be necessary to protect prime insurance carriers and reinsurers from institution-destroying loss levels under circumstances where adequate reserves have either not yet been accumulated or where the magnitudes of catastrophes exceed the levels estimated by computer simulation models (Atkisson et al., 1980; Atkisson and Petak, 1981a).

It is clear that there is resistance on the part of many elements of the American insurance industry to any mandated system of natural hazards insurance and a pervasive fear that any federal intrusion whatsoever into this field would lead either to unwanted federal regulation of the insurance industry or to requirements that might threaten the financial viability of the existing industry. Atkisson and Petak (1981a) also suggest that a variety of alternative schemes, such as federal development of an actuarially sound national catastrophe fund developed by payments from the federal government and the states, might accomplish essentially the same ends as a more conventional insurance system.

Action Possibility 8. A Comprehensive Study Should Be Conducted to Identify the Past, Present, and Probable Future Annual Costs of Natural Hazard Exposures That Have Been, or Will Be, Distributed to Non-Risk-Taking Parties. Utilizing comparatively gross assumptions, we have developed estimates concerning the current level of federal liability for natural hazard losses (see Chapter 5). A much more detailed study of this subject is needed, however, and much more information is required concerning other hazard loss distribution mechanisms at state and local levels of government.

Action Possibility 9. The Federal Government Should Require Parties Involved in the Sale, Transfer, Renting, or Leasing of Property to Engage in Full Disclosure of the Natural Hazard-Related Risks Associated With the Property and the Hazard Zones in Which it is Located. This action contemplates ultimate legislation that would require "full disclosure" on the part of those who sell or offer to sell and of those who offer for rent or lease any parcel of developed or undeveloped property within the United States, when such property falls within an area that the Federal Government has determined to be bounded by a federally designated hazard zone. It is further contemplated that potential purchasers, renters, and lessees of such property be advised by the seller, lessor, or any agents thereof of the nature and classification of the hazard zone within which the property is located, and that this assertion be verified (in the case of property sales) at time of title transfer by the title insuring agency. Although the adoption of a direct federal legislative requirement of this type may exceed the Constitutional authority of the Federal Government, it would appear that adoption of such legislation by state legislatures could well be "forced" by the Federal Government through insistence that the adoption of such legislation be a necessary precondition to the establishment of the eligibility of the state, any portion thereof, or any citizen thereof for any benefits under the terms of the Federal Disaster Relief of 1974; the Federal Flood Insurance Act, as amended; the various federal flood control acts; and those acts that authorize the activities of such agencies as the Federal Housing Administration and the Federal Home Loan Administration.

With respect to regulations issued under the authority of federal agencies engaged in mortgage transactions, it should be noted that Executive Order 11988 (May 24, 1977) now requires that "agencies which guarantee, approve, regulate, or insure any financial transaction which is related to an area located in a floodplain shall, prior to completing action on such transaction, inform any private parties participating in the transaction of the hazards of locating structures in the floodplain" (Federal Register, 1977).

Action Possibility 10. The Federal Government Should Continue to Sponsor and Fund Feasibility Studies Regarding Area Flood Protection Works, Should Continue to Engage in the Design of Such Works, but Should Alter Past Policies to Require the Participation of Beneficiary Parties in the Funding of the Capital and Operating Costs Associated With Such Facilities, and to Prohibit Federal Participation in Any Project in Which the Required Benefit–Cost Relationships Can Be Demonstrated Only Under Conditions of Increased Floodplain Occupancy. Past federal policies have usually involved an all-or-nothing approach to the provision of coastal and riverine flood control facilities to communities and states. If such projects were deemed necessary and qualifying under appropriate policies, then the Federal Government picked up the lion's share of the bills for the project. The recent study of the U.S. Water Resources Council found that only 11 to 20% of the mean, effective, composite cost of rural and urban flood damage reduction facilities and programs, respectively, is being borne by non-federal entities (U.S. Water Resources Council, 1975).

If this action possibility were to be implemented, project beneficiary states would be partners in the financing of future flood control projects. The action also contemplates the further requirement that precise boundaries be drawn by federal authorities to reveal the land areas and property owners who are the real beneficiaries of risk reductions associated with each flood control project, and that all, or a substantial fraction of, the capital and operating costs of such facilities be imposed on such land areas and parties. The intent of this action possibility is to limit past externalization of flood control costs and provide an opportunity for existing hazard-exposed populations to conveniently exercise the option of mitigating their local risks through use of flood control works, but only under conditions where the project beneficiaries internalize a large fraction of the associated costs. Such an amendment of policy could be linked to further requirements intended to prevent future buildups in floodplain exposures.

In its major thrust, this action possibility is consistent with the spirit and purpose of the several recommendations and declarations advanced by the Water Resources Council (1973, 1976), and previous national administrations. In his message to the Congress on May 23, 1977, the President observed that "it is essential to confine the public works efforts of the water development agencies to projects that can meet such defensible criteria as economic efficiency, safety, environmental protection, and fair distribution of project benefits."

This action statement contemplates that (a) the major fraction of the costs associated with construction of area hazard protection works (dams, levees, seawalls, etc.) should be internalized to those who receive the benefits from such expenditures; (b) the construction of such works should be justified in terms of the need to reduce the risks associated with the exposure of existing populations and property to the hazard, and should be managed to prevent unreasonable increases in the sizes of the populations at risk as well as increases in the absolute level of losses resulting from such exposures; (c) the opportunity for current natural hazard risk takers who employ this approach to reduce *their* levels of risk should be preserved, but in a context in which such risk takers must weigh the associated costs, benefits, and trade-offs. That local beneficiaries should share in the costs of federal flood control projects and activities was endorsed by several national groups during the 1975 public hearings conducted by the U.S. Water Resources Council (1975).

Action Possibility 11. Building Standards Enforced by Federal Lending and Mortgage Guarantee Agencies Should Be Amended to Require the Use of Building Strengthening Mitigations (For Wind, Flood, and Earthquake Hazards) on New Structures in Those Counties and Subcounty Areas in Which the Use of Such Mitigations is Cost Feasible. Under the terms of this action statement, federal lending and mortgage guarantee entities would be required to foster the use of loss-reducing building strengthening and/or floodproofing/elevating mitigations in those areas and under those conditions where the use of such mitigations is both cost feasible and contributory to the production of *net* reductions in the annual expected losses from area natural hazard exposures. This requirement could be extended to expansive soils and other hazard mitigations as rapidly as warranted by available evidence.

Action Possibility 12. Hazard Impact and Mitigation Statements Should Be Required Additions to the Environmental Impact Statements Currently Required in Respect to Federally Conducted or Federally Financed Projects. This recommendation intends that such statements might well be prepared for such critical capital facilities as schools, hospitals, nursing homes, community health centers, and other facilities whose location could increase the size of the involuntary population at risk of exposure to natural hazards. Such statements could further be employed to determine the extent to which federally funded education and other programs have been appropriately adapted to the natural hazard needs of specific areas. Thus, in respect to educational subventions to the states, such statements might well deal with questions relating to the extent to which public school curricula foster community use of life-saving and injury-avoiding adjustments to natural hazards.

Action Possibility 13. The Federal Disaster Relief Act of 1974 Should Be Amended so as to Accomplish the Following Purposes.

a. Clarification of the meaning of Section 406 and extension of its applicability to all forms of federal assistance. This section of the Act vests the administrators of the disaster relief program with the authority and responsibility to require the recipients of program benefits authorized under the Act to evaluate and mitigate the hazards within the area in which the federal funds are used. However, the section was added without clarifying interpretive committee comment during the process of adopting the Act and program administrators have therefore expressed doubt concerning the Congressional intent behind the Section and the degree of Congressional support for its rigorous enforcement. Although Section 406 authorizes the Executive Branch to require that state and local communities evaluate the natural hazards in the areas in which federal disaster relief benefits are provided under the specific terms of the Act, and although it further provides that the Executive Branch may require that appropriate mitigations be taken to reduce such hazards as a condition for award of the specified benefits, the section clearly does not apply to federal benefits authorized under other statutes (such as loans provided by the Small Business Administration). Its phraseology raises more questions than it answers.

In the past, administrators have debated whether the terms of Section 406 are discretionary or mandatory, whether Congress intended that a disaster caused by one hazard should result in the Federal Government requiring that all other hazards within the disaster area also be evaluated and mitigated, and whether or not Congress intended that a grant to a local government for reconstruction of one of its own facilities be conditioned upon a complete rewriting of that same jurisdiction's land use planning ordinance so as to secure mitigation of the risks faced by all other types of property, including those owned by private parties.

In the past, very narrow interpretations have been applied to the section by program administrators. Thus, in the case of the Houston floods, which caused much damage within the Texas Medical Center and resulted in the Presidential designation of Houston and Harris County as a disaster area, program administrators chose to define the term "area" in a very narrow sense; i.e., they invoked Section 406 requirements only within the boundaries of the Texas Medical Center, which is a multi-block component within the city.

b. Clearly define the criteria to be utilized in defining disasters and disaster areas and in qualifying applicants within such areas for specific federal benefits. Under current conditions, local politicians, state governors, and other actors within the American policy system face unusual pressures to secure the designation of a wide variety of hazardous occurrences as "disasters," and thereby to qualify the populations within broad areas of their jurisdictions for federal subsidies and services. At state and local levels there is little incentive to abridge the circumstances under which such declarations are made by the President. In addition, there is not much incentive for state and local communities to build actuarially sound funds which can be used to deal with the needs posed by future hazardous occurrences within their jurisdictions.

Once a disaster declaration has been issued, loss experiencing parties within the area qualify for federal assistance, whether or not their losses have been of a trivial or catastrophic nature to them. An appropriate amendment to the Act might therefore make explicit the guidelines which are to be followed by the administering agent(s) in promulgating regulations under which a potential benefit-receiving party may be entitled to one or more of the several forms of disaster relief benefits. These Congressionally-defined criteria could relate to such matters as the relationship of the party's loss to the party's annual income, annual operating budget, or total assets, and the extent to which the entitlements may be apportioned to a party (from 0 to 100%), based on the classification of the party within a group defined by loss as a function of income, annual operating budget, or net assets.

Although the Act requires that governmental and non-profit parties receiving assistance in disaster areas must thereafter procure insurance against future losses to damaged properties, this requirement has been limited to insurance against the amount of loss sustained in the existing disaster. Thus, federal liabilities for future large losses continue and disaster assistance recipients have little incentive to independently develop their own financial and hazard mitigation plans for possible future hazard-induced losses.

Current policies also permit federal loans, grants, and other assistance to be extended to owners of property which, prior to the occurrence of the loss-inducing disaster, would have met criteria conventionally used to describe a facility as a hazardous structure. Accordingly, it might be appropriate to amend the Disaster Relief Act so as to declare ineligible for any form of disaster relief any building or structure which meets the "hazardous structures" criteria. These criteria would be contained in regulations which the amendment would require be issued by an appropriate division of the Federal Government.

Considered as a whole, it seems fair to state that current federal disaster relief policies are riddled with inconsistencies, inequities and inefficiencies. Appropriate amendments to the Disaster Relief Act, including extension of its applicability to all forms of federal benefits, could partially correct this situation.

c. Define criteria for predisaster determination of state/local eligibility for disaster relief benefits. Within the United States, the typical Building and Safety Department is lightly staffed and ill equipped to deal with highly technical questions. Even some of the larger building and safety jurisdictions within the United States have encountered difficulties in checking plans and construction activities to assure that new construction conforms to natural hazard safety standards. Moreover, most of the nation's continuing growth in population and new structures is occurring outside the boundaries of those municipalities and counties which have both the largest and best-equipped building and safety, as well as land use planning and zoning departments. The result is a milieu in which prudent observers may well question the natural hazard resistance of much of the new construction which is occurring within the country, not to mention the quality of those structures built in earlier years. The suitability of both past and current site locations for whole community segments may also be questioned.

From state professional licensure laws, through constraints on the legal capacity of local governments to engage in the zoning of their land, to the sad state of new construction regulatory processes and land use standards in many communities, the problems which beset the design and implementation of effective loss-reducing natural hazard strategies and policies are legion.

Although the current Disaster Relief Act and its implementing regulations require predisaster preparation and federal approval of state emergency plans, neither the Act nor its implementing regulations require any predisaster determinations that state and local governments have enacted and are enforcing measures to assure cost-effective mitigation of possible future natural hazard damages and losses within their jurisdictions. Instead, federal disaster relief benefits are bestowed alike on the competent and incompetent, on the careful and the unwary, and on those who have taken effective steps to curb future losses as well as on those who have done comparatively little (Petak and Atkisson, 1981). We have documented elsewhere the comparatively low quality of state/local hazard mitigation efforts, particularly in respect to the earthquake hazard (Atkisson and Petak, 1981b).

d. Provide predisaster hazard evaluation and mitigation assistance to state and local governments. By and large, federal assistance to state and local communities is now limited to hazard evaluation and mitigation activities undertaken in the immediate wake of a catastrophe. Incentives and financial capacities therefore are not provided to communities to engage in those activities which are necessary in planning for and mitigating the future effects of such events.

This situation could be corrected through appropriate amendments, to the Federal Disaster Relief Act and annual appropriation bills, which would provide federal funds for assessment of local natural hazard risks by states, counties, and municipalities. Few U.S. communities are fully aware of the magnitude, characteristics and areal distribution of natural hazard risks within their boundaries, and even fewer are knowledgeable concerning the mix of benefits and costs associated with alternative risk reduction measures. Lacking motivation to concern themselves with these matters, caught up in local fiscal crises of their own, and typically short of competent in-house counsel concerning natural hazard safety matters, few U.S. communities have engaged in the kind of natural hazard risk assessment and mitigation planning efforts which are required in any effective community program. Accordingly, this action possibility contemplates enacting suitable legislative amendments to existing statutes which would authorize federal grants to a priority list of states and communities. These grants would promote locally initiated natural hazard risk assessment and mitigation planning projects which meet specified Legislative and Executive Branch criteria.

In addition, this action possibility contemplates appropriate amendments to the Disaster Relief Act to realize a federally-funded program of voluntary or mandatory professional audits, in order to monitor the hazard-diminishing power of state and local building codes and zoning regulations, as well as the adequacy of supporting professional licensure, agency staffing, and agency operating practices.

In this way, the action could deal with several characteristics of the setting in which current policies are made at local and state levels concerning natural hazard safety and building structural quality. Typically, the activities of building and safety departments and their sister land use planning and zoning agencies do not attract much attention until some calamity or significant malpractice prompts publicity within the community. The technical activities of such agencies often are not well understood by local elected officials; the contents of building codes are essentially inexplicable to all but the technically-trained. It is therefore difficult both to (1) escalate concern about these processes to the point that they receive attention on local policy-making agendas, and (2) provide local policy makers with informed counsel concerning the quality and relevance of their operations and of the substantive content of the codes and regulations they enforce.

In sum, this action alternative advocates that, on the basis of priorities determined by the level of natural hazard risk faced by communities, a federally-funded program of professional audits (voluntary or involuntary) of local codes and related enforcement operations be conducted. It is conceivable that such an activity could be completely voluntary and that audits would be conducted only in those jurisdictions whose governing councils specifically requested this service. The focus of an audit would be not on witch-hunting, but on an objective review of code content and departmental operations in terms of their compliance with reasonable professional models, and on providing local policy makers with counsel concerning deficiencies and means for improving operations.

Action Possibility 14. The Federal Internal Revenue Code Should Be Amended so as to Accomplish the Following Purposes.

a. Modify those casualty and business loss provisions of the code which are relevant to losses sustained during hazardous natural events. At present, many hazardous structures are maintained and used within the United States simply because it is economical to do so. Owners of properties which are offered for rent to multiple families or for business/industrial use may fail to apply needed retrofit measures to their structures. Their failure to do so may be based on narrowly conceived benefit/cost calculations, which consider the fact that some losses sustained during hazardous natural events will be partially recovered through reductions in tax payments that otherwise would be necessary. To some extent, then, the current provisions of the Internal Revenue Code provide subsidies to those who own and/or operate hazardous structures and may even provide those same parties with incentives for doing nothing to correct the vulnerabilities of their structures.

Moreover, business and corporate enterprises may deduct the cost of hazard-related insurance premium payments, but private owners of non-revenue-producing residential structures may not. Thus, a further inequity is created.

As noted earlier, current tax writeoff provisions in the Code tend to provide a greater level of reimbursement for natural hazard losses to the affluent than to the less affluent. This situation could be corrected through extension of tax credits for disaster losses to those parties within qualifying lower income ranges.

b. Modify current restrictions on buildup of reserves by prime insurance carriers and reinsurers. Where provided, insurance against property losses due to the occurrence of natural hazards typically involves premium rates which are escalated beyond levels justified by long-term annual expected loss rates. This escalation is due, in part, to the increased short-term risk faced by insurers due to the possibility of an occurrence of a high-consequence (albeit low-probability) loss-producing event, before adequate reserves can be accumulated to meet event payoff requirements. The problem is further exacerbated by the fact that current IRS regulations have the effect of limiting the total size of the reserve funds which may be accumulated by insurance companies to meet the liabilities associated with a rare but catastrophic occurrence. Liberalization of these regulations could therefore contribute to an escalation in the reserve capacity of the industry, improve its ability to enter more aggressively into the marketplace, and do so with policies which might permit lower premium payments.

Action Possibility 15. Legislation Should Be Adopted Providing for the Accumulation of Adequate Federal Financial Reserves to Meet the Large and Sudden Financial Requirements Associated with Probable Maximum Loss Events Induced by Hazardous Natural Occurrences. The occurrence of a natural hazard-induced PML event in a large metropolitan area could create federal liabilities of substantial magnitude. Federal borrowing to meet legal obligations for disaster relief, recovery, and reconstruction could disturb capital markets. Moreover, because of the true interest component of such loans, the need for federal borrowing could well exacerbate the long term economic burdens of the disaster on both disaster-exposed and nonexposed populations.

This action alternative therefore requires that appropriate studies be conducted of the annual expected and probable maximum losses associated with hazardous natural occurrences and that, based on such studies, one of several possible types of actuarially sound catastrophe reserve funds be created.

Type 1. A fund financed solely by federal tax receipts: This fund would provide a means for the federal government to accumulate reserves sufficient to meet predicted high losses from natural catastrophes in heavily populated areas. Since the occurrence of the "Great Galveston Flood" at the turn of the century, and the "Great San Francisco Earthquake" of 1906, no high magnitude occurrence has struck a major populated area within the United States. Thus, the average of federal disaster relief expenditures is nowhere near as impressive as the expenses from probable maximum loss events which may soon occur. Since the probability of a high magnitude earthquake in a major California metropolitan area is now generally believed to be strong, the exercise of this option would permit the federal government to build appropriate reserves for the eventual occurrence and to avoid the strains created by the large disaster relief demands which it will ultimately confront.

Type 2. Joint federal-state funds: Here, federal government would stimulate the formation and serve as the manager, of a set of actuarially sound catastrophe reserve funds in cooperation with each of the states. Each fund would be built from annual tax contributions by each participating state and by the federal government. To prompt fund formation, a variety of incentives might be utilized by the federal government, including high contributions in the initial years and lower contributions in later years.

Type 3. Federal-state funds linked to mitigation criteria: This type of fund is identical to the previous, except that it incorporates variable federal participation in each state reserve fund dependent on the compliance of the state and its local governmental entities with specific hazard evaluation and mitigation criteria.

Some have challenged the idea that a catastrophe reserve fund could ever be accumulated that would be secure from periodic raids by the Federal Government under pressure of short-term budgetary and political requirements (Reinsurance Association of America, 1980). Assuming that this criticism is valid, the problem might be solved through the establishment of a new organization, such as "Corporation for the National Catastrophe Fund," whose function would be to serve as the contracting agent for both the United States government and the participating states and as the manager of the fund. The development of the fund would be consummated through contracts between the several parties, the contracts explicitly indicating the sanctity of the contract payments and the integrity of the fund. The fund could be managed by board members appointed jointly by representatives of the federal and participating state governments. Creation of the organization and specification of the contracting procedures would require specific Congressional legislation. Thereafter, however, the corporation would have an independent status and would be legally able to modify disaster benefit formulas under conditions where contracting parties, including the federal government, take action inconsistent with the integrity of the fund or the ability of annual contract payments to meet the obligations associated with the established benefit formulas.

Action Possibility 16. Legislation Should Be Enacted to Facilitate the Improved "Free Market" Delivery of Natural Hazards Insurance by Private Companies and/or to Provide for Government Sponsored Insurance and Augmentation of That Available Through the Private Market. In a recent report, the U.S. Comptroller General expressed the belief that:

> . . . Insurance is the most efficient and equitable method of providing disaster assistance . . . insurance programs are inherently more consistent over time than loans, whose availability is triggered by, and whose terms are the result of, political decisions. Thus, the principles that should guide formulating disaster policy indicate that both in theory and in practice insurance is more appropriate than loans and grants. In practice, insurance produces lower differences in income and wealth distribution among victims and between victims and taxpayers (Comptroller General of the United States, 1980).

In theory, insurance systems have the capacity to assign the costs and benefits of risk exposure to those who experience the risk, to provide assurance to participants that their losses will be covered under known conditions and at known prices, to stimulate the policy holder's motivation to avoid risks through a differential rate structure which rewards prudence and penalizes imprudence, to provide equitable treatment to policy holders of a comparable class, and to provide rapid response when the need occurs–with minimum resort to excessive paper work, appeals, and qualifying checks. Appropriately structured insurance systems also provide for the development of reserve funds necessary to meet the large payout requirements associated with rare catastrophic occurrences. In practice, however, the ability of insurance systems to function in this fashion is dependent on the price that is charged for the service, the appeal of the service to various consuming groups, the willingness of the industry to provide the service, and the assessment by the industry of its capacity to meet payout requirements without sacrificing its own financial and economic viability. As the industry's assessment of its ability to accurately predict risk declines, the prices for the service rise, or may not be offered at all. When a solid actuarial base of information concerning the probability of loss to particular persons or properties under particular defined circumstances is missing, the price goes up or the industry's willingness to engage in the service goes down. When the industry's perception of its capacity to predict the coming and going of loss-producing events goes down, the price for its services goes up, and its willingness to extend the service goes down. When wild fluctuations, from year to year, can be anticipated in the volume of losses that will be sustained among policy holders, the requirements for reserve accumulations go up, the difficulties of the industry in coping with regulatory requirements and IRS policies are escalated, and the industry's interest in extending the service goes down–and, sometimes, the price of the service, when extended, goes up.

To the extent that the major public problems resulting from hazardous natural events may be ameliorated by widespread private purchase of insurance, and to the extent that the current low level of purchase of such insurance represent "failures" within the private marketplace, then federal activity might well be directed toward the correction of those failures. This action possibility therefore includes a federal role which is targeted, in part, on assisting the private insurance industry in overcoming those problems which now prevent insurance from meeting the larger social ends which are the target of federal disaster relief and tax policies. Such assistance could be provided by the federal government through the adoption of legislation and policies intended to (1) remove the existing disincentives to private purchase of natural hazard insurance which are caused by the current mix of federal disaster relief and tax policies; (2) assist the insurance industry in better understanding the characteristics of natural hazards and hazard mitigation processes, and in more accurately assessing the natural hazard risks of specific areas and types of structures; (3) assist the industry in accumulating reserves or otherwise acquiring a capacity adequate to meet the payout requirements associated with a natural hazard-induced

PML; (4) stimulate the market for private purchase of residential insurance; and (5) reduce the uncertainties which currently lead insurers to include large rate-escalating "risk factors" in their premium calculations. A more complete discussion of these alternatives appears in a special report to the Federal Insurance Administration which was primarily on the subject of earthquake insurance (Atkisson and Petak, 1981a).

In addition to action calculated to correct current "failures" within the private marketplace, the federal government could consider more direct forms of action involving the direct extension of federally-provided insurance to selected classes of potential purchasers. For example, in lieu of present forms of disaster relief to state and local governments, involving reimbursement of losses sustained by government-owned structures, the federal government could provide a program of insurance to meet these needs. Such a program could involve a sharing of premium payments by the federal government and state and local jurisdictions, and could involve either use of privately-provided insurance or federally-provided insurance. Similarly, the federal government could act as a direct reinsurer for prime carriers, and thereby assist the insurance industry in acquiring the capacity to meet the payout requirements associated with rare PML events. It could also act as the provider of "catastrophe caps" for insurance provided to owners of private residential structures and other facilities.

Further, of course, the federal government could provide research, technical assistance, and financial support to the establishment of state created insurance programs targeted on meeting the needs associated with the exposure of people and property to hazardous natural events.

Finally, the federal government and/or its several operating components could require the purchase of insurance by those parties who receive such federal benefits as disaster relief payments, federally subsidized loans, federally insured loans, and/or loans provided by federally regulated financial institutions. Referred to by some members of the insurance industry as "mandation," a federal requirement that natural hazard insurance by purchased as a condition for obtaining any federally sanctioned loan or any loan from a federally regulated financial institution would drastically raise the current level of natural hazard insurance coverage of residential units and other properties. Such a requirement, however, is much feared by the insurance industry. Unaccompanied by means for meeting the large payout requirements resulting from a PML or by other measures intended to protect the insurance industry from catastrophic losses, such a requirement would be both politically controversial and financially improvident. However, if combined with an appropriate mix of industry-supporting legislation, much of the sting of such mandated purchases would be removed.

Although industry fear of "mandation" is understandable, the fact is that much insurance that is now purchased by private households is the result of mandates by some group or another. The state may require the purchase of automobile insurance, a lender may require the purchase of fire insurance, the state government may require the purchase of workmen's compensation insurance, etc. The issue, therefore, is not simply one of "mandation," but of who issues the mandates, under what circumstances, and for what purposes.

Action Possibility 17. Legislation Should Be Enacted to Eliminate Most Existing Forms of Federal Disaster Relief. Two general forms of federal disaster relief now are provided to state and local governments, private indiviuals, and other groups. The first form involves the extension of emergency service in the immediate wake of a disaster and includes services such as the clearance of debris, the provision of emergency medical care, food, housing, and other similar needs. The second, and more expensive form of assistance, involves low cost loans, grants, and other forms of assistance intended to stimulate the economic recovery of the disaster area and to minimize the losses borne by exposed parties. This action possibility has to do with the second set of existing federal services.

As noted in Chapter 5, current federal disaster relief policies result in the imposition of net tax burdens in approximately 30 states to reduce net natural hazard costs in 21 others (counting the District of Columbia as a state). However, a major part of disaster-related interstate tax transfers are directed to such wealthy states as Florida, California, Washington, and Texas. Four states (Florida, California, Mississippi and Louisiana) collectively account for approximately three-fourths of disaster-related interstate tax transfers, with Florida and California alone accounting for nearly one-half (47.92%) of the total interstate transfers. Six states donate approximately 75% of the total amount, including New York, Michigan, Pennsylvania, New Jersey, Ohio, and West Virginia. What is surprising is that a comparatively few states receive interstate tax transfers of sufficient magnitude and comparative importance to be of high value to them, and that few, if any, states contribute sums in sufficiently large quantities and at sufficiently burdensome levels to be of much concern to them.

Under these circumstances, it is therefore appropriate to ask: Why continue this current mix of federal disaster relief policies at all? Why not adopt a system in which each state is solely responsible for its own losses, with the possible exception of emergency services required in the immediate wake of a disastrous occurrence?

Action Possibilities for State Governments

Policy makers in state governments face the same need as those at federal levels to determine which of the major strategic options discussed earlier should guide their policy-making activities, and which goal statements should be included in natural hazard policy plans for their own states. Once these questions are resolved, state-level policy makers face the need to determine which action possibilities should be selected in order to accomplish the ends specified in the goals contained in their state policy plans.

Many of the alternatives identified above as being appropriate for consideration by federal policy makers are also appropriate for consideration by state policy makers. Certainly, the alternatives related to government intervention in the insurance system and to governmental initiation of more comprehensive hazard evaluation and mitigation activities fall into this category. In addition, however, the following specific action alternatives seem worthy of consideration by state governments. As in the case of action possibilities identified for the federal level, these are options to be considered, rather than firm recommendations offered from one party to another.

Action Possibility 1. Coastal States Exposed to Tsunamis Should Promptly (a) Identify Tsunami Inundation Areas Within Their Borders; (b) Provide for Appropriate Marking of These Areas; and (c) Prohibit Any New Residential Development Within the Boundaries of Such Zones. The technical mitigations proposed for tsunami in this analysis are based on the assumption that (a) there are no effective building strengthening methods that might be employed to protect residents of nonengineered dwellings from the threat of injury and loss of life; and (b) the primary methods for reducing or preventing future injury and loss of life involve avoidance, hazard warning, and population evacuation systems. The only structures that should be located in hazardous areas should meet three tests: (1) they are economically necessary to such places; (2) they cannot be located in other areas; and (3) they are engineered structures.

Action Possibility 2. All States Abutting the Gulf of Mexico and the Atlantic Ocean Should Enact Legislation and/or Take Such Other Action as May Be Necessary to Secure the Mapping of Coastal Floodplains (by Frequency and Magnitude of Expected Floods), and the Early Use of Appropriate Risk-Reducing Measures Within Those Areas, Including (a) Avoidance of All New and Replacement Construction Within the More Hazardous of These Areas; (b) Avoidance of Net New Growth in All Areas In Which There Is a Moderate or High Risk of Property and/or Life Loss; and (c) the Use of Area Protection, Building Strengthening, and Building Elevation Safeguards for Those Portions of the Floodplains In Which Low to Moderate Risks are Faced, Where Such Actions Are Cost Feasible, and Where There Is Some Ascendant Need for Further Development and/or for the Protection of Existing Populations and Properties.

Action Possibility 3. All States Should Adopt Natural Hazard Management Acts That Require the Mapping of Natural Hazard Zones Within Their Boundaries, and Require the Parties Involved in the Sale, Transfer, Renting, or Leasing of Property to Engage in Full Disclosure of the Natural Hazard Related Risks Associated With Occupancy of Structures and Places Within Natural Hazard Zones. With respect to the second portion of this alternative, some states may wish to incorporate the requirement into their business and licensing codes governing the conditions under which business or professional licenses may be awarded or periodically reawarded to persons engaged in the development, sale, transfer, rent, or lease of property.

Action Possibility 4. All States Should Adopt Legislation That Either Imposes Minimum and Maximum Building and Housing Code Requirements in All Appropriately Populated Local Areas, or Requires State-Conducted Audits and Public Reports Concerning Such Enactments and the Effectiveness of Related Local Enforcement Efforts. Assuming the correctness of the findings offered elsewhere in this analysis concerning the quality and effectiveness of locally enacted building and housing codes and enforcement programs, such codes and enforcement activities may well be of uncertain quality in many areas of the nation and inconsistent with local needs for the mitigation of natural hazard risks.

This alternative assumes three major options available to state governments for dealing with this possible problem: (a) ignore the problem; (b) solve the problem by providing for a state-enacted and enforced system of building, housing, and subdivision codes; (c) take more limited action that preserves local choice over this important subject but that provides maximum assurance that decisions are based on the current state of related knowledge and are made with public participation.

Action Possibility 5. All States Should Adopt or Amend State Subdivision Statutes to Require That All Subdivision Applications Submitted to Any Local or State Entity Must Include a Hazard Impact or Mitigation Statement, the Contents of Which Must Also Be Provided to All Prospective Purchasers of Such Property Prior to Consummation of Title Transfer. Implementation of this alternative by state governments would provide an immediate mechanism for dealing with exposure problems that otherwise might arise until the natural hazard zones of the nation have been fully mapped and appropriately classified. It requires the subdividers of property to appropriately examine the major natural hazard features of all potential subdivision sites and to demonstrate that site and construction practices will conform to hazard-mitigating criteria. This approach stops short of suggesting that any specified set of mitigations should be required for sites of particular types, but does suggest that the information provided in the "impact and mitigation statement" would prompt appropriate regulatory activity by local and state groups responsible for reviewing and approving subdivision applications.

Action Possibilities for Local Units of Government

An old adage states that "the federal government has all the money, the states have all the power, and local government has all the problems!" So it appears to be in the field of natural hazards management. It is within the boundaries of local units of government that people and property are exposed to hazardous natural occurrences, and it is to these areas that loss-reducing efforts must be targeted.

Although the feasible range of natural hazard mitigating policy alternatives which are open to local units of government will inevitably be influenced by the characteristics of federal and state policies concerning this subject, most local governments are nonetheless now able to initiate a wide range of activities potentially capable of reducing both present and anticipated future natural hazards exposure losses. Among the many decision alternatives to be faced by local policy makers are the following:

Action Possibility 1. Each Local Unit of Government Should Identify the Natural Hazards To Which the Human or Building Populations Within Its Jurisdiction Are Exposed, the Boundaries of the Natural Hazard Zones, the Frequency With Which Hazardous Occurrences of Various Magnitudes Are Expected, and the Type and Extent of Damage, Injury, or Life Loss. This alternative requires each local unit of government, after technical counsel and public hearings, to determine the hazardous natural occurrences that are significant enough to that community to warrant public mitigating actions and/or the drawing of natural hazard zones for each such hazard. In those circumstances where such action is deemed appropriate, boundaries would be fixed for natural hazard zones in which such occurrences as expansive soil activity, riverine or coastal flooding, earthquake fault activity, landslides, local winds of high velocity, and other similar occurrences are expected.

A wide variety of data currently is available to support local activities of this sort. The National Oceanic and Atmospheric Administration (NOAA), the U.S. Geological Survey (USGS), the U.S. Army Corps of Engineers, and the U.S. Soil Conservation Service are now engaged in the identification and mapping of hazardous natural areas, including those susceptible to earthquakes, tsunamis, landslides, volcanic activity, coastal flooding, and riverine flooding. In addition, one or more of these same agencies engage in the preparation of maps and other data concerning local soil characteristics, which may assist local jurisdictions in identifying those areas in which landslides or expansive soil activity may be expected.

Action Possibility 2. Each Local Unit of Government Should Identify Those Buildings and Other Structures Which, Because of Their Location or Vulnerability to the Forces Exerted by Hazardous Natural Events, May Be Judged to Be Particularly Hazardous for Continued Future Use or Which May Require Retrofitting In Order to Qualify for Such Use. Because of the long lifetimes of buildings and other structures, many communities now face a significant risk of life and property loss as a result of continued use of hazardous structures. Buildings whose parapets may fall in the event of an earthquake or which may face the threat of collapse in the event of windstorms can be identified. Appropriate locally-determined means for reducing the risk can be required. In respect to the earthquake hazard, Lagorio has recently observed that ". . . despite all our talk about hazardous buildings, we have not solved the problem in real life terms; consequently, it remains a major unresolved policy issue." (Lagorio, 1979)

Action Possibility 3. The Master Land Use Plans Adopted by Local Units of Government Should Be Modified to Include an Identification of the Natural Hazard Zones For Which Boundaries Have Been Drawn as a Result of the Implementation of Action Possibility 1.

Action Possibility 4. Each Local Unit of Government Should Prepare and Adopt, Following Review by Local, State, and Federal Agencies, and Following Public Hearings, a Local Natural Hazards Management Plan and Program. This alternative requires that such a plan specify the periodicity and conditions under which outside audit of code-enforcing operations and local code content should be conducted. Few local policy-makers are currently aware of the level of technical efficacy of the content of their building safety and planning ordinances, and fewer still are aware of the competence of their local staff to adequately enforce such requirements.

Action Possibility 5. Local Subdivision Ordinances Should Be Amended to (a) Require Hazard Impact and Mitigation Statements To Be Filed With All Subdivision Applications; (b) Prohibit New Subdivisions in High-Hazard Zones; and (c) Require Soil Testing, Site Stabilization, Site Avoidance, or Other Site Modification in Specified Hazard Areas, Such as Those Associated With Potential Landslides or Expansive Soils. The suggestion that subdivisions be prohibited in high-hazard areas is a means to implement the technical mitigations identified for landslide, and to provide local units of government with the means for avoidance of new construction in such high-hazard areas as tsunami inundation zones and the high-frequency/magnitude segments of riverine and coastal floodplains.

Action Possibility 6. By Local Ordinance, Provision Should Be Made by Local Governing Boards for the Mandatory and Periodic Review of (a) the Natural and Other Hazards For Which Zonal Boundaries Have Been Set by the Jurisdiction; (b) the Accuracy of Current Boundaries; and (c) the Hazard-Mitigating Efficacy and Cost Feasibility of the Jurisdiction's Natural Hazards Management Plan.

Action Possibilities for Private Entities

James E. Anderson (1975) has observed how difficult it is for any subject to win a place on the agenda of any public policy-making body, and how important it is for outside constituencies to argue in favor of causes to be placed on such agendas and to engineer opinion climates within which policy-making bodies may dispose of those agenda items. A former state legislator and law professor has put it this way:

> Legislatures work almost exclusively as boards of review to judge proposals brought forward by various groups. Legislatures respond,

> they seldom lead. Those who want something from a legislature must ask for it . . . Legislators themselves seldom invent an idea, draft that idea into a bill, educate the press and public to a bill's merits, or lead a lobbying effort in both Houses of the Legislature and with the Executive Branch. It is unrealistic to expect them to. What actually happens is that new ideas in the form of bill drafts come to legislators from citizens, scholars, lawyers, bureaucrats, and lobbyists; these nonlegislators then help pass the bill by explaining its merits to legislators and to the public. A bill coming from outside the Legislature has a political legitimacy—a credential—that the few bills legislators think up on their own do not possess . . . the effective position is served on a silver platter as a soundly-conceived and well-drafted bill. It is accompanied by supporting advocacy which convinces legislators that the bill is sound and that they will not incur serious political vulnerabilities if they support it (Davies, 1975).

If changes are to be made in the current mix of federal, state and local policies concerned with the management of natural hazards it seems essential that nongovernmental groups concern themselves with the facts, issues, candidate public problems, strategic policy options, alternative goals, and action possibilities which have been presented in this book. It would be naive to imagine that the probability is high that legislator-initiated action will occur to modify the current mix of public policies.

Accordingly, more rapid and effective amelioration of problems posed by natural hazard exposures may be fostered if private organizations and associations take action to improve public and policy-maker understanding of exposure problems and the major problem-mitigating opportunities. Illustrative of the possibilities are the following:

Action Possibility 1. The American Association for the Advancement of Science Should (a) Appoint a Multidisciplinary Commission on Management of Environmental Hazard Zones; (b) Charge the Commission With Responsibility for Preparation of a Report on the Subject, Which Is of the Scope and Quality of the Report Issued by the AAAS Commission on Air Conservation.

Action Possibility 2. An Intersociety Committee on National Hazard Management Criteria and Research Should Be Sponsored by Appropriate National Professional Societies. The memberships of a number of professional societies are professionally involved in operations associated with the identification of (a) natural hazard areas; (b) natural hazard effects; and (c) natural hazard mitigations, costs, and consequences. These societies are the American Meteorological Society; American Society of Civil Engineers; American Institute of Architects; American Planners Association; American Public Works Association; National Association of Insurance Commissioners; International Conference of Building Officials; and others. Currently, the American Public Works Association has established a National Hazards Task Force.

A joint committee representing these societies could profitably be employed to review the need for and develop specific recommendations concerning the several policy alternatives identified above. Such a committee could continuously concern itself with identification of the types of criteria needed in the field of natural hazards management and with the type and scope of research necessary to the development of such criteria.

The Problem of Future Policy Making

The losses which are now being experienced, and which will be experienced in the future, as a result of the exposure of people and property to hazardous natural events are a direct result of the policy choices which have been made by state and local governments, by builders and subdividers, by architects, engineers, and other professionals, and by property purchasers and investors. The technological means are now at hand for identifying high-risk hazardous sites and for strengthening structures against forces from high intensity hazardous natural events; and appropriate knowledge is now available concerning the legal and administrative systems which can effectively implement such policies. Few communities, however, have fully utilized the range of methods which now are available in the arsenal of modern technology to cope with future hazardous natural occurrences. Low levels of concern for such occurrences are characteristic of state and local policy makers; it is difficult to secure the placement of natural hazard management issues on the policy agendas of state and local legislative bodies; and few local jurisdictions are equipped with technical staffs of the quality and strength necessary to adequately aid in natural hazard management policy-making and policy-implementation. Economic issues concerning the mix of benefits and costs of alternative natural hazard evaluation and mitigation policies are rife, and when such subjects finally find their way to the agendas of state and local policy-making bodies, the confusion is exacerbated by the arguments advanced by special interest groups representing builders and the purveyors of various types of construction materials and systems (Atkisson and Petak, 1981b).

Low levels of citizen concern for natural hazard risks feed a system in which apathy and long timelines are all too characteristic. Even in states such as California, which has led the nation in the invention and adoption of earthquake mitigation schemes, extraordinarily long timelines have been associated with important earthquake mitigation programs. For example, nearly half a century passed before all California school buildings were brought into compliance with the seismic safety standards which were demanded after the occurrence of the Long Beach Earthquake of 1933. In spite of the clear warnings provided by the Santa Barbara earthquake of 1925, the Long Beach earthquake of 1933, and the San Fernando earthquake of 1971, the Los Angeles City Council did not act to abate the seismic risk problems posed by hazardous buildings until January 1981. Few other county or municipal jurisdictions in the multicounty Southern

California area have yet acted to deal with this problem. Although ordinances intended to mitigate the risks associated with building parapets, cornices, and other overhangs have long been on the books of a few California cities, they have rarely been enforced with vigor and imagination. Similar accounts can be recited for other hazards and for other areas of the nation.

In those areas of the country where building and safety departments and land use planning and zoning entities have been created which are of a size and quality adequate to deal with natural hazard management problems, the great bulk of community growth seems to be occurring beyond boundary lines of their jurisdictions.

Although model building codes and recommended natural hazard mitigating standards have been developed by a number of reputable groups, these model standards and codes all too infrequently lead to timely modification of locally-enacted and legally-enforceable codes and regulations. Currently, there is no dearth of nationally-recommended hazard-related standards for buildings and other structures. However, the technical arguments now occurring over the content of such recommended codes and standards are not primarily focused on the level of safety they provide; rather, they question whether or not the standards have been pushed to such high levels as to escalate the cost of risk mitigation beyond the level of the benefits to be derived.

Even professional standards of practice have been brought into question in many states and are now attracting the attention of state professional licensure bodies. Charges have recently been made in California that builders and subdividers can buy almost any kind of geologic finding they wish in respect to the safety of proposed development sites. In Utah, any engineer, no matter what his or her specialty, may sign approval of building plans, and efforts now are being made by licensure boards within the state to tighten up the requirements as to the type of professional who may approve a building plan in terms of its seismic safety. One major development project was halted in Utah when it was found that the structure did not, in fact, conform to UBC seismic safety standards in spite of the fact that it had been certified by a licensed professional (Atkisson and Petak, 1981b).

Uncertainty as to the adequacy of natural hazard related structural standards and the effectiveness of their enforcement is pervasive throughout states at risk of experiencing high intensity hazardous natural events. Numerous national advisory bodies have decried the low quality of building plan review and construction inspection within the United States, dating back at least to the National Commission on Urban Problems (1968). Although excellence, dedication, and high levels of motivation can be identified in several communities, these attributes are all too rare among the total communities at risk within the United States.

At the federal level, a contradictory mix of disaster relief and tax policies is now in force. It is difficult to identify any truly comprehensive federal policy and management response to the potential problems posed by existing policies regarding current and future exposure of people and property to hazardous natural events. Nonetheless, it is clear that the occurrence of hazardous natural events produces far less loss of life and property within the United States than in other sections of the world, and that imprudently conceived policies targeted on reducing natural hazard losses may actually exacerbate the net cost burdens associated with natural hazard exposures (see Chapter 6). If current federal disaster relief policies are inconsistent and result in inefficiencies, they nevertheless have successfully mitigated the more acute problems that otherwise might have been faced by disaster-struck communities.

Thus, difficult questions of fact and value confront policy makers at all levels of government regarding present and future natural hazard exposures and the current mix of public policies targeted on these subjects. Like other subjects addressed by policy makers, the policy-framing and policy-amending which may take place in the future in respect to the management of natural hazards will occur in milieus in which "issues are confused, where you are never sure you're right, where good and bad are inextricably fused with the partly good and the partly bad, where often you can't do one worthy thing without endangering some other worthy thing" (Gardner, 1970).

References

Anderson, J. E. 1975. Public Policy Making. Praeger Publishers, New York.

Atkisson, A. A., and W. J. Petak. 1981a. Earthquake Insurance: A Public Policy Analysis. Report to the Federal Insurance Administration. J. H. Wiggins Company, Redondo Beach, California.

Atkisson, A. A., and W. J. Petak. 1981b. Seismic Safety Policies and Practices in U.S. Metropolitan Areas: A Three City Case Study. Report to the Federal Emergency Management Agency. J. H. Wiggins Company, Redondo Beach, California.

Atkisson, A. A., W. J. Petak, and D. R. Anderson. 1980. Earthquake Insurance Issues Workshop. Report to the Federal Insurance Administration. J. H. Wiggins Company, Redondo Beach, California.

Comptroller General of the United States. 1980. Federal Disaster Assistance: What Should the Policy Be? U.S. General Accounting Office, Washington, D.C.

Council on Environmental Quality, Executive Office of the President. 1977. Ninth Annual Report. U.S. Govt. Printing Office, Washington, D.C.

Davies, J. 1975. Legislative Law and Process in a Nutshell. West Publishing Company, St. Paul, Minnesota.

Fed. Regist. 1977. 42, No. 101, pp. 26951–26957.

Field, C. G., and S. R. Rivkin. 1975. The Building Code Burden. D. C. Heath, Lexington, Massachusetts.

Gardner, J. W. 1970. Recovery of Confidence. W. W. Norton, New York.

Lagorio, H. 1979. Failures in public policy research. *In*: S. Scott (Ed.), What Decision Makers Need to Know: Policy and Social Science Research on Seismic Safety. Institute of Governmental Studies, Berkeley, California, pp. 15–16.

National Commission on Urban Problems. 1968. Local Land and Building Regulation. Research Report No. 6. U.S. Govt. Printing Office, Washington, D.C.

Petak, W. J., and A. A. Atkisson. 1981. Alternative Policies and Procedures for Implementing Section 406 of the Disaster Relief Act of 1974. Report to the Federal Emergency Management Agency. J. H. Wiggins Company, Redondo Beach, California.

Reinsurance Association of America. 1980. Background Information and Summary of the Federal Catastrophe Loan Plan. *In*: D. Anderson, W. J. Petak, and C. E. Widell. Earthquake Insurance Practices. Prepared for the Federal Insurance Administration. J. H. Wiggins Company, Redondo Beach, California, Appendix F.

Rettger, M. J., and R. Boisvert. 1979. Economics of federal flood insurance and loan programs. Search 9(2):1–39.

U.S. Water Resources Council. 1973. Principles and Standards for Planning Water and Related Land Resources. U.S. Govt. Printing Office, Washington, D.C.

U.S. Water Resources Council. 1975. Options for Cost Sharing. U.S. Govt. Printing Office, Part 5A, Table 5A-1.

U.S. Water Resources Council. 1976. A Unified National Program for Flood Plain Management. U.S. Govt. Printing Office, Washington, D.C.

Index

Springer Series on Environmental Management
Robert S. DeSanto, Series Editor

Gradient Modeling
Resource and Fire Management
***by* Stephen R. Kessell**

Gradient modeling is a promising new approach to resource management simulation and land management planning—a method that applies the tools of remote sensing, computer technology, and gradient analysis to resource management problems.

"The author has done much research in ecosystem modeling for resource management. Most of the text is devoted to Kessell's study in Glacier National Park, and covers field methods and modeling methods in that study. The author also includes methods for model implementation and applications of gradient modeling systems. A lengthy bibliography and index provide an excellent reference resource. Two appendixes give species lists for the studied area, and gradient model nomograms for Glacier Park. Advanced students and professionals in resource management should welcome this work."

—*Choice*

". . . deserves to be read and understood by all those concerned with the wise management of land and resources."

—*BioScience*

1979/xv, 432 pp./175 illus./27 tables/cloth
ISBN 0-387-**90379**-8

Springer Series on Environmental Management
Robert S. DeSanto, Series Editor

Disaster Planning
The Preservation of Life and Property
***by* Harold D. Foster**

Almost daily, stories of tragic occurrences—disasters—appear in the news. However, many communities still ignore the possible risks to life and property imposed by an ever-increasing range of hazards, both natural and manmade. **Disaster Planning** offers local administrators, institutions, and organizations a blueprint for increasing community safety through a comprehensive risk management program. Dr. Foster presents a step-by-step approach to delineating safety goals in which urban and regional planners are introduced to numerous innovative techniques and concepts applicable to existing staff and equipment, thereby minimizing expense while maximizing safety. Also, a new safety index based on stress factors and a methodology for deciding the necessity of evacuation are introduced.

1980/x, 275 pp./48 illus./cloth
ISBN 0-387-**90498**-0

Air Pollution and Forests
Interactions between Air Contaminants and Forest Ecosystems
***by* William H. Smith**

"A definitive book on the complex relationship between forest ecosystems and atmospheric deposition . . . long needed . . . a thorough and objective review and analysis."

—*Journal of Forestry*

"It presents a tremendously useful framework for thinking about the problems it treats."

—*Atmospheric Environment*

1981/xv, 379 pp./60 illus./cloth
ISBN 0-387-**90501**-4

MIX
Papier aus verantwortungsvollen Quellen
Paper from responsible sources
FSC® C105338

If you have any concerns about our products,
you can contact us on
ProductSafety@springernature.com

In case Publisher is established outside the EU,
the EU authorized representative is:
Springer Nature Customer Service Center GmbH
Europaplatz 3, 69115 Heidelberg, Germany

Printed by Libri Plureos GmbH
in Hamburg, Germany